This is an advanced text for higher degree materials science students and researchers concerned with the strength of highly brittle covalent–ionic solids, principally ceramics. It is a reconstructed and greatly expanded edition of a book first published in 1975.

The book presents a unified continuum, microstructural and atomistic treatment of modern-day fracture mechanics from a materials perspective. Particular attention is directed to the basic elements of bonding and microstructure that govern the intrinsic toughness of ceramics. These elements hold the key to the future of ceramics as high-technology materials – to make brittle solids strong, we must first understand what makes them weak. The underlying theme of the book is the fundamental Griffith energy-balance concept of crack propagation. The early chapters develop fracture mechanics from the traditional continuum perspective, with attention to linear and nonlinear crack-tip fields, equilibrium and non-equilibrium crack states. It then describes the atomic structure of sharp cracks, the topical subject of crack–microstructure interactions in ceramics, with special focus on the concepts of crack-tip shielding and crack-resistance curves, and finally deals with indentation fracture, flaws, and structural reliability.

Brittle fracture crosses the boundaries between materials science, structural engineering, and physics and chemistry. This book develops a cohesive account by emphasising basic principles rather than detailed factual information. Due regard is given to model brittle materials such as silicate glass and polycrystalline alumina, as essential groundwork for ultimate extension of the subject matter to more complex engineering materials.

This book will be used by advanced undergraduates, beginning graduate students and research workers in materials science, mechanical engineering, physics and earth science departments interested in the brittle fracture of ceramic materials.

T0328176

Fracture of brittle solids

Cambridge Solid State Science Series

EDITORS:
Professor E. A. Davis
Department of Physics, University of Leicester
Professor I. M. Ward FRS
Department of Physics, University of Leeds

Titles in print in this series

Polymer Surfaces
B. W. Cherry

An Introduction to Composite Materials
D. Hull

Thermoluminescence of Solids
S. W. S. McKeever

Modern Techniques of Surface Science
D. P. Woodruff and T. A. Delchar

New Directions in Solid State Chemistry
C. N. R. Rao and J. Gopalakrishnan

The Electrical Resistivity of Metals and Alloys
P. L. Rossiter

The Vibrational Spectroscopy of Polymers
D. I. Bower and W. F. Maddams

Fatigue of Materials
S. Suresh

Glasses and the Vitreous State
J. Zarzycki

Hydrogenated Amorphous Silicon
R. A. Street

Microstructural Design of Fiber Composites
T.-W. Chou

Liquid Crystalline Polymers
A. M. Donald and A. H. Windle

Fracture of Brittle Solids – Second Edition
B. R. Lawn

An Introduction to Metal Matrix Composites
T. W. Clyne and P. J. Withers

BRIAN LAWN

NIST Fellow

Fracture of brittle solids

SECOND EDITION

Published by the Press Syndicate of the University of Cambridge
The Pitt Building, Trumpington Street, Cambridge CB2 1RP
40 West 20th Street, New York, NY 10011-4211, USA
10 Stamford Road, Oakleigh, Melbourne 3166, Australia

First published 1975
Second edition 1993

A catalogue record of this book is available from the British Library

Library of Congress cataloguing in publication data

Lawn, Brian R.
    Fracture of brittle solids/Brian Lawn. – 2nd edn
      p.   cm. – (Cambridge solid state science series)
    Includes bibliographical references and index.
    ISBN 0 521 40176 3. – ISBN 0 521 40972 1 (pbk.)
    1. Fracture mechanics. 2. Brittleness. I. Title. II. Series.
TA409.L37 1993
620.1′126–dc20   91-26191 CIP

ISBN 0 521 40176 3 hardback
ISBN 0 521 40972 1 paperback

Transferred to digital printing 2004

Contents

Preface   x

Glossary of symbols and abbreviations   xiii

1 The Griffith concept   **1**
1.1 Stress concentrators   2
1.2 Griffith energy-balance concept: equilibrium fracture   5
1.3 Crack in uniform tension   7
1.4 Obreimoff's experiment   9
1.5 Molecular theory of strength   12
1.6 Griffith flaws   13
1.7 Further considerations   14

2 Continuum aspects of crack propagation I: linear elastic crack-tip field   **16**
2.1 Continuum approach to crack equilibrium: crack system as thermodynamic cycle   17
2.2 Mechanical-energy-release rate, G 20
2.3 Crack-tip field and stress-intensity factor, K 23
2.4 Equivalence of G and K parameters   29
2.5 G and K for specific crack systems   30
2.6 Condition for equilibrium fracture: incorporation of the Griffith concept   39
2.7 Crack stability and additivity of K-fields   41
2.8 Crack paths   44

3 Continuum aspects of crack propagation II: nonlinear crack-tip field   **51**
3.1 Nonlinearity and irreversibility of crack-tip processes   52
3.2 Irwin–Orowan extension of the Griffith concept   56
3.3 Barenblatt cohesion-zone model   59

3.4 Path-independent integrals about crack tip   66
3.5 Equivalence of energy-balance and cohesion-zone
       approaches   70
3.6 Crack-tip shielding: the R-curve or T-curve   72
3.7 Specific shielding configurations: bridged interfaces and
       frontal zones   80

4 Unstable crack propagation: dynamic fracture   **86**
4.1 Mott extension of the Griffith concept   87
4.2 Running crack in tensile specimen   88
4.3 Dynamical effects near terminal velocity   93
4.4 Dynamical loading   99
4.5 Fracto-emission   103

5 Chemical processes in crack propagation: kinetic fracture **106**
5.1 Orowan generalisation of the Griffith equilibrium
       concept: work of adhesion   108
5.2 Rice generalisation of the Griffith concept   112
5.3 Crack-tip chemistry and shielding   117
5.4 Crack velocity data   119
5.5 Models of kinetic crack propagation   128
5.6 Evaluation of crack velocity parameters   138
5.7 Thresholds and hysteresis in crack
       healing–repropagation   139

6 Atomic aspects of fracture   **143**
6.1 Cohesive strength model   144
6.2 Lattice models and crack trapping: intrinsic bond
       rupture   149
6.3 Computer-simulation models   162
6.4 Chemistry: concentrated crack-tip reactions   165
6.5 Chemistry: surface forces and metastable crack-interface
       states   175
6.6 Crack-tip plasticity   185
6.7 Fundamental atomic sharpness of brittle cracks: direct
       observations by transmission electron microscopy   188

7 Microstructure and toughness   **194**
7.1 Geometrical crack-front perturbations   195
7.2 Toughening by crack-tip shielding: general
       considerations   208
7.3 Frontal-zone shielding: dislocation and microcrack
       clouds   211

7.4 Frontal-zone shielding: phase transformations in
    zirconia   221
7.5 Shielding by crack-interface bridging: monophase
    ceramics   230
7.6 Ceramic composites   242

8 Indentation fracture   **249**
8.1 Crack propagation in contact fields: blunt and sharp
    indenters   250
8.2 Indentation cracks as controlled flaws: inert strength,
    toughness, and T-curves   263
8.3 Indentation cracks as controlled flaws: time-dependent
    strength and fatigue   276
8.4 Subthreshold indentations: crack initiation   282
8.5 Subthreshold indentations: strength   293
8.6 Special applications of the indentation method   296
8.7 Contact damage: strength degradation, erosion and wear   300
8.8 Surface forces and contact adhesion   304

9 Crack initiation: flaws   **307**
9.1 Crack nucleation at microcontacts   309
9.2 Crack nucleation at dislocation pile-ups   314
9.3 Flaws from chemical, thermal, and radiant fields   319
9.4 Processing flaws in ceramics   325
9.5 Stability of flaws: size effects in crack initiation   328
9.6 Stability of flaws: effect of grain size on strength   332

10 Strength and reliability   **335**
10.1 Strength and flaw statistics   337
10.2 Flaw statistics and lifetime   343
10.3 Flaw elimination   347
10.4 Flaw tolerance   350
10.5 Other design factors   357

References and reading list   **363**
Index   **372**

Preface

This book is a restructured version of a first edition published in 1975. As before, the objective is a text for higher degree students in materials science and researchers concerned with the strength and toughness of brittle solids. More specifically, the aim is to present fracture mechanics in the context of the 'materials revolution', particularly in ceramics, that is now upon us. Thus whereas some chapters from the original are barely changed, most are drastically rewritten, and still others are entirely new.

Our focus, therefore, is 'brittle ceramics'. By brittle, we mean cracks of atomic sharpness that propagate essentially by bond rupture. By ceramics, we mean covalent–ionic materials of various persuasions, including glasses, polycrystalline aggregates, minerals, and even composites. Since 1975, our knowledge of structural ceramics has equalled, some would insist surpassed, that of metals and polymers. But it is brittleness that remains the singular limiting factor in the design of ceramic components. If one is to overcome this limitation, it is necessary first to understand the underlying mechanics and micromechanics of crack initiation and propagation. Prominent among improvements in this understanding have been a continuing evolution in the theories of continuum fracture mechanics and new conceptions of fundamental crack-tip laws. Most significant, however, is the advent of 'microstructural shielding' processes, as manifested in the so-called crack-resistance- or toughness-curve, with far-reaching consequences in relation to strength and toughness. This developing area promises to revolutionise traditional attitudes toward properties design and processing strategies for ceramics.

The unifying theme of the book is the thermodynamic energy-balance concept expounded by Griffith in his classic 1920 paper. Griffith's concept leads naturally to classifications of crack systems as equilibrium or dynamic, stable or unstable, reversible or irreversible. His concept survives because of its inherent generality: in proceeding to more complex systems one needs only to modify existing terms, or add new ones, in the expression for the total energy of the crack system. All soundly-based fracture theories derive either directly from the Griffith concept or from some alternative concept with underlying equivalence, such as Irwin's stress-intensity factor.

In attempting to construct an integrated picture of fracture, one becomes aware of widely diverse perspectives on brittle cracks. Most traditional is the 'global' perspective of the engineer, who sees cracks in terms of a slit continuum, treating the tip and its surrounds as a singular (black box) zone. At the opposite end of the spectrum is the crack-tip 'enclave' perspective of the physicist–chemist, who defines the processes of discrete bond rupture in terms of intersurface force functions. Both viewpoints are valuable: the first gives us general parameters such as mechanical-energy-release rate G and stress-intensity factor K for quantifying the 'motive' for fracture in terms of extraneous variables like applied loads, specimen geometry, environmental concentration, etc.; the second provides us with a basis for describing the fundamental structure of atomically sharp cracks and thereby defining laws of extension. And now we must add a relatively new perspective, that of the materials scientist, who seeks to incorporate discrete dissipative elements into ceramic microstructures in order to overcome the intrinsic brittleness. It is at this level that the concept of shielding emerges, in the form of an intervening dissipative zone which screens the crack-tip enclave from the external applied loads. Innovations in microstructural shielding processes hold the key to the next generation of strong and tough brittle materials.

As with any attempt to tie these disparate perspectives into a cohesive description, it is inevitable that conflicts in notation will arise. In seeking compromise I have leant toward materials terminology. Among the more conspicuous symbols is the Griffith c rather than the solid mechanics a for crack size. Also notable are the symbols for toughness, R and T, in place of the engineering parameters G_R and K_R; the former serve to emphasise that the intrinsic resistance to crack propagation is an equilibrium material property, ultimately expressible as an integral of a constitutive stress–displacement relation without reference to fracture at all.

The layout of the book follows a loose progression from scientific fundamentals at one end to engineering design at the other. Historical and conceptual foundations are laid in chapter 1, with a review of the energy-balance concept and flaw hypothesis of Griffith. Chapters 2 and 3 develop a theoretical description of crack propagation in terms of continuum fracture mechanics, with an emphasis on equilibrium configurations.

Chapters 4 and 5 extend these considerations to moving cracks, dynamic ('fast') and kinetic ('slow'), with special attention in the latter case to environmental chemistry. In chapter 6 we analyse crack-tip processes at the atomic level, again with provision to include chemistry in the fundamental crack laws. Chapter 7 considers the influence of microstructure on the fracture mechanics, with accent on some of the promising shielding mechanisms that are emerging in the toughness description. One of the most powerful and widespread methodologies for evaluating ceramic materials, indentation fracture, is surveyed in chapter 8. In chapter 9 we deal with the issue of flaws and crack initiation. Finally, in chapter 10, strength and reliability are addressed.

An understanding of fracture mechanics is best obtained by concentrating on basic principles rather than on factual information. Consequently, our attention to 'model' materials like homogeneous glass and polycrystalline alumina should be seen as essential groundwork for ultimate extension to more complex engineering materials. That philosophy extends to the literature citations. We have not sought to provide an extensive reference list, but rather a selective bibliography. It is a hope that, in an age where the published word is fast becoming a lost forum of communication, the reader will be persuaded to consult the open literature.

Many colleagues and students have contributed greatly to this venture. Special mention is due to Rodney Wilshaw, former co-author and old friend, with whom the first edition was conceived and produced. Soon after publication of that earlier version Rod turned from academic endeavours to a life on the land. He gracefully withdrew his name from the cover of this edition. His spirit is nevertheless still to be found in the ensuing pages. Other major contributors over the years include: S. J. Bennison, L. M. Braun, S. J. Burns, H. M. Chan, P. Chantikul, R. F. Cook, T. P. Dabbs, F. C. Frank, E. R. Fuller, B. J. Hockey, R. G. Horn, S. Lathabai, Y.-W. Mai, D. B. Marshall, N. P. Padture, D. H. Roach, J. Rödel, J. E. Sinclair, M. V. Swain, R. M. Thomson, K.-T. Wan and S. M. Wiederhorn. I also thank R. W. Cahn for his encouragement to embark on this second edition, and his perseverance during its completion. Finally, to my wife Valerie, my heartfelt appreciation for enduring it all.

Brian Lawn

Glossary of symbols and abbreviations

SI units are used throughout, with the following prefixes:

k	kilo	10^3		m	milli	10^{-3}
M	mega	10^6		μ	micro	10^{-6}
G	giga	10^9		n	nano	10^{-9}
T	tera	10^{12}		p	pico	10^{-12}
				f	femto	10^{-15}
				a	atto	10^{-18}

Symbols (with units)

a inclusion or pore radius (μm); characteristic contact radius (μm)

a_c critical contact size (μm)

a_0 atomic spacing (nm)

A cross-sectional area (mm^2); Auerbach constant

b minor axis in Inglis elliptical cavity (μm); magnitude of Burgers vector (nm)

b_0 lattice spacing (nm)

c characteristic crack size (μm)

c_B crack size at branching (μm)

c_C critical crack size (μm)

c_f flaw size (μm)

c_F crack size at failure (μm)

c_I crack size at pop-in (μm)

c_M crack size at activated failure (μm)

c_0 starter crack (notch) size (mm)

C crack area (μm^2)

d beam thickness (mm); characteristic spacing between microstructural elements (μm)

E Young's modulus (GPa)

E' E, plane stress; $E/(1-v^2)$, plane strain (GPa)

F line force (force per unit length) (N m^{-1})

F_B force on stretched atomic bond (nN)

F_n lattice-modified force (nN)

ΔF activation free energy (aJ molec^{-1})

f_i angular function in crack-tip displacement field

f_{ij} angular function in crack-tip stress field

$\mathscr{G}$ net crack-extension force, or 'motive' (J m^{-2})

G mechanical-energy-release rate (J m^{-2})

G_A global mechanical-energy-release rate (J m^{-2})

G_C critical mechanical-energy-release rate (J m^{-2})

G_R G_A in material with shielding (J m^{-2})

G_* crack-tip enclave mechanical-energy-release rate (J m^{-2})

G_μ shielding-zone mechanical-energy-release rate (J m^{-2})

G_0 cohesion-zone mechanical-energy-release rate (J m^{-2})

h cantilever-beam crack-opening displacement (μm)

$\mathbf{h}$ Planck constant (6.6256×10^{-34} J s)

H indentation hardness (GPa)

J Rice line integral (J m^{-2})

k elastic coefficient for Hertzian contact

$\mathbf{k}$ Boltzmann constant (1.3805×10^{-23} J K^{-1})

ℓ net K-field at singular tip (MPa m$^{1/2}$)

K stress-intensity factor (MPa m$^{1/2}$)

K_A global stress-intensity factor (MPa m$^{1/2}$)

K_B stress-intensity factor at crack branching (MPa m$^{1/2}$)

K_C critical stress-intensity factor (MPa m$^{1/2}$)

K_R residual stress-intensity factor (MPa m$^{1/2}$)

K_R K_A in material with shielding (MPa m$^{1/2}$)

K_μ shielding-zone stress-intensity factor (MPa m$^{1/2}$)

K_* crack-tip enclave stress-intensity factor (MPa m$^{1/2}$)

K_0	cohesion-zone stress-intensity factor (MPa m$^{1/2}$)
K_I, K_{II}, K_{III}	mode I, II, III stress-intensity factors (MPa m$^{1/2}$)
l	beam span in flexure specimen (mm); grain size (μm)
l_C	critical grain size for spontaneous microcracking (μm)
L	bridging zone length (mm); specimen dimension (mm)
m	molecular mass (10^{-27} kg)
n	crack velocity power-law exponent; number of atoms in lattice-crack chain
p_C^B	critical bridging stress (MPa)
p_C^P	critical fibre pullout stress (MPa)
p^D	fibre debonding stress (MPa)
p_E	environmental gas pressure (kPa)
p_{Th}	theoretical cohesive stress (GPa)
p_γ	cohesive surface stress at crack interface (GPa)
p_μ	microstructural shielding tractions at crack interface (MPa)
p_0	mean contact pressure (MPa)
P	applied point load, contact load (N)
P_C	critical contact load (N)
P_+, P_-	applied load extremes for lattice trapping (N)
$\mathbf{P}$	probability of failure
Q	heat input (J)
r	radial crack-tip coordinate (μm); fibre or sphere radius (μm)
R	crack-resistance energy per unit area (J m^{-2})
R-curve	resistance-curve
R_E	crack-resistance energy in interactive environment (J m^{-2})
R_μ	microstructural shielding component of resistance energy (J m^{-2})
R_0	crack-resistance energy in a vacuum (J m^{-2})
R_∞	steady-state crack-resistance energy (J m^{-2})
R^+, R^-	crack-resistance trapping range (J m^{-2})
R'	quasi-equilibrium crack-resistance energy (J m^{-2})
s	arc length (m)
S	entropy (J K^{-1})

t	time (s)
t_F	time to failure (lifetime) (s)
T	toughness (MPa m$^{1/2}$)
T-curve	toughness-curve
T_E	toughness in interactive environment (MPa m$^{1/2}$)
T_μ	microstructural shielding component of toughness (MPa m$^{1/2}$)
T_0	toughness in a vacuum (MP m$^{1/2}$)
T_∞	steady-state toughness (MPa m$^{1/2}$)
$\mathbf{T}$	absolute temperature (K)
$\mathscr{T}$	traction vector in J-integral (MPa)
u	crack-opening displacement (μm); load-point displacement (μm)
$\mathbf{u}$	displacement vector (μm)
u_i	component of displacement vector (μm)
u_Z	crack-opening displacement at edge of traction zone (μm)
u_y	crack-opening displacement in cohesion zone (nm)
U	system internal energy (J)
U_A	energy of applied loading system (J)
U_{AA}	cohesion energy of molecule A–A (fJ)
U_{AB}	energy of terminal bond A–B– (fJ)
U_B	energy of stretched cohesive bond (fJ)
U_{BB}	cohesion energy of bond –B–B– (fJ)
U_E	elastic strain energy (J)
U_i, U_f	initial, final energy states (J)
U_K	kinetic energy (J)
U_M	mechanical energy (J)
U_S	surface energy of crack area (J)
ΔU_{Ad}	adsorption energy (J m^{-2})
$\mathscr{U}$	strain energy density in *J*-integral (J m^{-3})
v	crack velocity (m s^{-1})
v_l	longitudinal wave velocity (km s^{-1})
v_T	terminal velocity (km s^{-1})
v_I, v_{II}, v_{III}	velocities in regions I, II, III
V_f	volume fraction

w	specimen width (mm)
w_{C}	critical width of frontal-zone wake (μm)
W	Dupré work of adhesion (J m^{-2})
$^{\mathrm{h}}W$	same, for crack growth through healed interface (J m^{-2})
$^{\mathrm{v}}W$	same, for crack growth through virgin solid (J m^{-2})
W_{AB}	work to separate unlike bodies A–B in a vacuum (J m^{-2})
W_{BB}	work to separate like bodies B–B in a vacuum (J m^{-2})
W_{BEB}	work to separate like bodies B–B in environment E (J m^{-2})
x, y, z	Cartesian coordinates for crack system (m)
X	crack-interface coordinate measured from crack tip (mm)
X_{Z}	crack-interface coordinate at edge of traction shielding zone (mm)
α	specimen geometry edge correction factor; activation area (nm^{2} molec^{-1}); lattice spring constant (nN nm^{-1}); thermal expansion coefficient (K^{-1})
α_{0}	contact geometry coefficient
β	gas pressure coefficient in crack velocity equation; lattice spring constant (nN nm^{-1}); normalised radial coordinate of contact crack initiation
γ	surface or interface energy per unit area (J m^{-2})
γ_{B}	intrinsic ('inert') surface energy of solid body B (J m^{-2})
γ_{BE}	interfacial energy for body B in environmental medium E (mJ m^{-2})
γ_{GB}	grain boundary energy (mJ m^{-2})
γ_{hE}	fault energy for interface healed in environment (mJ m^{-2})
γ_{IB}	interphase boundary energy (mJ m^{-2})
Γ	Gibbs surface excess (nm^{-2})
Γ_{B}	lattice-trapping modulation factor in cohesion energy (J m^{-2})
δ	Barenblatt crack-opening displacement (nm)
ε	strain
ε^{B}	bridge rupture strain
ε^{M}	constrained microcrack-zone dilational strain
ε^{T}	constrained transformation-zone dilational strain

ε^{Y} rupture strain for plastic bridge

ε_{μ} dilational strain in frontal-zone shielding field

ζ kink coordinate (nm)

η order of chemical interaction

θ, ϕ polar coordinates for crack system

θ fractional surface adsorption coverage

κ Knudsen attenuation factor for free molecular flow

λ elastic compliance (m N^{-1}); Barenblatt zone length (nm)

Λ entropy production rate (J s^{-1})

μ friction coefficient; shear modulus (GPa)

ν Poisson's ratio

ν_0 lattice frequency (Hz)

ξ critical range for stress cutoff at edge of closure zone (μm)

ξ^{B} critical cutoff range for bridge disengagement (μm)

ξ^{P} critical cutoff range for fibre pullout (μm)

ρ tip radius of elliptical cavity (nm); density (kg m^{-3}); radial coordinate (m)

σ stress (MPa)

σ_{A} applied uniform stress (MPa)

σ_{C} stress at tip of elliptical cavity (GPa)

σ_{C}^{D} critical activation stress for dislocation motion (MPa)

σ_{C}^{M} critical activation stress for microcracking (MPa)

σ_{C}^{T} critical activation stress for transformation (MPa)

σ_{C}^{Y} yield stress (MPa)

σ_{F} failure stress (MPa)

σ_{ij} component of stress tensor (MPa)

σ_{I} inert strength (MPa)

σ_{M} activated failure stress (MPa)

σ_{P} proof stress (MPa)

σ_{R} residual stress (MPa)

σ_{S} surface stress (MPa)

σ_{T} tensile stress at Hertzian contact circle (MPa)

σ_{TS} thermal shock stress (MPa)

σ_{μ} dilational stress in frontal-zone shielding field

σ_{0} Weibull scaling stress (MPa)

τ interfacial friction stress (MPa)

Φ indenter half-angle

χ indentation residual-contact coefficient

ψ crack-geometry factor

Abbreviations

CT	compact tension specimen
DCB	double-cantilever beam specimen
DT	double-torsion specimen
NDE	non-destructive evaluation
PSZ	partially stabilised zirconia
SENB	single-edge notched beam specimen
TEM	transmission electron microscope

1
The Griffith concept

Most materials show a tendency to fracture when stressed beyond some critical level. This fact was appreciated well enough by nineteenth century structural engineers, and to them it must have seemed reasonable to suppose strength to be a material property. After all, it had long been established that the stress response of materials within the elastic limit could be specified completely in terms of characteristic elastic constants. Thus arose the premise of a 'critical applied stress', and this provided the basis of the first theories of fracture. The idea of a well-defined stress limit was (and remains) particularly attractive in engineering design; one simply had to ensure that the maximum stress level in a given structural component did not exceed this limit.

However, as knowledge from structural failures accumulated, the universal validity of the critical applied stress thesis became more suspect. The fracture strength of a given material was not, in general, highly reproducible, in the more brittle materials fluctuating by as much as an order of magnitude. Changes in test conditions, e.g. temperature, chemical environment, load rate, etc., resulted in further, systematic variations in strengths. Moreover, different material types appeared to fracture in radically different ways: for instance, glasses behaved elastically up to the critical point, there to fail suddenly under the action of a tensile stress component, while many metallic solids deformed extensively by plastic flow prior to rupture under shear. The existing theories were simply incapable of accounting for such disparity in fracture behaviour.

This, then, was the state of the subject in the first years of the present century. It is easy to see now, in retrospect, that the inadequacy of the critical stress criterion lay in its empirical nature: for the notion that a solid should break at a characteristic stress level, however intuitively appealing, is not based on sound physical principles. There was a need to take a closer look at events within the boundaries of a critically loaded solid. How, for example, are the applied stresses transmitted to the inner regions where fracture actually takes place? What is the nature of the fracture mechanism itself? The answers to such questions were to hold the key to an understanding of all fracture phenomena.

The breakthrough came in 1920 with a classic paper by A. A. Griffith. Griffith considered an isolated crack in a solid subjected to an applied stress, and formulated a criterion for its extension from the fundamental energy theorems of classical mechanics and thermodynamics. The principles laid down in that pioneering work, and the implications drawn from those principles, effectively foreshadowed the entire field of present-day fracture mechanics. In our introductory chapter we critically analyse the contributions of Griffith and some of his contemporaries. This serves to introduce the reader to many of the basic concepts of fracture theory, and thus to set the scene for the remainder of the book.

1.1 Stress concentrators

An important precursor to the Griffith study was the stress analysis by Inglis (1913) of an elliptical cavity in a uniformly stressed plate. His analysis showed that the local stresses about a sharp notch or corner could rise to a level several times that of the applied stress. It thus became apparent that even submicroscopic flaws might be potential sources of weakness in solids. More importantly, the Inglis equations provided the first real insight into the mechanics of fracture; the limiting case of an infinitesimally narrow ellipse might be considered to represent a *crack*.

Let us summarise briefly the essential results of the Inglis analysis. We consider in fig. 1.1 a plate containing an elliptical cavity of semi-axes b, c, subjected to a uniform applied tension σ_A along the Y-axis. The objective is to examine the modifying effect of the hole on the distribution of stress in the solid. If it is assumed that Hooke's law holds everywhere in the plate, that the boundary of the hole is stress-free, and that b and c are small in comparison with the plate dimensions, the problem reduces to a relatively straightforward exercise in linear elasticity theory. Although the mathematical treatment becomes somewhat unwieldy, involving as it does the use of elliptical coordinates, some basic results of striking simplicity emerge from the analysis.

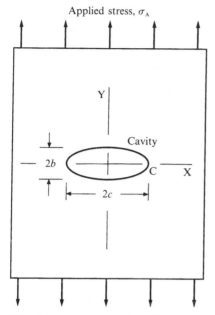

Fig. 1.1. Plate containing elliptical cavity, semi-axes b, c, subjected to uniform applied tension σ_A. C denotes 'notch tip'.

Beginning with the equation of the ellipse,

$$x^2/c^2 + y^2/b^2 = 1, \tag{1.1}$$

one may readily show the radius of curvature to have a minimum value

$$\rho = b^2/c, \quad (b < c) \tag{1.2}$$

at C. It is at C that the greatest concentration of stress occurs:

$$\begin{aligned} \sigma_C &= \sigma_A (1 + 2c/b) \\ &= \sigma_A [1 + 2(c/\rho)^{1/2}]. \end{aligned} \tag{1.3}$$

For the interesting case $b \ll c$ this equation reduces to

$$\sigma_C/\sigma_A \simeq 2c/b = 2(c/\rho)^{1/2}. \tag{1.4}$$

The ratio in (1.4) is an elastic *stress-concentration factor*. It is immediately evident that this factor can take on values much larger than unity for narrow holes. We note that the stress concentration depends on the *shape* of the hole rather than the *size*.

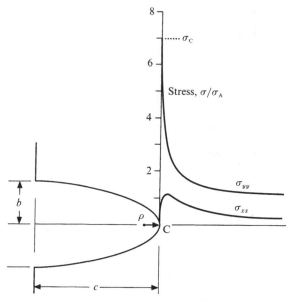

Fig. 1.2. Stress concentration at elliptical cavity, $c = 3b$. Note that concentrated stress field is localised within $\approx c$ from tip, highest gradients within $\approx \rho$.

The variation of the local stresses along the X-axis is also of interest. Fig. 1.2 illustrates the particular case $c = 3b$. The stress σ_{yy} drops from its maximum value $\sigma_C = 7\sigma_A$ at C and approaches σ_A asymptotically at large x, while σ_{xx} rises to a sharp peak within a small distance from the stress-free surface and subsequently drops toward zero with the same tendency as σ_{yy}. The example of fig. 1.2 reflects the general result that significant perturbations to the applied stress field occur only within a distance $\approx c$ from the boundary of the hole, with the greatest gradients confined to a highly localised region of dimension $\approx \rho$ surrounding the position of maximum concentration.

Inglis went on to consider a number of stress-raising configurations, and concluded that the only geometrical feature that had a marked influence on the concentrating power was the highly curved region where the stresses were actually focussed. Thus (1.4) could be used to estimate the stress-concentration factors of such systems as the surface notch and surface step in fig. 1.3, with ρ interpreted as a characteristic radius of curvature and c as a characteristic notch length. A tool was now available for appraising the potential weakening effect of a wide range of structural irregularities, including, presumably, a real crack.

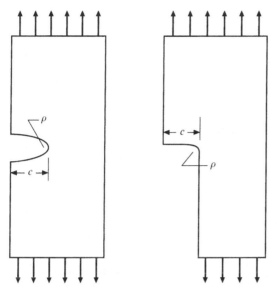

Fig. 1.3. Stress concentration half-systems: surface cavity and surface step of characteristic length c and notch radius ρ.

Despite this step forward the fundamental nature of the fracture mechanism remained obscure. If the Inglis analysis were indeed to be applicable to a crack system, then why in practice did large cracks tend to propagate more easily than small ones? Did not such behaviour violate the size-independence property of the stress-concentration factor? What is the physical significance of the radius of curvature at the tip of a real crack? These were some of the obstacles which stood between the Inglis approach and a fundamental criterion for fracture.

1.2 Griffith energy-balance concept: equilibrium fracture

Griffith's idea was to model a static crack as a reversible thermodynamic system. The important elements of the system are defined in fig. 1.4: an elastic body B containing a plane-crack surface S of length c is subjected to loads applied at the outer boundary A. Griffith simply sought the configuration that minimised the total free energy of the system; the crack would then be in a state of equilibrium, and thus on the verge of extension.

The first step in the treatment is to write down an expression for the total energy U of the system. To do this we consider the individual energy terms that are subject to change as the crack is allowed to undergo virtual

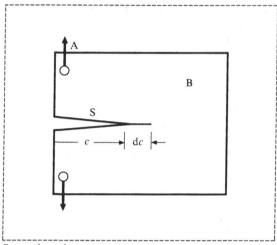

System boundary

Fig. 1.4. Static plane-crack system, showing incremental extension of crack length c through dc: B, elastic body; S, crack surface; A, applied loading.

extension. Generally, the system energy associated with crack formation may be partitioned into *mechanical* or *surface* terms. The mechanical energy itself consists of two terms, $U_M = U_E + U_A$: U_E is the strain potential energy stored in the elastic medium; U_A is the potential energy of the outer applied loading system, expressible as the negative of the work associated with any displacements of the loading points. The term U_S is the free energy expended in creating the new crack surfaces. We may therefore write

$$U = U_M + U_S. \tag{1.5}$$

Thermodynamic equilibrium is then attained by balancing the mechanical and surface energy terms over a virtual crack extension dc (fig. 1.4). It is not difficult to see that the mechanical energy will generally *decrease* as the crack extends ($dU_M/dc < 0$). For if the restraining tractions across the incremental crack boundary dc were suddenly to relax, the crack walls would, in the general case, accelerate outward and ultimately come to rest in a new configuration of lower energy. On the other hand, the surface energy term will generally *increase* with crack extension, since cohesive forces of molecular attraction across dc must be overcome during the creation of the new fracture surfaces ($dU_S/dc > 0$). Thus the first term in (1.5) favours crack extension, while the second opposes it. This is the

Griffith energy-balance concept, a formal statement of which is given by the *equilibrium* requirement

$$dU/dc = 0. \tag{1.6}$$

Here then was a criterion for predicting the fracture behaviour of a body, firmly rooted in the laws of energy conservation. A crack would extend or retract reversibly for small displacements from the equilibrium length, according to whether the left-hand side of (1.6) were negative or positive. This criterion remains the building block for all brittle fracture theory.

1.3 Crack in uniform tension

The Griffith concept provided a fundamental starting point for any fracture problem in which the operative forces could be considered to be conservative. Griffith sought to confirm his theory by applying it to a real crack configuration. First he needed an elastic model for a crack, in order to calculate the energy terms in (1.5). For this he took advantage of the Inglis analysis, considering the case of an infinitely narrow elliptical cavity ($b \to 0$, fig. 1.1) of length $2c$ in a remote, uniform tensile stress field σ_A. Then, for experimental verification, he had to find a well-behaved, 'model' material, isotropic and closely obeying Hooke's law at all stresses prior to fracture. Glass was selected as the most easily accessible material satisfying these requirements.

In evaluating the mechanical energy of his model crack system Griffith invoked a result from linear elasticity theory (cf. sect. 2.2), namely that for any body under constant applied stress during crack formation,

$$U_A = -2U_E, \quad \text{(constant load)} \tag{1.7}$$

so that $U_M = -U_E$. The negative sign indicates a mechanical energy *reduction* on crack formation. Then from the Inglis solution of the stress and strain fields the strain energy density is readily computed for each volume element about the crack. Integrating over dimensions large compared with the length of the crack then gives, for unit width along the crack front,

$$U_E = \pi c^2 \sigma_A^2 / E' \tag{1.8}$$

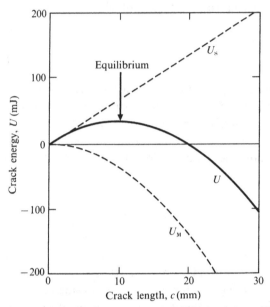

Fig. 1.5. Energetics of Griffith crack in uniform tension, plane stress. Data for glass from Griffith: $\gamma = 1.75\,\mathrm{J\,m^{-2}}$, $E = 62\,\mathrm{GPa}$, $\sigma_A = 2.63\,\mathrm{MPa}$ (chosen to give equilibrium at $c_0 = 10$ mm).

where E' identifies with Young's modulus E in plane stress ('thin' plates) and $E/(1-v^2)$ in plane strain ('thick' plates), with v Poisson's ratio. The application of additional loading parallel to the crack plane has negligible effect on the strain energy terms in (1.8). For the surface energy of the crack system Griffith wrote, again for unit width of front,

$$U_S = 4c\gamma \tag{1.9}$$

with γ the free surface energy per unit area. The total system energy (1.5) becomes

$$U(c) = -\pi c^2 \sigma_A^2 / E' + 4c\gamma. \tag{1.10}$$

Fig. 1.5 shows plots of the mechanical energy $U_M(c)$, surface energy $U_S(c)$, and total energy $U(c)$. Observe that, according to the Inglis treatment, an edge crack of length c (limiting case of surface notch, $b \to 0$, fig. 1.2) may be considered to possess very nearly one-half the energy of an internal crack of length $2c$.

The Griffith equilibrium condition (1.6) may now be applied to (1.10).

We thereby calculate the critical conditions at which 'failure' occurs, $\sigma_A = \sigma_F$, $c = c_0$, say:

$$\sigma_F = (2E'\gamma/\pi c_0)^{1/2}. \tag{1.11}$$

As we see from fig. 1.5, or from the negative value of d^2U/dc^2, the system energy is a maximum at equilibrium, so the configuration is *unstable*. That is, at $\sigma_A < \sigma_F$ the crack remains stationary at its original size c_0; at $\sigma_A > \sigma_F$ it propagates spontaneously without limit. Equation (1.11) is the famous Griffith strength relation.

For experimental confirmation, Griffith prepared glass fracture specimens from thin round tubes and spherical bulbs. Cracks of length 4–23 mm were introduced with a glass cutter and the specimens annealed prior to testing. The hollow tubes and bulbs were then burst by pumping in a fluid, and the critical stresses determined from the internal fluid pressure. As predicted, only the stress component normal to the crack plane was found to be important; the application of end loads to tubes containing longitudinal cracks had no detectable effect on the critical conditions. The results could be represented by the relation

$$\sigma_F c_0^{1/2} = 0.26 \text{ MPa m}^{1/2}$$

with a scatter $\approx 5\%$, thus verifying the essential *form* of $\sigma_F(c_0)$ in (1.11).

If we now take this result, along with Griffith's measured value of Young's modulus, $E = 62$ GPa, and insert into (1.11) at plane stress, we obtain $\gamma = 1.75$ J m^{-2} as an estimate of the surface energy of glass. Griffith attempted to substantiate his model by obtaining an independent estimate of γ. He measured the surface tension within the temperature range 1020–1383 K, where the glass flows easily, and extrapolated linearly back to room temperature to find $\gamma = 0.54$ J m^{-2}. Considering that even present-day techniques are barely capable of measuring surface energies of solids to very much better than a factor of two, this 'agreement' between measured values is an impressive vindication of the Griffith theory.

1.4 Obreimoff's experiment

Plane cracks in uniform tension represent just one application of the energy-balance equation (1.6). To emphasise the generality of the Griffith concept we digress briefly to discuss an important experiment carried out

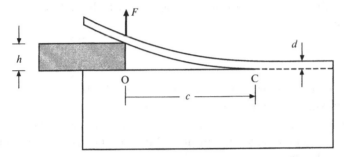

Fig. 1.6. Obreimoff's experiment on mica. Wedge of thickness h inserted to peel off cleavage flake of thickness d and width unity. In this configuration both crack origin O and tip C translate with wedge.

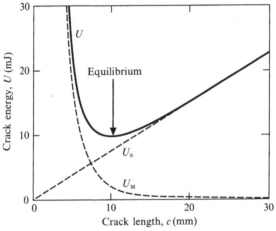

Fig. 1.7. Energetics of Obreimoff crack. Data for mica from Obreimoff: $\gamma = 0.38$ J m^{-2} (air), $E = 200$ GPa, $h = 0.48$ mm, $d = 75$ μm (chosen to give equilibrium at $c_0 = 10$ mm).

by Obreimoff (1930) on the cleavage of mica. This second example provides an interesting contrast to the one treated by Griffith, in that the equilibrium configuration is *stable*.

The basic arrangement used by Obreimoff is shown in fig. 1.6. A glass wedge of thickness h is inserted beneath a thin flake of mica attached to a parent block, and is made to drive a crack along the cleavage plane. In this case we may determine the energy of the crack system by treating the cleavage lamina as a freely loaded cantilever, of thickness d and width unity, built-in at the crack front distant c from the point of application of the wedge. We note that on allowing the crack to form under constant wedging conditions the bending (line) force F suffers no displacement, so the net work done by this force is zero, i.e.

$$U_{\text{A}} = 0. \tag{1.12}$$

At the same time we have, from simple beam theory, the elastic strain energy in the cantilever arm,

$$U_{\text{E}} = Ed^3h^2/8c^3. \tag{1.13}$$

The surface energy is

$$U_{\text{S}} = 2c\gamma. \tag{1.14}$$

The total system energy $U(c)$ in (1.5) now follows, and application of the Griffith condition (1.6) leads finally to the equilibrium crack length

$$c_0 = (3Ed^3h^2/16\gamma)^{1/4}. \tag{1.15}$$

The energy terms $U_{\text{M}}(c)$, $U_{\text{S}}(c)$, and $U(c)$ are plotted in fig. 1.7. It is evident from the minimum at $U(c_0)$ that (1.15) corresponds to a stable configuration. In this instance the fracture is 'controlled': the crack advances into the material at the same rate as that of the wedge.

Equation (1.15) indicates that, as in Griffith's uniform tension example, a knowledge of equilibrium crack geometry uniquely determines the surface energy. Obreimoff proceeded thus to evaluate the surface energy of mica under different test conditions, and found a dramatic increase from $\gamma = 0.38$ J m^{-2} at normal atmosphere (100 kPa pressure) to $\gamma = 5.0$ J m^{-2} in a vacuum (100 μPa). The test *environment* was clearly an important factor to be considered in evaluating material strength. Moreover, Obreimoff noticed that on insertion of the glass wedge the crack did not grow immediately to its equilibrium length: in air equilibrium was reached within seconds, whereas in a vacuum the crack continued to creep for several days. Thus the time element was another complicating factor to be considered. These observations provided the first indication of the role of *chemical kinetics* in fracture processes.

Obreimoff also observed phenomena that raised the question of *reversibility* in crack growth. Propagation of the crack was often erratic, with an accompanying visible electrostatic discharge ('triboluminescence'), especially in a vacuum. On partial withdrawal of the glass wedge the crack was observed to retreat and apparently 'heal', but re-insertion of the wedge revealed a perceptible reduction in cleavage strength. These results imply the existence in the energy balance of *dissipative* elements.

1.5 Molecular theory of strength

Although Griffith formulated his criterion for fracture in terms of macroscopic thermodynamical quantities, he was aware that a complete description required an evaluation of events at the molecular level. He argued that the maximum stress at the tip of an equilibrium crack must correspond to the *theoretical cohesive strength* of the solid; that is, the largest possible stress level that the molecular structure can sustain by virtue of its intrinsic bond strength. Griffith accordingly estimated the theoretical strength of his glass from the stress-concentration formula (1.4), inserting $\rho \approx 0.5$ nm (molecular dimensions), as a 'reasonable' tip radius for a crack growing by sequential bond rupture, together with his measured value $\sigma_A c^{1/2} = \sigma_F c_0^{1/2}$ at instability (sect. 1.3). The value obtained, $\sigma_C \approx 23$ GPa, is an appreciable fraction of Young's modulus for glass, representing a bond strain of some 0.3–0.4. Griffith appreciated that Hooke's law could hardly be assumed to hold at such strain levels, for the force–separation relationship for interatomic bonds surely becomes *nonlinear* immediately prior to rupture. Nor could (1.4), based on the continuum concept of matter, be relied upon to give accurate results on the molecular scale. With due allowance for these factors, Griffith concluded that the limiting cohesive strain was probably in the vicinity of 0.1.

By way of confirmation of his estimate, Griffith consulted the literature for values of the 'intrinsic pressure' of solids (as determined, for instance, from the heat of vaporisation or equation of state). Since both the theoretical strength and intrinsic pressure essentially measure the molecular cohesion, their magnitudes should be comparable, at least for nearly isotropic solids. Griffith determined this to be the case. He thus inferred that the theoretical strength should be a material constant, closely related to the energy of cohesive bonds, with a value of order $E/10$ for all solids.

Thus with both ρ and σ_C effectively predetermined by the molecular structure of the solid, the critical applied tension in the Inglis equation (1.4) becomes dependent on the crack *size*. There is an implication here of an invariant crack-tip structure. The last obstacle to a basic fracture criterion (sect. 1.1) is thereby removed.

1.6 Griffith flaws

The argument in the previous section gave an indication of the strength that could be achieved by an ideal solid, an ultimate target in the fabrication of strong solids. Griffith was intrigued by the fact that the strengths of 'real materials' fell well short of this level, typically by two orders of magnitude, despite great care in maintaining specimen perfection on an optical scale. A further discrepancy was also evident. If a solid were to fail at its theoretical strength the applied stress would reach a maximum at rupture, implying a zero elastic modulus at this point: at such a rupture point a sudden release of stored elastic strain energy, equivalent approximately to the heat of vaporisation, would be expected to manifest itself as an explosive separation of the constituent atoms. Again, real materials behaved differently, parting instead with relatively little kinetic energy on a more or less well-defined separation plane.

Griffith concluded that the typical brittle solid must contain a profusion of submicroscopic *flaws*, microcracks or other centres of heterogeneity too small to be detected by ordinary means. The 'effective length', $c_0 = c_f$, of these so-called 'Griffith flaws' was calculated by inserting the tensile strength of the strongest as-received glass specimen tested (sect. 1.3), $\sigma_F = 170$ MPa, along with the previously measured values of E and γ, into the critical condition (1.11): this gave $c_f \approx 2 \, \mu m$. We may deduce from (1.2) that a molecularly sharp microcrack of this length has a wall separation $2b \simeq 0.05 \, \mu m$, which is about one-tenth of the wavelength of visible light and therefore barely on the limit of optical delectability. The theoretical stress-concentration factor (1.4) is of order 100 in this instance, emphasising the potential weakening power of even the most minute of flaws.

To test his flaw hypothesis Griffith ran a series of experiments on the strength of glass fibres. The fibres were drawn from the same glass as used in the previous tests (sect. 1.3), and were broken either in tension or in bending under a monotonically increasing dead weight. Well-prepared, pristine fibres showed unusually high strengths, shattering in the explosive manner expected of ideal, flawless solids. However, on exposure to laboratory atmosphere all fibres declined steadily in strength, reaching after a few hours a 'steady state' value more typical of ordinary glass specimens. Griffith next tested a large number of such 'aged' fibres with diameters ranging from 1 mm down to 3 μm, and found an apparent *size effect*; the thinner specimens showed a tendency to greater strength. Arguing that a single chain of molecules must possess the theoretical strength (since such a chain could hardly sustain a flaw), he extrapolated his data to molecular dimensions, and once again arrived at a value close to one-tenth of the elastic modulus. Thus in the one series of tests Griffith had demonstrated convincingly not only that sources of weakness exist in the average specimen, but also that these could be avoided if sufficient care and skill were to be exercised in preparation. The production of ultra-high strength optical fibres, in which freshly drawn glass filaments are coated with a protective resin, is a modern exploitation of this principle.

It remained only for Griffith to speculate on the *genesis* of these flaws. He actually rejected the possibility that the flaws might be real microcracks, since the observed decrease in fibre strength with time would require the system energy to increase spontaneously by the amount of surface energy of the crack faces. He also rejected the possibility that the flaws might generate spontaneously by stress-assisted thermal fluctuations, regarding as highly improbable the synchronised rupture of a large number (say 10^8) of neighbouring bonds, except perhaps at temperatures close to the melting point. Griffith considered that the most likely explanation lay in a highly localised rearrangement of molecules within the glass network, with transformations from the metastable, amorphous state into a higher density, crystalline phase (devitrification). He envisaged sheet-like units with an associated internal field capable of nucleating full-scale fractures. As we shall see later, Griffith's speculations on the origin and nature of flaws have largely been superseded. The basic notion of the flaw as a source of weakness in a solid has, nevertheless, played a vital part in the historical development of the present-day theory of strength.

1.7 Further considerations

With his energy-balance concept (pertaining to crack *propagation*) and flaw hypothesis (pertaining to crack *initiation*), Griffith had laid a solid foundation for a general theory of fracture. In a second paper in 1924 he developed his ideas still further, giving explicit consideration to the effect of applied *stress state* on the critical fracture conditions, and discoursing on the factors which determine *brittleness*. With regard to stress state, Griffith extended his analysis of sect. 1.3 to the case of a biaxial applied stress field, in which the crack plane is subjected to both normal (tensile *or* compressive) stress and shear stress. Referring once more to the Inglis stress analysis of an elliptical cavity, he argued that the location of the local tensile stress at the near-tip contour, hence the direction of crack extension, will rotate away from the major axis of the ellipse as the shear component increases. Conclusions concerning the crack path and critical applied loading could then be drawn. A somewhat surprising result of the analysis is that the crack tip may develop high tensile stresses even when both principal stresses of the applied field are compressive, *provided* these principal stresses are unequal. This concept has been developed most strongly in rock mechanics, where compressive stress states are the norm.

As to the question of brittleness, Griffith could but touch on the complications that were apparent in the fracture of many different material types. In many structural steels, for instance, the incidence of plastic flow prior to or during rupture was known to have a profound effect on the strength, but there seemed no way of reconciling this essentially irreversible behaviour with the energy-balance model. It will be recalled that Griffith had based his original model on the notion of an 'ideally' brittle solid in which the creation of new fracture surface by the conservative rupture of cohesive bonds constitutes the sole mode of mechanical energy absorption. In 'real materials', however, irreversible processes inevitably accompany crack growth, and a substantially greater amount of mechanical energy may be consumed in the process of separating the material. Thus it was recognised that different materials might exhibit different 'degrees of brittleness'. A theoretical understanding of this factor remained an important and difficult problem for future researchers.

What follows in the subsequent chapters is the logical extension of the theory of brittle fracture from the fundamental concepts expounded by Griffith.

2

Continuum aspects of crack propagation I: linear elastic crack-tip field

The Griffith study usefully identifies two distinct stages in crack evolution, *initiation* and *propagation*. Of these, initiation is by far the less amenable to systematic analysis, governed as it invariably is by complex (and often ill-defined) local nucleation forces that describe the flaw state. Accordingly, we defer investigation of crack initiation to chapter 9. A crack is deemed to have entered the propagation stage when it has outgrown the zone of influence of its nucleating forces. The term 'propagation' is not necessarily to imply departure from an equilibrium state: indeed, for the present we shall concern ourselves *exclusively* with equilibrium crack propagation. Usually (although not always), a single 'well-developed' crack, by relieving the stress field on neighbouring nucleation centres, propagates from a 'dominant flaw' at the expense of its potential competitors. In the construction of experimental test specimens for studying propagation mechanics such a well-developed crack may be artificially induced, e.g. by machining a surface notch. This pervasive notion of a well-developed crack, taken in conjunction with the fundamental Griffith energy-balance concept, provides us with the starting point for a powerful analytical tool called *fracture mechanics*, the many facets of which will become manifest in the remaining chapters.

The formulation of fracture mechanics began with Irwin and his associates round about 1950. The impetus for the development of this discipline originally came from the increasing demand for more reliable safety criteria in engineering design. In more recent times there has been a growing trend toward a 'materials science' perspective, where fracture mechanics is used to provide insight into the fundamental processes of fracture themselves, at the microstructural and atomic levels. This trend has been especially evident in the current surge toward stronger and tougher *ceramic* materials. The Irwin formulation, couched in the

16

continuum view of matter, retains the macroscopic, or thermodynamic, view of crack propagation. It embraces two major needs:

(i) For the routine analysis of a wide range of crack-loading geometries the Griffith concept needs to be placed within a more general theoretical framework. The requirement is for functional quantities that characterise the driving force for fracture. Of these quantities, *mechanical-energy-release rate G* and *stress-intensity factor K*, with certain properties of linear superposability, stand pre-eminent in present-day formalisms.

(ii) A methodology for dealing with the complexity of *stability* conditions that define the nature of crack equilibria is required. We have seen in chapter 1 how equilibrium cracks can be energetically stable as well as unstable. Many important crack systems pass through a sequence of different equilibrium states in their propagation to ultimate failure. A complete description of stability includes consideration of path, in addition to energetics, of fracture.

The basic principles underlying the above two elements of fracture mechanics have received insufficient attention from the materials community.

Accordingly, in the present chapter we shall outline these principles, in the strict thermodynamical context of linearity and reversibility laid down by Griffith. In so doing we shall bypass detailed consideration of the nonlinear, dissipative terms that inevitably come into play when dealing with fundamental crack-tip separation processes in 'real materials'. The means for incorporating such material-specific terms into appropriate fracture resistance parameters (analogous to Griffith's surface energy) will be discussed in chapter 3.

2.1 Continuum approach to crack equilibrium: crack system as thermodynamic cycle

Let us begin by restating Griffith's thermodynamic concept of crack equilibrium in broader terms. Reconsider the plane-crack system of fig. 1.4. The solid is an isotropic linear elastic continuum, loaded arbitrarily at its outer boundary, and the crack is formed from an infinitesimally narrow slit. For a specified crack length the problem reduces to a formal exercise

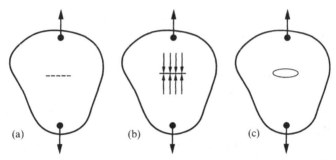

Fig. 2.1. Reversible crack cycle, (a) → (b) → (c) → (b) → (a). Mechanical energy released in crack formation is determined by prior stresses on separation plane.

in elasticity theory, in which solutions may be found for the stress and strain (or displacement) fields in the loaded solid. The question then arises as to how these fields, particularly in the vicinity of the crack tip, determine the energetics of crack propagation. Here the Inglis analysis (sect. 1.1) provides some foresight: the *intensity* of the field is largely determined by the outer boundary conditions (applied loading configuration), the *distribution* by the inner boundary conditions (stress-free crack walls).

Our approach is to treat the energetics of crack propagation in terms of an operational, hypothetical opening and closing cycle. There are two ways in which such a cycle may be conceived. One is to consider the formation of the entire crack from the initially intact body (as done effectively by Griffith and Obreimoff in sects. 1.3 and 1.4). The other way is to consider an incremental extension of an existing crack. It will be implicit in our constructions, consistent with the Griffith thesis, that the processes which determine the mechanical and surface energies operate independently of each other. While this may appear to be a trivial point, we will find cause in later chapters to question the decoupling of energy terms.

The first kind of opening and closing cycle, although not explicitly part of the Irwin scheme, deserves attention if only because of its insistence that the mechanical energy term U_M in (1.5) is determined uniquely by the stresses in the loaded solid *prior* to cracking. At first sight this insistence may seem untenable, for it certainly can be argued that the progress of a crack must be determined by the highly modified stress state at the instant of extension. But the correspondence between crack energetics and prior stresses can be unequivocally demonstrated by the sequence in fig. 2.1. We start with the crack-free state (a), for which it is presumed the elastic field is known. Suppose now that we make an infinitesimally narrow cut along the ultimate crack plane, and impose tractions equal and opposite to the prior stresses there to maintain the system in equilibrium. This operation takes us to state (b), and the only energy involved thus far is the amount U_s supplied by the cutting process in creating new fracture surfaces. We now relax the imposed tractions to zero (slowly, to avoid kinetic energy terms), applying constraints at the crack ends to prevent further extension. The resulting configuration is the equilibrium crack (c), and the mechanical energy released in achieving this state is precisely U_M. At this point the process is reversed. The tractions are re-applied, starting from zero and increasing linearly until the crack is closed again over its whole area. Since the elastic system is conservative the *final* stress state must be identical to the *prior* stress state (b). Thus the mechanical energy decrease associated with crack formation may, within the limits of Hooke's law, be expressed as an integral over the crack area of prior stresses multiplied by crack-wall displacements. Since the displacements are themselves related linearly through the elasticity equations to the crack-surface tractions, the prior stress distribution must uniquely determine the crack energetics. The final stage of the cycle merely involves a healing operation to recover the surface energy, and the removal of the imposed tractions to restore state (a).

It is worth emphasising once more the implications of the above result: the entire propagation history of a crack is *predestined* by the existing stress state before fracture has even begun. Thus in many cases all that is needed to describe the fracture behaviour of an apparently complex system is a standard stress analysis of the system in its uncracked state. This principle will prove useful when we consider specific crack systems in sect. 2.5.

The second kind of cycle, that involving the extension and closure over a small *increment* of slit-crack area, makes use of detailed linear elasticity solutions for the field at the tip of an existing crack. The presence of the crack assuredly complicates the elasticity analysis, but there is a certain universality in the near-tip solutions (foreshadowed in our allusion above to the Inglis analysis) that makes this an especially attractive route. It is the element of universality that is the key to the innate power of Irwin's fracture mechanics.

We shall return later (sect. 2.4) to this second application of the reversibility argument to incorporate the Griffith concept into our generalised description. At this point we turn our attention to specific details of fracture mechanics terminology.

2.2 Mechanical-energy-release rate, *G*

Consider now the elemental crack system of fig. 2.2. The body contains a slit of length *c*, the walls of which are traction-free. Consider the lower end to be rigidly fixed, the upper end to be loaded with a tensile point force *P*. If workless constraints are imposed at the ends of the slit to prevent extension the specimen will behave as an equilibrium elastic spring in accordance with Hooke's law

$$u_0 = \lambda P \tag{2.1}$$

where u_0 is the load-point displacement and $\lambda = \lambda(c)$ is the elastic compliance. The strain energy in the system is equal to the work of elastic loading;

$$U_E = \int_0^{u_0} P(u_0)\,du_0$$
$$= \tfrac{1}{2}Pu_0 = \tfrac{1}{2}P^2\lambda = \tfrac{1}{2}u_0^2/\lambda. \tag{2.2}$$

Now suppose, with the body maintained in a loaded configuration, we release our end constraints on the slit and allow incremental extensions through d*c*. We should expect the compliance to increase. To show this formally we differentiate (2.1), thus;

$$du_0 = \lambda\,dP + d\lambda\,P, \tag{2.3}$$

so that for $du_0 \geqslant 0$, $dP \leqslant 0$ (general loading conditions for d*c* > 0) we have $d\lambda \geqslant 0$ always. At the same time we should expect the composite mechanical energy term $U_M = U_E + U_A$ to decrease (sect. 1.2). It is convenient to consider two extreme loading configurations:

(i) *Constant force* ('dead-weight' loading). The applied force remains constant as the crack extends. At *P* = const the change in potential energy of the loading system, i.e. the negative of the work associated with the load-point displacement, is determinable from (2.3) as

$$dU_A = -P\,du_0 = -P^2\,d\lambda, \tag{2.4a}$$

and the corresponding change in elastic strain energy from (2.2) and (2.3) as

$$dU_E = \tfrac{1}{2}P^2\,d\lambda. \tag{2.4b}$$

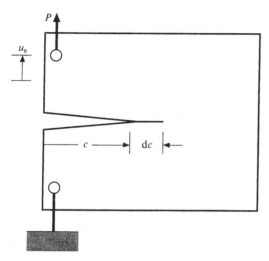

Fig. 2.2. Simple specimen for defining mechanical-energy-release rate. Applied point load P displaces through u_0 during crack formation c, increasing system compliance.

The total mechanical energy change $dU_M = dU_E + dU_A$ is therefore

$$dU_M = -\tfrac{1}{2}P^2\,d\lambda. \tag{2.5}$$

(ii) *Constant displacement* ('fixed-grips' loading). The applied loading system suffers zero displacement as the crack extends. At $u_0 = $ const the energy changes are

$$dU_A = 0 \tag{2.6a}$$

$$dU_E = -\tfrac{1}{2}(u_0^2/\lambda^2)\,d\lambda = -\tfrac{1}{2}P^2\,d\lambda, \tag{2.6b}$$

again using (2.2) to compute the strain energy term. This gives

$$dU_M = -\tfrac{1}{2}P^2\,d\lambda. \tag{2.7}$$

We see that (2.5) and (2.7) are identical: that is, *the mechanical energy released during incremental crack extension is independent of loading configuration.* We leave it to the reader to prove this result for the more complex case in which neither P nor u_0 are held constant.

We have considered only one particular specimen configuration here, that of loading at a point, but a more rigorous analysis shows our conclusion to be quite general. It is accordingly convenient to define a quantity called the *mechanical-energy-release rate*,[1]

$$G = -dU_M/dC \qquad (2.8a)$$

with C the crack interfacial *area*. Observe that G has the dimensions of energy per unit area, as befits our ultimate goal of reconciling the crack energetics with surface energy. For the special case of a straight crack, where length c is sufficient to define the crack area, (2.8a) may be reduced to an alternative, more common (but more restrictive) form

$$G = -dU_M/dc \qquad (2.8b)$$

per unit width of crack front. Thus, G may also be regarded as a generalised line force, in analogy to a surface tension. Since G does not depend on the loading type, we may confine our attention to the constant-displacement configuration without loss of generality. Equation (2.8b) then reduces to

$$G = -(\partial U_E/\partial c)_{u_0}, \qquad (2.9)$$

which defines the (fixed-grips) *strain-energy-release rate* per unit width of crack front. It should be noted that, notwithstanding our references to surface energy and tension, the definitions (2.8) and (2.9) have been made without specifying any criterion for crack extension.

   The above analysis also provides us with a means for determining G experimentally. We can write

$$G = \tfrac{1}{2}P^2\,d\lambda/dc, \quad (P = \text{const}) \qquad (2.10a)$$

$$G = \tfrac{1}{2}(u_0^2/\lambda^2)\,d\lambda/dc, \quad (u_0 = \text{const}). \qquad (2.10b)$$

Given a suitable load–displacement (P–u_0) monitor, we may use (2.1) to obtain an empirical compliance calibration $\lambda(c)$ over any specified range of crack size. It is interesting that while the elastic strain energy *decreases* with crack extension at $u_0 = $ const in (2.6b), it actually *increases* at $P = $ const in (2.4b): whereas the release of strain energy drives the crack in fixed-grips loading, it is the reduction in potential energy of the applied loading that drives the crack in dead-weight loading. This is manifested in (2.10) by a divergence in crack response once the system is disturbed by more than an incremental amount from an initial equilibrium loading:

---

[1] *Rate* relative to spatial crack coordinate, area or length, *not* time.

recalling from (2.3) that $\lambda(c)$ is always an increasing function, we conclude that $G(u_0 = \text{const})$ will generally diminish relative to $G(P = \text{const})$ as the crack extends. Fixed-grips loading will thus always produce the more stable configuration.

However, more powerful, analytical methods are available for evaluating G. Many of these follow from the crack-field stress analysis below.

2.3 Crack-tip field and stress-intensity factor, *K*

2.3.1 Modes of crack propagation

In proceeding to a continuum stress analysis for plane-cracks it is useful to distinguish three basic 'modes' of crack-surface displacement, as in fig. 2.3. Mode I (*opening* mode) corresponds to normal separation of the crack walls under the action of tensile stresses; mode II (*sliding* mode) corresponds to longitudinal shearing of the crack walls in a direction normal to the crack front; mode III (*tearing* mode) corresponds to lateral shearing parallel to the crack front. Extensions in the shear modes II and III bear a certain analogy to the respective glide motions of edge and screw dislocations.

Of the three modes, the first is by far the most pertinent to crack propagation in highly brittle solids. As we shall see in sect. 2.8, there is always a tendency for a brittle crack to seek an orientation that minimises the shear loading. This would appear to be consistent with the picture of crack extension by progressive stretching and rupture of cohesive bonds across the crack plane. Genuine shear fractures do occur, for instance in the constrained propagation of cracks along weak interfaces (e.g. cleavage planes in monocrystals, grain or interphase boundaries in polycrystals) inclined to a major tensile axis, in the rupture of metals and polymers where ductile tearing is favoured, and in rocks where large geological pressures suppress the tensile mode. In most (not all) of the fracture processes to be discussed in this book the role of the shear modes will be subordinate to that of the tensile mode. The use of any fracture mechanics parameter without qualification may accordingly be taken to imply pure mode I loading.

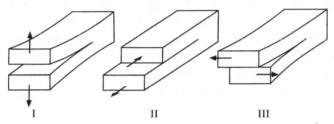

Fig. 2.3. The three modes of fracture: I, opening mode; II, sliding mode; III, tearing mode.

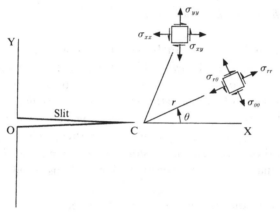

Fig. 2.4. Stress field at Irwin slit-crack tip C, showing rectangular and polar-coordinate components.

2.3.2 Linear elastic crack-tip field

Let us now examine analytical solutions for the stress and displacement fields around the tip of a slit-like plane crack in an ideal Hookean continuum solid. The classic approach to any linear elasticity problem of this sort involves the search for a suitable 'stress function' that satisfies the so-called biharmonic equation (fourth-order differential equation embodying the condition for equilibrium, strain compatibility, and Hooke's law), in accordance with appropriate boundary conditions. The components of stress and displacement are then determined directly from the stress function. For internal cavities of general shape the analysis can be formidable, but for an infinitesimally narrow slit the solutions take on a particularly simple, polar-coordinate form. The first stress-function analyses for such cracks evolved from the work of elasticians like Westergaard and Muskhelishvili, leading to the now-familiar Irwin 'near-field' solutions (for reviews see Irwin 1958, Paris & Sih 1965). It is important to re-emphasise a key assumption here, that the crack walls behind the tip remain free of tractions at all stages of loading.

The Irwin crack-tip solutions are given below for each of the three modes in relation to the coordinate system of fig. 2.4. The K terms are the *stress-intensity factors*, E is Young's modulus, v is Poisson's ratio, and

$$\kappa = (3-v)/(1+v), \quad v' = 0, \quad v'' = v, \quad \text{(plane stress)}$$
$$\kappa = (3-4v), \quad\quad\quad v' = v, \quad v'' = 0, \quad \text{(plane strain)}.$$

Mode I:

$$\begin{Bmatrix} \sigma_{xx} \\ \sigma_{yy} \\ \sigma_{xy} \end{Bmatrix} = \frac{K_{\mathrm{I}}}{(2\pi r)^{1/2}} \begin{Bmatrix} \cos(\theta/2)\,[1 - \sin(\theta/2)\sin(3\theta/2)] \\ \cos(\theta/2)\,[1 + \sin(\theta/2)\sin(3\theta/2)] \\ \sin(\theta/2)\cos(\theta/2)\cos(3\theta/2) \end{Bmatrix}$$

$$\begin{Bmatrix} \sigma_{rr} \\ \sigma_{\theta\theta} \\ \sigma_{r\theta} \end{Bmatrix} = \frac{K_{\mathrm{I}}}{(2\pi r)^{1/2}} \begin{Bmatrix} \cos(\theta/2)\,[1 + \sin^2(\theta/2)] \\ \cos^3(\theta/2) \\ \sin(\theta/2)\cos^2(\theta/2) \end{Bmatrix}$$

$$\sigma_{zz} = v'(\sigma_{xx} + \sigma_{yy}) = v'(\sigma_{rr} + \sigma_{\theta\theta})$$
$$\sigma_{xz} = \sigma_{yz} = \sigma_{rz} = \sigma_{\theta z} = 0$$

$$\begin{Bmatrix} u_x \\ u_y \end{Bmatrix} = \frac{K_{\mathrm{I}}}{2E} \left\{ \frac{r}{2\pi} \right\}^{1/2} \begin{Bmatrix} (1+v)\,[(2\kappa-1)\cos(\theta/2) - \cos(3\theta/2)] \\ (1+v)\,[(2\kappa+1)\sin(\theta/2) - \sin(3\theta/2)] \end{Bmatrix}$$

$$\begin{Bmatrix} u_r \\ u_\theta \end{Bmatrix} = \frac{K_{\mathrm{I}}}{2E} \left\{ \frac{r}{2\pi} \right\}^{1/2} \begin{Bmatrix} (1+v)\,[(2\kappa-1)\cos(\theta/2) - \cos(3\theta/2)] \\ (1+v)\,[-(2\kappa+1)\sin(\theta/2) + \sin(3\theta/2)] \end{Bmatrix}$$

$$u_z = -(v''z/E)(\sigma_{xx} + \sigma_{yy}) = -(v''z/E)(\sigma_{rr} + \sigma_{\theta\theta}). \qquad (2.11)$$

Mode II:

$$\begin{Bmatrix} \sigma_{xx} \\ \sigma_{yy} \\ \sigma_{xy} \end{Bmatrix} = \frac{K_{\mathrm{II}}}{(2\pi r)^{1/2}} \begin{Bmatrix} -\sin(\theta/2)\,[2 + \cos(\theta/2)\cos(3\theta/2)] \\ \sin(\theta/2)\cos(\theta/2)\cos(3\theta/2) \\ \cos(\theta/2)\,[1 - \sin(\theta/2)\sin(3\theta/2)] \end{Bmatrix}$$

$$\begin{Bmatrix} \sigma_{rr} \\ \sigma_{\theta\theta} \\ \sigma_{r\theta} \end{Bmatrix} = \frac{K_{\mathrm{II}}}{(2\pi r)^{1/2}} \begin{Bmatrix} \sin(\theta/2)\,[1 - 3\sin^2(\theta/2)] \\ -3\sin(\theta/2)\cos^2(\theta/2) \\ \cos(\theta/2)\,[1 - 3\sin^2(\theta/2)] \end{Bmatrix}$$

$$\sigma_{zz} = v'(\sigma_{xx} + \sigma_{yy}) = v'(\sigma_{rr} + \sigma_{\theta\theta})$$
$$\sigma_{xz} = \sigma_{yz} = \sigma_{rz} = \sigma_{\theta z} = 0$$

$$\begin{Bmatrix} u_x \\ u_y \end{Bmatrix} = \frac{K_{\mathrm{II}}}{2E} \left\{ \frac{r}{2\pi} \right\}^{1/2} \begin{Bmatrix} (1+v)\,[(2\kappa+3)\sin(\theta/2) + \sin(3\theta/2)] \\ -(1+v)\,[(2\kappa-3)\cos(\theta/2) + \cos(3\theta/2)] \end{Bmatrix}$$

$$\begin{Bmatrix} u_r \\ u_\theta \end{Bmatrix} = \frac{K_{II}}{2E} \left(\frac{r}{2\pi}\right)^{1/2} \begin{Bmatrix} (1+v)[-(2\kappa-1)\sin(\theta/2)+3\sin(3\theta/2)] \\ (1+v)[-(2\kappa+1)\cos(\theta/2)+3\cos(3\theta/2)] \end{Bmatrix}$$

$$u_z = -(v''z/E)(\sigma_{xx}+\sigma_{yy}) = -(v''z/E)(\sigma_{rr}+\sigma_{\theta\theta}). \tag{2.12}$$

Mode III:

$$\sigma_{xx} = \sigma_{yy} = \sigma_{rr} = \sigma_{\theta\theta} = \sigma_{zz} = 0$$

$$\sigma_{xy} = \sigma_{r\theta} = 0$$

$$\begin{Bmatrix} \sigma_{xz} \\ \sigma_{yz} \end{Bmatrix} = \frac{K_{III}}{(2\pi r)^{1/2}} \begin{Bmatrix} -\sin(\theta/2) \\ \cos(\theta/2) \end{Bmatrix}$$

$$\begin{Bmatrix} \sigma_{rz} \\ \sigma_{\theta z} \end{Bmatrix} = \frac{K_{III}}{(2\pi r)^{1/2}} \begin{Bmatrix} \sin(\theta/2) \\ \cos(\theta/2) \end{Bmatrix}$$

$$u_x = u_y = u_r = u_\theta = 0$$

$$u_z = (4K_{III}/E)(r/2\pi)^{1/2}[(1+v)\sin(\theta/2)]. \tag{2.13}$$

There are several corollary points from these solutions that highlight the power of the stress-intensity factor K as a fracture parameter:

(i) The stress and displacement formulas in (2.11)–(2.13) may be reduced to particularly *simple forms*,

$$\sigma_{ij} = K(2\pi r)^{-1/2} f_{ij}(\theta) \tag{2.14a}$$

$$u_i = (K/2E)(r/2\pi)^{1/2} f_i(\theta) \tag{2.14b}$$

in which the cardinal elements of the field appear as separable factors. The K factors depend only on the *outer boundary conditions*, i.e. on the applied loading and specimen geometry (see sect. 2.5), and consequently determine the *intensity* of the local field. The remaining factors depend only on the spatial coordinates about the tip, and determine the *distribution* of the field: these coordinate factors consist of a *radial* component (characteristic $r^{-1/2}$ dependence in the stresses) and an *angular* component (fig. 2.5).

(ii) Details of the applied loading enter *only* through the multiplicative K terms. Thus for any given mode there is an intrinsic (spatial) *invariance* in the near field. This invariance reflects the existence of a *singularity* in the linear stresses and strains at $r = 0$, unavoidably introduced by requiring the continuum slit to be perfectly sharp. We shall take up this anomaly in full in the next chapter. Moreover, higher-order terms need to be included

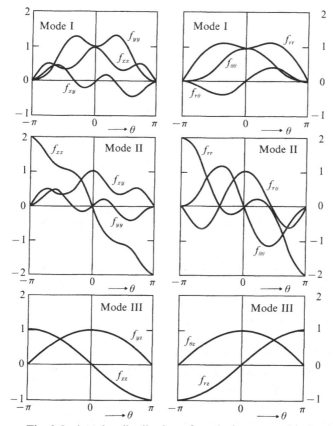

Fig. 2.5. Angular distribution of crack-tip stresses for the three modes. Rectangular components (left) and polar components (right). Note comparable magnitudes of normal and shear components in modes I and II, absence of normal components in mode III.

in the near-field equations if the stresses and displacements are to match the outer boundary conditions. Hence we must be careful not to apply (2.14) at very small or very large distances r from the tip.

(iii) Since the principle of superposition applies to all linear elastic deformations at a point, the invariance in (ii) means that, *for a given mode*, K terms from superposed loadings are *additive*. This result is of far-reaching importance in the analysis of crack systems with complex loadings (sect. 2.7).

(iv) In pure tensile loading, the Irwin crack-opening displacement in the near field is *parabolic* in the crack-interface coordinate $X = c - x$, fig. 2.6.

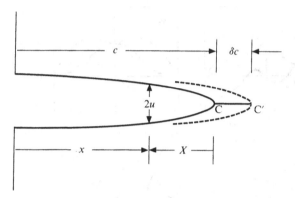

Fig. 2.6. Opening and closure of crack increment CC′ in specimen of unit thickness. Open crack has parabolic profile, in accordance with (2.15).

This is shown by inserting $\theta = \pm\pi$ and $r = X$, $u = u_y$ and $K_{\mathrm{I}} = K$ (mode I), into (2.11);

$$u(X) = (K/E')(8X/\pi)^{1/2}, \quad (X > 0). \tag{2.15}$$

Again, following the point made above in (ii) concerning the singularity at $r = 0$, (2.15) can *not* be taken as physically representative of the crack profile at the actual *tip* (see sect. 3.3).

As with G in sect. 2.2, we point out that our definition of K has been made without recourse to any criterion for crack extension.

A brief word may be added concerning potential complications in the K-field solutions due to inhomogeneity, anisotropy, etc. in the material system. In particular, for systems in which the elastic properties on opposing sides of a plane-crack interface are non-symmetrical, the crack-tip fields will reflect that non-symmetry. Thus, for example, a crack interface between two unlike materials subjected to pure tensile loading will exhibit not only mode I stresses and displacements, but some mode II and III as well, depending on the degree of elastic mismatch. Specific details of such 'cross-term' K-fields have been the subject of much debate over the last decade or so, but it is now unequivocally resolved that the $r^{-1/2}$ singularity remains essentially intact, thereby preserving the super-posability property of stress-intensity factors (Hutchinson 1990).

2.4 Equivalence of *G* and *K* parameters

We are now placed to resume the reversibility argument of sect. 2.1, with special attention to the extension and closure of the crack increment CC' in fig. 2.6. In analogy to fig. 2.1, we may identify the mechanical energy released during the extension half-cycle C → C' with the work done by hypothetically imposed surface tractions during the closure half-cycle C' → C. Evaluation of (2.9) for straight cracks at fixed grips (u = const) is sufficient to provide a general result.

The strain-energy release may therefore be expressed as the following integral over the interfacial crack area immediately behind the tip,

$$\delta U_{\rm E} = 2 \int_{c+\delta c}^{c} \tfrac{1}{2} (\sigma_{yy} u_y + \sigma_{xy} u_x + \sigma_{zy} u_z) \, {\rm d}x, \quad (u = {\rm const}) \qquad (2.16)$$

per unit width of front, the factor 2 arising because of the displacement of the two opposing crack surfaces, the factor $\tfrac{1}{2}$ because of the proportionality between tractions and corresponding displacements. The relevant stresses σ_{ij} are those across CC' prior to extension, i.e. those corresponding to $r = x - c \, (c \leqslant x \leqslant c + \delta c)$, $\theta = 0$; the displacements u_i are those across CC' prior to closure, i.e. those corresponding to $r = c + \delta c - x$, $\theta = \pi$. Making the appropriate substitutions into the field equations (2.11)–(2.13), and proceeding to the limit $\delta c \to 0$, (2.16) reduces to

$$G = -(\partial U_{\rm E}/\partial c)_u = G_{\rm I}(K_{\rm I}) + G_{\rm II}(K_{\rm II}) + G_{\rm III}(K_{\rm III}). \qquad (2.17)$$

Integration then gives

$$G = K_{\rm I}^2/E' + K_{\rm II}^2/E' + K_{\rm III}^2(1+v)/E, \qquad (2.18)$$

recalling (sect. 1.3) that $E' = E$ in plane stress and $E' = E/(1 - v^2)$ in plane strain, with v Poisson's ratio. We see that G terms from superposed loadings *in different modes* are additive.

With G and K thus defined we have a powerful base for quantifying the *driving* forces for the crack. We have yet to consider the cutting and healing sequence which is needed to complete our thermodynamic cycle, and which embodies the *resisting* forces for the crack. This last stage in the formulation will be dealt with in sect. 2.6. We simply note here that the above opening–closing and cutting–healing operations may be effected in a mutually independent manner, i.e. the mechanical and surface energies in the Irwin (as in the Griffith–Inglis) crack are truly decoupled.

This is then an opportune time to examine applications to specific crack systems.

2.5 *G* and *K* for specific crack systems

The crack systems that find common usage in fracture testing are many and varied. Several factors influence the design of a test specimen, among them the nature of the fracture property to be studied, but the features that distinguish one system from another are basically geometrical. In general, the test procedure involves monitoring the response of a well-defined, pre-formed planar crack to controlled applied loading. It is the aim of the fracture mechanics approach to describe the crack response in terms of *G* or *K*, or some other equivalent parameter.

It is not our intent to provide detailed analyses for given specimen geometries here. That is the business of theoretical solid mechanics. We have already alluded to some of the more common approaches: direct measurement, using a compliance calibration for *G* (sect. 2.2); elasticity theory, either (a) inserting an expression for mechanical energy directly into (2.8) or (2.9) for *G* (as in sects. 1.3 and 1.4), or (b) determining *K* factors by the stress-function method used to solve the crack-tip field (sect. 2.3). These and other analytical techniques are adequately described in specialist fracture mechanics handbooks and engineering texts (e.g. Rooke & Cartwright 1976; Tada, Paris & Irwin 1985; Atkins & Mai 1985).

Some of the more important crack systems used in the testing of brittle solids are summarised below. In keeping with our goal of emphasising principles rather than details we present only basic *G* and *K* solutions, ignoring for the most part higher-order terms associated with departures from idealised specimen geometries. The serious practitioner is advised to consult the appropriate testing literature before adopting any specific test geometry for materials fracture evaluation.

2.5.1 *Cracks in uniform applied loading*

The simplest stress state in a continuous elastic body occurs under conditions of uniformly applied loading. Examples of this type of loading for specimens into which planar cracks have been introduced are shown in fig. 2.7 (straight-fronted cracks) and fig. 2.8 (curved-front cracks). The thickness of the specimen is taken to be unity in the two-dimensional configurations, infinity in the three-dimensional configurations.

Example (a) in fig. 2.7 shows an infinite specimen containing an embedded, double-ended straight crack of length $2c$, with uniform applied stresses σ_A in modes I, II and III. The stress-intensity factors are

$$\left.\begin{array}{l} K_I = \psi\sigma_A^I c^{1/2} \\ K_{II} = \psi\sigma_A^{II} c^{1/2} \\ K_{III} = \psi\sigma_A^{III} c^{1/2} \end{array}\right\}. \tag{2.19}$$

Usually, we confine our attention to mode I, in which case (2.19) condenses to a more familiar form,

$$K = \psi\sigma_A c^{1/2}. \tag{2.20}$$

The dimensionless *geometry term* ψ in (2.19) and (2.20) is

$$\psi = \pi^{1/2}, \quad \text{(straight crack, infinite specimen)}. \tag{2.21a}$$

A consistency check may be made by evaluating G from (2.8b), inserting Griffith's energy terms (1.7) and (1.8) (allowing a factor of two for the double-ended crack), and using (2.18) in mode I to determine K_I.

Other crack systems in uniform loading differ only in the numerical factor ψ. In the straight-crack examples (b) and (c) in fig. 2.7, this factor is modified by the presence of outer free surfaces. For the single-ended crack in a semi-infinite specimen, example (b), one obtains, in direct analogy to (2.21a),

$$\psi = \alpha\pi^{1/2}, \quad \text{(edge crack)}, \tag{2.21b}$$

with $\alpha \simeq 1.12$ a simple edge correction factor. For the double-ended crack in a long specimen of finite width $2w$, example (c),

$$\psi(c/w) = [(2w/c)\tan(\pi c/2w)]^{1/2}, \quad \text{(finite specimen)}. \tag{2.21c}$$

Observe in the limit $c \ll w$, $\psi(c/w) \to \pi^{1/2}$, i.e. (2.21c) reverts to (2.21a).

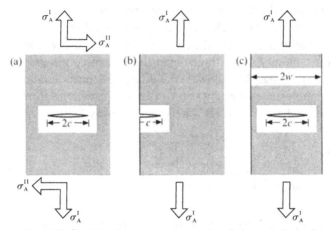

Fig. 2.7. Straight-fronted plane cracks of characteristic dimension c subjected to uniform stresses σ_A: (a) internal crack in infinite specimen (three modes operating, mode III out of plane of diagram), (b) edge crack in semi-infinite specimen (mode I), (c) internal crack in specimen of finite width $2w$ (mode I). The specimen thickness is unity in all cases.

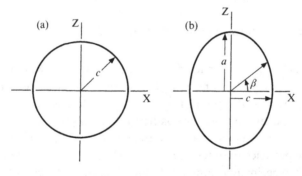

Fig. 2.8. Curved-front plane cracks in infinite body subjected to uniform tension σ_A along Y-axis: (a) penny crack of radius c; (b) crack with elliptical front, semi-axes a, c.

Now consider curved-front cracks, fig. 2.8, in an infinite body. The simplest case is that of a penny crack with radius c, example (a), for which

$$\psi = 2/\pi^{1/2}, \quad \text{(penny crack).} \tag{2.21d}$$

The elliptical crack with semi-axes c and a, example (b), is interesting for the way it highlights the effect of crack-front curvature. K then varies with angular coordinate β according to

$$\psi(a/c, \beta) = \pi^{1/2}[\cos^2\beta + (c/a)^2\sin^2\beta]^{1/2}/E(a/c), \quad \text{(ellipse)},$$
(2.21e)

where $E(a/c)$ is the elliptic integral

$$E(a/c) = \int_0^{\pi/2}[1 - (1 - c^2/a^2)\sin^2\Phi]^{1/2}\,\mathrm{d}\Phi$$

with Φ a dummy variable. A plot of $\psi(a/c, 0)$ is given in fig. 2.9. We note the following special cases: (i) $a/c \to \infty$, corresponding to the straight crack of fig. 2.7(a), $\psi(\infty, 0) = \pi^{1/2}$; (ii) $a/c = 1$, corresponding to the penny crack of fig. 2.8(a), $\psi(1, 0) = 2/\pi^{1/2}$. We may further note from (2.21e) that $\psi(a/c, \pi/2)/\psi(a/c, 0) = (c/a)^{1/2} < 1$ for all $a/c > 1$, from which it can be inferred that the stress-intensity factor will always be greatest at the point where the elliptical crack front intersects the minor axis. The implication here is that a crack free of interference from outer boundaries will tend to extend on a circular front, as in fig. 2.8(a): if, on the other hand, an ever-expanding crack does intersect the free surfaces bounding a specimen of finite thickness, its front will straighten and ultimately tend to the line geometry of fig. 2.7(a).

2.5.2 Cracks in distributed internal loading

Let us turn now to another important class of crack geometry, that of loading at the inner walls. Suppose first that the loads are continuously distributed as internal mirror-symmetric stresses $\sigma_I(x, 0) = \sigma_I(x)$ at the crack plane for *straight* cracks, axial-symmetric stresses $\sigma_I(r, 0) = \sigma_I(r)$ for *penny* cracks, fig. 2.10. For infinite bodies the solutions are:

$$K = 2(c/\pi)^{1/2}\int_0^c [\sigma_I(x)/(c^2 - x^2)^{1/2}]\,\mathrm{d}x, \quad \text{(straight crack)} \quad (2.22a)$$

$$K = [2/(\pi c)^{1/2}]\int_0^c [r\sigma_I(r)/(c^2 - r^2)^{1/2}]\,\mathrm{d}r, \quad \text{(penny crack)}. \quad (2.22b)$$

The quantities $(c^2 - x^2)^{-\frac{1}{2}}$ and $r(c^2 - r^2)^{-\frac{1}{2}}$ are Green's functions, which 'weight' the integrals in favour of the stresses closest to the crack tip.

A special case is that of homogeneously loaded faces, $\sigma_I = \sigma_A = $ const, whence (2.22a) and (2.22b) reduce to $K = \psi\sigma_A c^{1/2}$, i.e. the result of (2.20), with geometrical factors ψ defined as in (2.21a) and (2.21d). Thus the

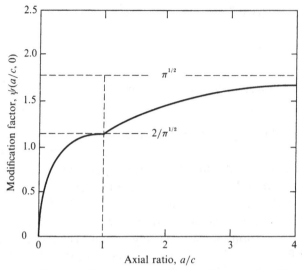

Fig. 2.9. Geometry modification factor at $\beta = 0$ in (2.21e) for elliptical crack as function of ellipticity.

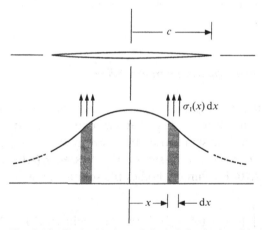

Fig. 2.10. Embedded crack in infinite body with distributed internal normal stresses $\sigma_I(x)$ at crack plane.

driving force for a crack under internal hydrostatic pressure is the same as that for a crack in equivalent external uniform tension. Using the argument of sect. 2.1, this line of reasoning may be extended to infinite bodies in more complex remote tensile loading, whereby K solutions may be written down immediately by identifying σ_I in (2.22) with the associated *prior* distributed stresses across the prospective crack plane.

Now consider another special case of (2.22), that of *concentrated* forces

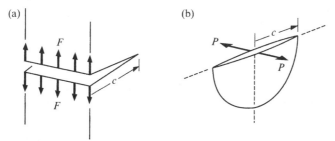

Fig. 2.11. Semi-infinite body with concentrated loading: (a) straight edge crack, line force F per unit length at mouth; (b) surface half-penny crack, point force P at centre.

at the crack walls. In particular, consider the half-crack systems in fig. 2.11: (a) mouth-loaded straight cracks, i.e. line forces $F = \sigma_1(x)\,dx$ per unit length at $x = 0$; (b) centre-loaded penny cracks, i.e. point forces $P = \pi r \sigma_1(r)\,dr$ at $r = 0$. We see that

$$K = 2\alpha F/(\pi c)^{1/2}, \quad \text{(straight edge crack)} \tag{2.23a}$$

$$K = 2\alpha P/(\pi c)^{3/2}, \quad \text{(half-penny crack)} \tag{2.23b}$$

where α is the same kind of edge correction factor as in (2.21b).

2.5.3 Practical crack test geometries

In figs. 2.12–2.14 we illustrate some of the most commonly used fracture test specimens. We re-iterate that departures from the idealised geometries may occur in many if not most instances, so that appropriate higher-order correction factors involving some characteristic specimen dimension will generally need to be incorporated into the solutions.

(i) *Flexure specimens – single-edge-notched beam (SENB) and biaxial flat-on-ring*, fig. 2.12. A single sharp pre-crack of characteristic depth c is cut in a specimen centre face. For *bars and rods*, fig. 2.12(a), loading is preferably in flexible-support four-point flexure, with straight edge (or other) notched pre-crack oriented for maximum tension. The applied stress σ_A in the outer surface (constant within the inner span) is given by thin-beam elasticity theory,

$$\sigma_A = 3Pl/4wd^2. \tag{2.24a}$$

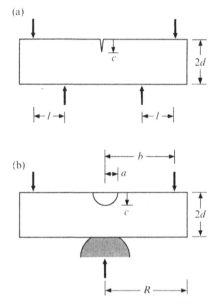

Fig. 2.12. Flexure specimens of thickness $2d$ under load P. (a) Single-edge-notched beam (SENB), four-point support, l outer support span. Straight crack of depth c through specimen width w (not shown, out of plane of diagram). (b) Biaxial flexure disc, upper ring (or distributed point) load on lower ring (or circular flat) support; a and b are inner and outer support radii, R specimen radius. Half-penny crack of radius c located at disc centre.

At $c \ll d$, (2.24a) may be combined with $K = \psi \sigma_A c^{1/2}$ from (2.20), $\psi = \alpha \pi^{1/2}$ in (2.21b), for the K-field. In the limit of zero inner span the system reduces to three-point loading. In more restrictive cases, stress-gradient and notch-radius factors $\psi = \psi(c/d)$ and $\psi = \psi(c/\rho)$ need to be included.

For discs, the geometrical analogue is the biaxial flexure of fig. 2.12(b), circular ring-on-ring support, with half-penny pre-crack. From thin-plate theory,

$$\sigma_A = (3P/16\pi d^2)\{(1+v)[2\ln(b/a)+1] + (1-v)(2b^2-a^2)/2R^2\} \quad (a \gg d). \tag{2.24b}$$

The K-field at $c \ll d$ again derives from (2.20), with $\psi = 2\alpha/\pi^{1/2}$ in (2.21d). Equation (2.24) is the basis of present-day strength testing.

(ii) *Double-cantilever beam specimens* (*DCB*), fig. 2.13. A versatile system is that of the double-cantilever, obtained by symmetrically pre-cracking

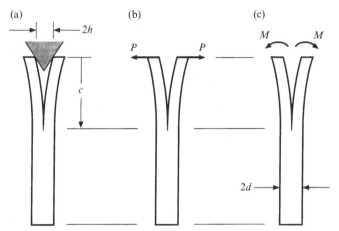

Fig. 2.13. Double-cantilever beam (DCB) test specimens with cracks of length c, width w (not shown, out of plane of diagram) and thickness $2d$, at (a) constant wedging displacement h, (b) constant point-force load P, and (c) constant moment M.

a beam specimen. This specimen may be seen as an elaboration of the Obreimoff arrangement (fig. 1.6). Indeed, a result for constant-displacement loading follows immediately by inserting Obreimoff's energy terms (1.12) and (1.13) into the definition (2.8b) for G (with a factor of two included for the double system). The theory of simple elastic beams ($d \ll c$) allows us to determine the analogous result for constant-force and constant-moment loading. Then

$$
\left.
\begin{aligned}
G &= 3Eh^2 d^3/4c^4, & (h\ \text{const}) \\
&= 12P^2 c^2/Ew^2 d^3, & (P\ \text{const}) \\
&= 12M^2/Ew^2 d^3, & (M\ \text{const})
\end{aligned}
\right\}.
\tag{2.25}
$$

A most interesting aspect of (2.25) is the divergent dependence of G on crack length: at $h = $ const, $G(c)$ is a decreasing function of c (stable); at $P = $ const, $G(c)$ is an increasing function of c (unstable); at $M = $ const, $G(c)$ is invariant (neutral). The relative merits of the first two cases rest with specific test requirements: generally, constant-displacement loading gives greater control over the crack propagation through the length of the specimen. The third configuration is also attractive to test designers: the independence of crack length at $M = $ const eliminates, in principle, all need to monitor the crack length, a distinct advantage in the testing of many brittle materials where the crack front may be difficult to locate. Actually, by suitably tapering the cantilever arms one can achieve constant

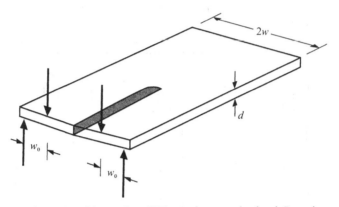

Fig. 2.14. Double-torsion (DT) specimen under load P. w_0 is support span, $2w$ and d beam width and thickness.

G conditions at either fixed h or fixed P, but then at the expense of simplicity in specimen fabrication. Again, it becomes important to pay attention to higher-order effects; end effects can severely restrict the range of crack size over which (2.25) remains valid. For asymmetric beams of non-similar materials, it is necessary only to substitute $1/Ed^3 = \frac{1}{2}(1/E_1 d_1^3 + 1/E_2 d_2^3)$ in (2.25); however, the attendant K solutions now require complex numerical analysis, because the asymmetry generates a component of mode II (Hutchinson 1990; Hutchinson & Suo 1991).

(iii) *Double-torsion specimen (DT)*, fig. 2.14. This is a useful configuration for thin slab specimens. In the approximation of simple plate theory

$$K = [12(1+v)]^{1/2} Pw_0/w^{1/2} d^2. \tag{2.26}$$

Note the absence of crack size in this relationship; the system has the same neutral stability characteristics at $P = \text{const}$ as the double cantilever in constant-moment loading. Once more, end effects restrict the range of crack size over which (2.26) may be applied.

(iv) *Bi-material interfacial-crack specimen*, fig. 2.15. A notch is machined into the thinner member of a bi-material bar, and a crack then made to run along the interface by loading in flexure. The specimen is intriguing because planes parallel to the surfaces and within the inner span of the *uncracked* specimen are, in the approximation of thin-beam theory, stress free. The crack loading arises exclusively from the elastic asymmetry,

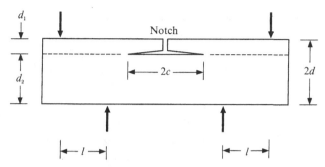

Fig. 2.15. Bi-layer specimen under load P, for measuring fracture properties of interface. l is outer support span. Specimen is of width w (not shown, out of plane of diagram) and composite beam thickness $2d = d_1 + d_2$.

resulting in a mixed-mode I + II field. For cracks well within the inner span, yet larger than the beam thickness, G may be determined directly via calculation of the mechanical energy: thus

$$G = 12\beta P^2 l^2 / E_2 w^2 d^3 \tag{2.27}$$

with coefficient $\beta(d_1, d_2, E_1/E_2, \nu_1/\nu_2)$

$$\beta = \{1/\eta_2^3 + 1/[\eta_2^3 + \eta_1^3/\lambda + 3\eta_1\eta_2/(\eta_1 + \lambda\eta_2)]\}$$
$$\eta_1 = d_1/2d, \quad \eta_2 = d_2/2d, \quad \lambda = (1 - \nu_1^2) E_2 / (1 - \nu_2^2) E_1.$$

Equation (2.27) is independent of c, so fig. 2.15 is another specimen with neutral stability.

Other popular test geometries include compact-tension (CT, compliance-calibrated hybrid of SENB and DCB) and chevron notch (Atkins & Mai 1985). Several new test geometries are being developed for mixed-mode crack geometries in interlayer structures (Hutchinson & Suo 1991). Indentation fracture techniques are discussed in chapter 8.

2.6 Condition for equilibrium fracture: incorporation of the Griffith concept

We turn now to incorporation of the Griffith energy-balance concept into the Irwin fracture mechanics. The extension and closure sequence in the thermodynamic operation of sect. 2.4 provided a formulation of the mechanical energy variation in the crack system. We need only obtain an analogous formulation of the surface energy variation to account for all the terms in the Griffith energy expression (1.5). This brings us to the final stage in the thermodynamic operation, the cutting and healing sequence required to complete the cycle.

Consider once more the crack element CC′ in fig. 2.6. Relative to crack area *C*, the energetics of the cutting–healing sequence is determined as the reversible work to separate the interfacial element against the fundamental intersurface forces,

$$dU_S = R_0 \, dC \tag{2.28}$$

where R_0 is the intrinsic work rate per unit area,

$$R_0 = +dU_S/dC, \tag{2.29a}$$

which *opposes* extension (i.e. R_0 positive). For straight cracks (2.29a) may be alternatively written

$$R_0 = +dU_S/dc \tag{2.29b}$$

per unit width of crack front. In analogy to *G* in (2.8b), R_0 may be interpreted as a surface tension.

The advantage of the above formulation lies in its applicability to a wide range of surface and interfacial fracture configurations. Griffith only treated the fracture of a virgin homogeneous solid. We remove this restriction by identifying R_0 with the *Dupré work of adhesion W* per unit interfacial area (Adamson 1982, Maugis 1985). Thus, for *in-vacuo* separation against cohesive interplanar forces between two *like* half-bodies (B), the Dupré work term identifies with Griffith's intrinsic surface energy,

$$R_0 = W_{BB} = 2\gamma_B, \quad \text{(cohesion)}. \tag{2.30a}$$

For the corresponding separation against interplanar forces between two *unlike*, but *coherent*, half-bodies (A–B), it identifies with

$$R_0 = W_{AB} = \gamma_A + \gamma_B - \gamma_{AB}, \quad \text{(adhesion)} \tag{2.30b}$$

recalling that all γ quantities are defined relative to initial cohesion (B–B or A–A) states. We shall encounter further *W* terms when we consider chemistry in chapters 5, 6 and 8.

Writing (1.5) in differential form then leads to an energy-balance criterion for crack extension:

$$\begin{aligned} dU &= dU_{\mathrm{M}} + dU_{\mathrm{s}} \\ &= -G\,dC + R_0\,dC \\ &= -\mathscr{g}\,dC, \end{aligned} \tag{2.31}$$

using (2.8a) and (2.29a). The quantity $\mathscr{g} = -dU/dC = G - R_0$ defines the net *crack-extension force*, or 'motive'. At Griffith equilibrium in (1.6), we put $\mathscr{g} = 0$, $G = G_{\mathrm{c}}$ (or $K = K_{\mathrm{c}}$), and use (2.18) for mode I to obtain

$$G_{\mathrm{c}} = K_{\mathrm{C}}^2/E' = R_0. \tag{2.32}$$

Then at $G_{\mathrm{c}} > R_0$ the crack extends, at $G_{\mathrm{c}} < R_0$ it retracts.

The above formulation establishes the framework for a general approach to the mechanics of fracture. However, (2.32) should be applied with caution. First, in identifying a measured value of G_{c} or K_{c} with an inherent surface or interface energy, one must ensure that the critical test configuration represents true equilibrium. An apparently static crack may in fact be far removed from an equilibrium state, and may simply be approaching this state at an immeasurably slow rate (chapter 5). Again, in our treatment we have neglected material anisotropy; many classes of brittle solids, e.g. single crystals, exhibit a high degree of anisotropy. A more exact formulation would require the anisotropy in both elastic constants (thus in G and K) and surface energies to be taken into account. We have also neglected microstructure which, as we shall see (chapter 7), can profoundly influence the crack resistance. Finally, there is the ever-present problem of the singularity in the crack field solutions of sect. 2.3: we have seen how the stresses and strains approach infinity at the crack tip, the very region where the critical bond-rupture processes are assumed to operate. In its present form the Irwin theory is incapable of accounting for the actual *mechanisms* of fracture.

2.7 Crack stability and additivity of *K*-fields

Several allusions have been made earlier in this and the first chapter to the different kinds of crack equilibrium, i.e. unstable or stable (or even neutral). The issue of energetic *stability* is important, for it determines whether the ensuing crack propagation is 'catastrophic' or 'controlled'.

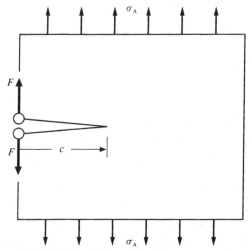

Fig. 2.16. Straight crack with superposed remote uniform stresses σ_A and mouth-opening line forces F.

It is a widely held misconception that $G = G_c$ or $K = K_c$ constitutes a condition for 'failure'. This condition is certainly necessary, but it is not sufficient. For instability, it is not enough that the energy $U(C)$ has an extreme value $(\mathrm{d}U/\mathrm{d}C = 0)$; that value must also be a maximum $(\mathrm{d}^2U/\mathrm{d}C^2 < 0)$. Thus, insofar as R_0 in (2.31) remains a material constant, the stability requirements may be conveniently expressed in terms of $G(C)$ or, from (2.18), $K(C)$:

$$\mathrm{d}G/\mathrm{d}C > 0, \quad \mathrm{d}K/\mathrm{d}C > 0, \quad \text{(unstable)} \tag{2.33a}$$

$$\mathrm{d}G/\mathrm{d}C < 0, \quad \mathrm{d}K/\mathrm{d}C < 0, \quad \text{(stable)}. \tag{2.33b}$$

Barenblatt (1962) and Gurney (Gurney & Hunt 1967) were among the first to emphasise the importance of such relations in the description of equilibrium crack propagation. We shall have occasion to make extensive use of stability criteria in chapters 7–10 (although in a form suitably modified in chapter 3 to allow for shielding contributions to the crack resistance).

In fact, not even (2.33a) guarantees complete failure: an unstable crack may ultimately 'arrest' at some remote, positive-gradient branch of the $U(C)$ curve ('pop-in'). Further increase in the applied loading may then lead to a second, catastrophic instability configuration. Such is often the case with cracks around contacts (chapter 8) and inclusions (chapter 9).

It is instructive to illustrate the principles stated in (2.33) with a contrived crack configuration that serves also to demonstrate the *additivity of K-fields*. Consider the mode I straight half-crack of length c subjected to combined uniform applied stresses σ_A and mouth-opening line forces F in fig. 2.16. Then from (2.20), (2.21b) and (2.23a) we may write

$$K = K_A + K_F = \alpha\sigma_A (\pi c)^{1/2} + 2\alpha F/(\pi c)^{1/2}. \tag{2.34}$$

The function $K(c)$ is plotted in fig. 2.17 for two values of σ_A at fixed F; the use of logarithmic coordinates is simply to highlight the dominance of each power-law term (asymptotic dashed lines of slopes $\frac{1}{2}$ and $-\frac{1}{2}$) in the large and small crack-size regions. The curve thus has two branches, with an intervening minimum.

Now let us investigate the requirements for failure. The equilibrium condition (2.32) is represented in fig. 2.17 as the horizontal dashed line $K = K_C$. Suppose the crack is initially stationary at $\sigma_A = 0$, $c_I = (1/\pi) (2\alpha F/K_C)^2$. For $\sigma_M > \sigma_A > 0$ (lower curve) the $K = K_C$ line intersects the solid curve at two points: at $c = c'_I > c_I$, corresponding to *stable* equilibrium ($dK/dc < 0$); and at $c = c_F$, corresponding to *unstable* equilibrium ($dK/dc > 0$). Further increments in σ_A cause c'_I to expand stably at $K = K_C$, and c_F simultaneously to contract, until ultimately $c = c'_I = c_F = c_M$. At this point the minimum in $K(c)$ is coincident with the equilibrium line, so the system is on the verge of instability (upper curve). The critical load $\sigma_A = \sigma_M$ at $F = \text{const}$ is then evaluated by invoking $dK/dc = 0$ at $K = K_C$:

$$c_M = (4\alpha F/\pi^{1/2}K_C)^2 \tag{2.35a}$$

$$\sigma_M = 2F/\pi c_M. \tag{2.35b}$$

We see that the K_F term in (2.34), by virtue of its inverse c dependence, stabilises the crack growth. An interesting consequence is that the critical configuration condition retains no 'memory' of the initial crack size: c_I does not appear in (2.35). We shall see later that such stabilising influences can have profound implications in flaw mechanics.

A final point may be raised concerning the calculation of G for systems in superposed loading. For the mode I crack system described by (2.34) we have, from (2.18),

$$G = (K_A^2 + K_F^2 + 2K_A K_F)/E'. \tag{2.36}$$

Thus, as implied in sect. 2.4, the mechanical-energy-release rates that would obtain if the applied or mouth-opening loads were to operate

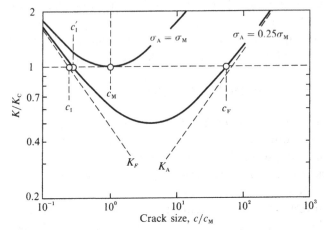

Fig. 2.17. Reduced plot of $K(c)$ in (2.34) for system of fig. 2.16. Inclined dashed lines are asymptotic K_F and K_A fields, horizontal dashed line is equilibrium condition $K = K_C$. Function plotted at fixed F, for $\sigma_A = 0.25\sigma_M$ and $\sigma_A = \sigma_M$ (failure configuration).

individually ($G_A = K_A^2/E'$ and $G_F = K_F^2/E'$) are *not* additive: (2.36) contains cross terms. The cross terms appear because the displacements resulting from each loading system contribute to the mechanical work done by the other loading system.

2.8 Crack paths

We turn now from the question of stability in the crack energetics to stability in the crack *path*. Thus far it has been tacitly assumed that cracks tend to extend in their own plane; and, further, that this plane tends to be that which experiences the maximum tensile component in the applied loading. These presumptions have long formed an intuitive basis for explaining the path of fracture in simple crack systems. In practice, however, cracks can deviate from their original growth plane, sometimes quite abruptly, and can also propagate through regions of substantial applied shear loading.

The crack-path problem can be considered within the framework of the thermodynamic fracture criterion outlined in sect. 2.6. Suppose we start with a simple plane-strain crack of length c in an isotropic homogeneous solid and then subject this crack to mixed-mode loading, as in fig. 2.18. The question arises as to the orientation of the incremental extension dc. We

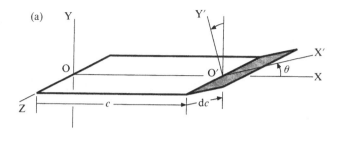

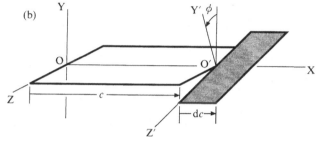

Fig. 2.18. Model for out-of-plane crack extension: (a) 'tilt' configuration, (b) 'twist' configuration.

propose that the favoured orientation be that which maximises the decrease in total system free energy. In the context of (2.31), this corresponds to a maximisation of the motive force, $\mathscr{G} = G - R_0$. For an isotropic system (R_0 independent of θ, ϕ) the problem reduces to one of seeking a maximum in G.

We should be aware at the outset that our formulation of the crack-deflection problem in relation to an initially planar system is mathematically restrictive. For once a crack extends beyond the first out-of-plane increment we are faced with a complex linked configuration for which there are no simple analytical solutions (Cotterell & Rice 1980). Nevertheless, valuable physical insight into the general behaviour can be gained from the plane-crack solutions, so we confine ourselves to the simplistic examples of fig. 2.18.

It is accordingly sufficient for our present purposes to treat only the two special cases illustrated: (a) a rotation θ of the crack plane about OZ ('tilt' configuration); and (b) a rotation ϕ of the crack plane about OX ('twist' configuration).[2] The stresses acting on the proposed plane of incremental prolongation can be determined from the standard solutions (2.11)–(2.13)

---

[2] A third angular rotation, about OY, leaves the crack plane unchanged.

by an appropriate tensor transformation, and a conventional fracture mechanics analysis thereby carried out. We demonstrate the procedure, after Gell & Smith (1967), for the following load states:

(i) *Pure mode I loading.* Consider first the θ-rotation of fig. 2.18(a). The pertinent normal and shear stress components on the new plane are given by the transformation relations

$$
\left.
\begin{aligned}
\sigma_{y'y'} &= \sigma^{\mathrm{I}}_{\theta\theta} = [K_{\mathrm{I}}/(2\pi r)^{1/2}]f^{\mathrm{I}}_{\theta\theta} = K'_{\mathrm{I}}(\theta)/(2\pi r)^{1/2} \\
\sigma_{x'y'} &= \sigma^{\mathrm{I}}_{r\theta} = [K_{\mathrm{I}}/(2\pi r)^{1/2}]f^{\mathrm{I}}_{r\theta} = K'_{\mathrm{II}}(\theta)/(2\pi r)^{1/2} \\
\sigma_{x'z'} &= \sigma^{\mathrm{I}}_{rz} = 0 = K'_{\mathrm{III}}(\theta)/(2\pi r)^{1/2}
\end{aligned}
\right\}
\tag{2.37}
$$

where the f^{I}_{ij} terms are directly obtainable from (2.11), and the 'transformed stress-intensity factors'

$$
\left.
\begin{aligned}
K'_{\mathrm{I}}(\theta) &= K_{\mathrm{I}}f^{\mathrm{I}}_{\theta\theta} \\
K'_{\mathrm{II}}(\theta) &= K_{\mathrm{I}}f^{\mathrm{I}}_{r\theta} \\
K'_{\mathrm{III}}(\theta) &= 0
\end{aligned}
\right\}
\tag{2.38}
$$

define the field for the modified crack. The angular variation of the mechanical-energy-release rate may now be determined from the plane-strain version of (2.18);

$$
G(\theta) = K'^{2}_{\mathrm{I}}(\theta)(1-v^2)/E + K'^{2}_{\mathrm{II}}(\theta)(1-v^2)/E.
\tag{2.39}
$$

Similarly, for the ϕ rotation of fig. 2.18(b), the transformation relations are

$$
\left.
\begin{aligned}
\sigma_{y'y'} &= \sigma^{\mathrm{I}}_{\phi\phi} = [K_{\mathrm{I}}/(2\pi r)^{1/2}]g^{\mathrm{I}}_{\phi\phi} = K'_{\mathrm{I}}(\phi)/(2\pi r)^{1/2} \\
\sigma_{x'y'} &= \sigma^{\mathrm{I}}_{x'\phi} = 0 = K'_{\mathrm{II}}(\phi)/(2\pi r)^{1/2} \\
\sigma_{x'z'} &= \sigma^{\mathrm{I}}_{z'\phi} = [K_{\mathrm{I}}/(2\pi r)^{1/2}]g^{\mathrm{I}}_{z'\phi} = K'_{\mathrm{III}}(\phi)/(2\pi r)^{1/2}
\end{aligned}
\right\}
\tag{2.40}
$$

where the g^{I}_{ij} are calculable from (2.11), and

$$
\left.
\begin{aligned}
K'_{\mathrm{I}}(\phi) &= K_{\mathrm{I}}g^{\mathrm{I}}_{\phi\phi} \\
K'_{\mathrm{II}}(\phi) &= 0 \\
K'_{\mathrm{III}}(\phi) &= K_{\mathrm{I}}g^{\mathrm{I}}_{z'\phi}
\end{aligned}
\right\}.
\tag{2.41}
$$

Again, from (2.18),

$$
G(\phi) = K'^{2}_{\mathrm{I}}(\phi)(1-v^2)/E + K'^{2}_{\mathrm{III}}(\phi)(1+v)/E.
\tag{2.42}
$$

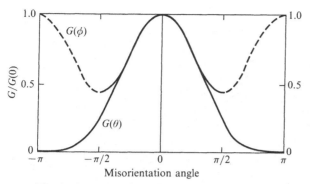

Fig. 2.19. Reduced mechanical-energy-release rate as a function of the misorientation angles θ and ϕ in fig. 2.18, for pure mode I loading. Plane strain assumed, with $\nu = \frac{1}{3}$.

The functions $G(\theta)$ and $G(\phi)$, normalised to values for straight-ahead extension, are plotted in fig. 2.19. From the maxima at $\theta = 0$, $\phi = 0$, we conclude that the crack is favoured to continue in its original plane. Thus the plane crack in mode I loading may be said to have 'directional stability'.

(ii) *Shear mode superposed onto mode I.* Now consider the superposition of mode II or mode III onto the mode I loading of the original crack. The procedure is identical to that above, except that there are additional contributions to the stress components $\sigma_{y'y'}, \sigma_{x'y'}, \sigma_{x'z'}$, in (2.37) from the field of mode II (see (2.12)), and in (2.40) from the field of mode III (see (2.13)). The transformed stress-intensity factors accordingly become

$$
\left.
\begin{aligned}
K'_{\text{I}}(\theta) &= K_{\text{I}} f^{\text{I}}_{\theta\theta} + K_{\text{II}} f^{\text{II}}_{\theta\theta} \\
K'_{\text{II}}(\theta) &= K_{\text{I}} f^{\text{I}}_{r\theta} + K_{\text{II}} f^{\text{II}}_{r\theta} \\
K'_{\text{III}}(\theta) &= 0
\end{aligned}
\right\} \quad \text{(modes I + II)} \tag{2.43}
$$

$$
\left.
\begin{aligned}
K'_{\text{I}}(\phi) &= K_{\text{I}} g^{\text{I}}_{\phi\phi} + K_{\text{III}} g^{\text{III}}_{\phi\phi} \\
K'_{\text{II}}(\phi) &= 0 \\
K'_{\text{III}}(\phi) &= K_{\text{I}} g^{\text{I}}_{z'\phi} + K_{\text{III}} g^{\text{III}}_{z'\phi}
\end{aligned}
\right\} \quad \text{(modes I + III).} \tag{2.44}
$$

The mechanical-energy-release rates $G(\theta)$ and $G(\phi)$ then follow, as before, from (2.39) and (2.42): normalised plots are shown in fig. 2.20. In both cases the action of the imposed shear is to deflect the crack away from plane geometry, i.e. 'directional instability'. Moreover, this deflection always tends toward the orientation of minimum shear loading: in this

(a)

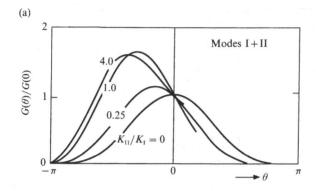

(b)

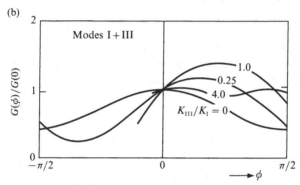

Fig. 2.20. Reduced mechanical-energy-release rate as a function of misorientation angles θ and ϕ, for (a) superposed modes I+II, (b) superposed modes I+III. Plane strain, $\nu = \frac{1}{3}$. Plots cover the range of misorientations over which the local normal stress is tensile.

sense the shear stresses may be seen as 'corrective', restoring deviant cracks to a stable path of orthogonality to the greatest principal tensile stresses in the applied field.

It is instructive to examine some practical manifestations of the above analysis. We take first the observed crack path from the tips of a pre-existing inclined slit in a tensile plate, fig. 2.21. In the orientation shown the original components of mode I and II loading on the slit are comparable. The initial crack extension, at $\theta \approx -50°$, conforms closely to the maximum in $G(\theta)$ in fig. 2.20(a). As the new crack grows out of the near field of the slit ($dc \ll c$) and into the far field of the externally applied loads ($\Delta c > c$) the crack system undergoes a continuous transition from mixed mode to pure mode I. If we were to present fig. 2.21 as a working model of crack

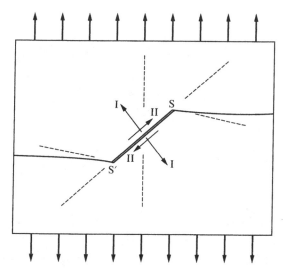

Fig. 2.21. Observed crack paths in brittle epoxy plate, loaded in uniform tension. Arrows indicate resolved components of loading on pre-existing slit SS'. Note that mode I ≈ mode II loading in initial extension, but that mode I ≫ mode II in subsequent (well-developed) extension. (After Erdogan, F. & Sih, G. C. (1963) *J. Basic Eng.* **85** 519.)

development from an incipient flaw, we would conclude that flaw orientation is indeed an important factor in the fracture mechanics of initial extension, but not of subsequent extension beyond the zone of influence of the flaw.

Our second practical example concerns the physical mechanism by which a crack in combined modes I and III loading changes its plane. A consideration of fig. 2.18 shows that while a θ-rotation of the crack plane can be accommodated by a continuous adjustment of the front, a ϕ-rotation can not. In the latter case the crack overcomes the accommodation problem by segmenting into 'partial fronts', separated by steps. This is shown to striking effect in the micrograph in fig. 2.22. Note that the shear disturbance need not be great: in fig. 2.22 the step height is ≈ 1 μm relative to a step separation of ≈ 50 μm (i.e. a rotation of ≈ 1°), corresponding to a shear-to-tension ratio ≈ 2%.

One final point may be made here. We refer to the tendency for brittle cracks in single crystals to propagate along favoured cleavage planes, or in layered structures along weak interfaces. Under such conditions it is strictly the quantity $\mathscr{g}$ that should be plotted in figs. 2.19 and 2.20: the orientation dependence of R_0, as well as of G, must be taken into account.

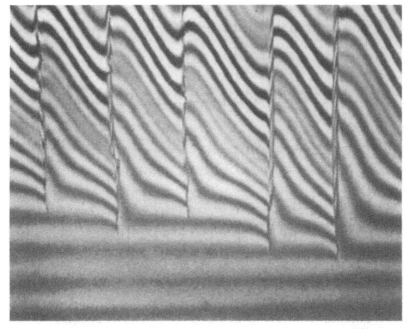

Fig. 2.22. Fracture surface of glass, showing rotation of crack plane, with step formation, for crack experiencing abrupt increase in mode III/mode I ratio. Main crack front propagates from bottom to top. Interference microscopy, white light. Width of field 350 μm. (After Sommer, E. (1969) *Eng. Fract. Mech.* **1** 539.)

In this context it is seen that $G(\theta)$ and $G(\phi)$ in fig. 2.20 do not have sharp maxima (although anisotropy in the elastic constants could enhance these maxima). For single crystals, the issue becomes one of degree of anisotropy in the surface energy: in covalent structures, where the crystallographic dependence of γ is typically modest (e.g. diamond structure, extreme range $\gamma_{111}/\gamma_{100} = 1/\sqrt{3}$), the cleavage tendency is usually slight; in ionic structures, where the crystallographic anisotropy is usually much more pronounced (e.g. mica, sharp minimum at γ_{0001}), the cleavage tendency can be very strong indeed. For polycrystals or layered composites it is the relative values of the interface (grain or interphase boundary) and bulk surface energies that determine the tendency for crack deflection (sect. 7.1).

3

Continuum aspects of crack propagation II: nonlinear crack-tip field

Irwin's generalisation of the Griffith concept, as outlined in the previous chapter, provides us with a powerful tool for handling the mechanics of fracture. In particular, by balancing mechanical energy released against surface energy gained we have a thermodynamically sound criterion for predicting *when* an ideally brittle crack will extend. But the Irwin mechanics tell us nothing as to *how* that crack extends. Whenever we have had occasion to refer to events at the tip the description has proved totally inadequate. The singularities in the *continuum linear elastic* solutions simply cannot be reconciled with any physically realistic local rupture process. Assuming the first law of thermodynamics to be beyond question, it is clear that some vital element is missing in the fracture mechanics: it is not the Griffith energy balance that is at issue here, but rather the mechanism by which this balance is effected.

In this chapter we question the adequacy of the linear continuum representation of matter. Real solids tend to a maximum in the intrinsic stress–strain characteristic. This is true even of perfectly brittle solids which fracture by bond rupture. In addition, crack growth can be accompanied by deformation processes in the near-tip field. Such deformation can be highly dissipative. If there is to be any proper basis for understanding the behaviour of 'tough' ceramics and other brittle materials for structural applications we must include provision for nonlinear and irreversible elements in the equilibrium fracture mechanics.

In acknowledging the potentially large influence of near-field irreversible deformation we make a strong qualifying statement, one that sets the tone for several later topics. It is implicit in our interpretation of a 'brittle solid' that bond rupture always remains the operative mechanism of crack extension. *Near-field deformations may interact with the crack tip via their influence on the stress field, but not alter the fundamental growth criteria defined in the preceding chapter.* This leads us to the important concept of crack-tip *shielding.* Exceptions to this sharp-crack picture most certainly exist, e.g. ductile metals and polymers, in which the tip deformation plays a primary role in the rupture process. We exclude such material types from explicit consideration here simply to avoid being drawn into the complex solid mechanics arena. The interested reader is referred to the engineering fracture literature for further details on this subject, summarised in several texts (Knott 1973; Broek 1982; Atkins & Mai 1985; Hertzberg 1988).

In principle, there seems nothing to preclude extension of the energy-balance concept to nonlinear crack systems. However, nonlinear problems are notoriously difficult to handle conceptually and mathematically. With this in mind we limit ourselves to the most rudimentary treatments. We begin with the so-called *small-scale zone* concept, incorporating dissipative work terms into a composite fracture-surface energy. The models that derive from this concept retain much of the simplicity of continuum-based linear elastic fracture mechanics, yet provide useful new physical insights into the innate nonlinear nature of the brittle crack-tip structure. We highlight the Barenblatt cohesion-zone model for the natural way it removes the singularity problem. Then we introduce a powerful line integral formalism due to Rice as a means of unifying brittle fracture descriptions at the macroscopic and microscopic levels. The treatment in this chapter will lead us to derivative equilibrium parameters, crack-resistance energy R and toughness T, which we shall use to quantify the material resistance to crack propagation.

3.1 Nonlinearity and irreversibility of crack-tip processes

3.1.1 Origin of crack-tip singularity: breakdown of linear-continuum mechanics

The nature of the crack-tip singularity is rooted in two sources: *Hooke's law* and the *continuum approximation.* We touch on the fundamental inadequacies of both here.

The problem stems from our assumption that the initially closed slit-like crack is infinitesimally sharp. Recall from sect. 2.3 that, mathematically, the Irwin crack tip widens into a rounded, parabolic contour (fig. 2.6). It is readily demonstrated (e.g. by differentiating (2.15) to obtain $\mathrm{d}u/\mathrm{d}X$ and proceeding to the limit $X \to 0$) that the strains must tend to infinity as one

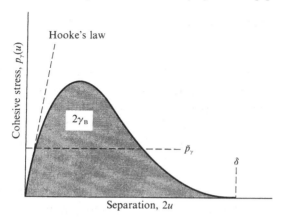

Fig. 3.1. Cohesive stress–separation function for two atom planes in brittle solid. Work to separate planes beyond range $2u = \delta$ defines work of cohesion $R_0 = W_{BB} = 2\gamma_B$ for body B. Horizontal dashed line is an 'averaged' stress, $\bar{p}_\gamma = R_0/\delta$.

approaches the tip. This is simply a manifestation of a law applied beyond its limits of validity: the idealised Hookean solid knows no bounds to elastic strain, i.e. it has 'infinite strength'.

In reality, there are maximum levels of stress that a solid can withstand. Picture our ideal brittle material as a homogeneous, defect-free solid with a regular atomic lattice structure. Consider an elemental volume of material at the crack plane, bounded by adjacent interatomic rows. The stress–extension response of such an element as the crack walls are displaced outward is represented in fig. 3.1 by an interplanar cohesion function, $p_\gamma(u)$, where the subscript γ serves to indicate that we are dealing with intrinsic surface forces. And now a sign convention: we shall here (and hereinafter) regard positive cohesive stress p as indicative of a state of *attraction*, equal and opposite to the countervailing stresses transmitted to the element from the remote applied loading. The stress–separation function is linear initially, but exhibits a maximum as the crack approaches and passes by, ultimately tailing off to zero as the material separates. It is apparent from fig. 3.1 that some kind of 'range' parameter, δ, is necessary to characterise the nonlinear stress–extension response fully. Strictly, this range is infinite, but for practical purposes it cuts off at a few atomic diameters. Observe that even the oversimplistic 'averaged cohesive stress' indicated as the horizontal dashed line of height $\bar{p}_\gamma$ in the figure has better inbuilt provision to describe the essential cutoff feature than does the unbounded Hooke's law.

According to this description the critical region of mechanical energy absorption for an extending, perfectly brittle crack is confined, via the underlying mechanism of bond rupture, to the two interatomic layers that bound the crack plane. Atom layers further from the crack plane deform along the same stress–separation curve, but only partially, and (not-withstanding certain relaxation processes that occur when atomic structures on newly created surfaces rearrange themselves) reversibly, restoring to their intact configuration when the crack moves ahead. The extensions in fig. 3.1 are therefore to be considered relative to a characteristic dimension of one interplanar spacing. Survival of the Irwin–Griffith fracture mechanics of chapter 2 in the context of crack-tip micromechanics ultimately rests with our ability to incorporate this essential atomic-scale range characteristic into the formalism.

The fundamentally nonlinear stress–separation function in fig. 3.1 thereby holds the key to the singularity problem for brittle cracks. It is to be noted that this function can be defined without any reference to a crack system. In principle, we might obtain it from conventional tensile tests on perfect (flaw-free) specimens (e.g. sect. 1.5) or from interatomic potential calculations. Such an independent determination would open up the possibility of a priori predictions of fracture properties, a suggestion explored further in chapter 6.

3.1.2 Extraneous energy dissipation in the crack-tip zone

We have just seen how the stress–extension curve for surfaces initially separated by a lattice spacing determines the intrinsic fracture resistance. The area under this curve defines the Dupré work of adhesion,

$$R_0 = W_{BB} = \int_0^\delta p_\gamma(u)\, d(2u)$$

$$= 2\int_0^{\delta/2} p_\gamma(u)\, du = 2\gamma_B, \tag{3.1}$$

consistent with our definition in (2.30). For the present we confine our interpretation of R_0 to the intrinsic *in-vacuo* cohesion of the bulk solid as defined in (2.30a), but this should not be seen as restrictive (sect. 2.6): more general interpretations will be developed when we consider the modifying influences of chemistry and microstructure in chapters 5–7. Note that R_0 integrates all the essential features of the nonlinear crack-tip response in fig. 3.1, i.e. cutoff stress $\bar{p}_\gamma$ and range δ.

Table 3.1. *Fracture parameters for selected brittle materials.* Group 1, monocrystals (covalent-to-ionic down table, indices denote cleavage plane); 2, polycrystals (pc); 3, ceramic composites; 4, metallics. Data include: experimental Young's modulus E, toughness $K_C = T$ and crack-resistance energy $G_C = R$ (*in vacuo*); theoretical surface energy $W_{BB} = 2\gamma_B$.

Material	Formula	E (GPa)	T (MPa m$^{1/2}$)	R (J m^{-2})	$2\gamma_B$ (J m^{-2})
Diamond	C (111)	1000	4	15	12
Silicon	Si (111)	170	0.7	3.0	2.4
Silicon carbide	SiC (basal)	400	2.5	15	8
Silica	SiO$_2$ (glass)	70	0.75	8.0	2
Sapphire	Al$_2$O$_3$ ($10\bar{1}0$)	400	3	25	8
Mica	KAl$_2$(AlSi$_3$O$_{10}$)OH$_2$ (basal)	170	1.3	10	
Magnesium oxide	MgO (100)	250	0.9	3	3
Lithium fluoride	LiF (100)	90	0.3	0.8	0.6
Alumina	Al$_2$O$_3$ (pc)	400	2–10	10–250	
Zirconia	ZrO$_2$ (pc)	250	3–10	30–400	
Silicon carbide	SiC (pc)	400	3–7	25–125	
Silicon nitride	Si$_3$N$_4$ (pc)	350	4–12	45–400	
Alumina composites		300–400	4–12	40–500	
Zirconia composites		100–250	3–20	30–3000	
Fibre-reinforced ceramic composites		200–400	20–25	1000–3000	
Ductile-dispersion ceramic composites		200–400	10–20	250–2000	
Cement paste		20	0.5	10	
Concrete		30	1–1.5	30–80	
Tungsten carbide	WC/Co	500–600	10–25	300–1000	
Steel	Fe + additives	200	20–100	50–50000	

Values for polycrystalline and composite materials embrace range of T-curve (R-curve).

But this idealistic picture does not always appear to be in accord with the fracture surface energies measured for real solids. Compare in table 3.1 values of experimental mechanical-energy-release rates $G_C = R$ with theoretical surface energies $2\gamma_B$ for a range of materials. One should not attach too much significance to discrepancies of up to a factor of two: experimental uncertainties in G_C arise from failure to account for the complications of specimen fabrication, crystal anisotropy, microstructure, strain-rate and environment; likewise, inadequacies exist in the theoretical calculations of γ_B. The data are nevertheless sufficiently accurate to suggest that some form of energy expenditure other than that associated with mere surface creation must occur in some cases, especially in polycrystals, composites and metallics. Those materials that deviate strongly from the brittle ideal ($G_C \gg 2\gamma_B$) are said to be 'tough'. It hardly bears stating that toughness is a desirable quality for structural materials.

Notwithstanding this additional form of energy expenditure, we re-assert that most of the materials listed in table 3.1 are indeed intrinsically brittle in the sense alluded to in the introductory remarks to this chapter. That is, the cracks remain essentially sharp but may activate secondary energy-absorbing sources within the near field. While such sources may have a strong influence on the crack driving force by 'screening' the tip from the applied loading, they play no part in the fundamental surface separation process. This *shielding* concept, first proposed by Thomson, forms the underlying basis of much of the current fracture mechanics modelling for 'tough' ceramics (chapter 7). Thus as we proceed from group to group down table 3.1 the trend toward increasing toughness may be seen as reflecting a greater activity of secondary dissipative processes; until, for ductile metals, these processes become so pervasive that the cracks lose their intrinsic sharpness and propagate by an altogether different, tearing mode.

3.2 Irwin–Orowan extension of the Griffith concept

A major problem facing early workers in fracture mechanics was how to accommodate the essential elements of nonlinearity and irreversibility into the linear elastic fracture mechanics framework. A significant advance was made by Irwin and Orowan (see Irwin 1958), who independently proposed the mathematical contrivance of dividing the crack system into two *zones*, as in fig. 3.2: the *outer* zone, linear elastic, transmits the applied loadings

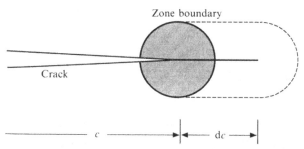

Fig. 3.2. Irwin–Orowan small-scale zone model. Surface-separation processes are confined to a frontal zone (shaded) small compared to crack dimensions. On extension from c (area C) through dc (area dC) the crack traces out a 'wake' deformation layer (region within dashed line).

to the *inner* zone (shaded), where all energy absorption processes (including intrinsic bond rupture) operate. With one major assumption, that the inner zone be negligibly small relative to the outer zone (*small-scale zone* approximation), the critical energy terms may be mathematically decoupled, as before:

(i) The intrinsic, inner zone work of separation is expended within an encompassing K-field, and is therefore insensitive to the nature of the remote loading. In a steady-state extension (where the inner zone configuration simply translates with the tip) this work term is a characteristic material energy parameter.

(ii) The mechanical-energy-release rate for the system is governed by the elastic configuration in the surrounding outer zone, where the bulk of the reversible strain energy resides, and is therefore insensitive to events within the (invariant) inner zone. Under these circumstances linear elastic fracture mechanics may be retained.

Given these features we may modify the Griffith concept in a simple way. Let the crack in fig. 3.2, total surface area C, undergo an incremental extension dC. First, excluding the inner zone as negligibly small, we specify the mechanical work rate dU_M exactly as for the reversible system of sect. 2.2. Second, excluding the outer zone, we specify a surface work rate dU_s associated with separation of the crack walls, where now the dissipative elements contribute to the energetics. Then, in direct analogy to (2.31), the total system energy increment may be written

$$\begin{aligned}
dU &= dU_M + dU_s \\
&= -GdC + RdC \\
&= -\mathscr{g}dC
\end{aligned} \tag{3.2}$$

where we define the mechanical-energy-release rate exactly as in (2.8a),

$$G = -dU_M/dC, \tag{3.3}$$

but replace R_0 in (2.29a) with a generalised *crack-resistance energy*,

$$R = +dU_s/dC. \tag{3.4}$$

Like R_0, R is positive ($dU_s > 0$ for $dC > 0$). The condition for equilibrium is that the crack-extension force $\mathscr{g} = -dU/dC = G - R$ should vanish: i.e. $\mathscr{g} = 0$, $G = G_C$, giving us a slightly modified version of (2.32),

$$G_C = K_C^2/E' = R. \tag{3.5}$$

The crack-resistance energy (fracture-surface energy) R is one of the most commonly used indicators of material toughness. Again, in the limit of zero dissipative component, R identifies with $R_0 = W_{BB} = 2\gamma_B$ for ideally brittle solids.

While Irwin–Orowan is a useful generalisation, it has major limitations. First, the model is strictly applicable only in the extreme of small inner zone size. This requirement is not always met in practice, even in ceramic materials. Second, while establishing a basis for defining a resistance R, it provides no insight into what value R should take. Unlike R_0, R can not be specified a priori; the dissipative component is not well-defined.

Moreover, contrary to widespread assumption, the dissipative energy may *not* be truly independent of the intrinsic surface energy. In this context we note a common expression of the Irwin–Orowan model, $R = 2\gamma_B + R_P$, where R_P is a 'plastic work' term. It is often presumed (specifically for tough metallic and polymeric materials) that R_P is so much larger than $2\gamma_B$ that the latter can be considered to be a disposable quantity. However, such presumption is dangerous: the two terms may in fact be interdependent, so that γ_B appears as a multiplicative rather than additive term in R, thus retaining a controlling influence in the crack energetics. Such coupling is in fact typical of the shielding processes that characterise ideally brittle materials with intrinsically sharp crack tips, as will be seen in chapter 7.

3.3 Barenblatt cohesion-zone model

The most important of the small-scale zone models for intrinsic brittle fracture is due to Barenblatt (1962). It recognises the underlying atomic nature of the fracture process by specifying the resistance in terms of a nonlinear cohesive force function of the type shown in fig. 3.1, yet preserves the continuum basis of the linear fracture mechanics by assuming the forces to be distributed over a sufficiently large zone (relative to atomic dimensions) along the crack plane instead of infinitesimally concentrated along a line (the 'crack front'). It is assumed that the cohesive forces are confined to the crack plane, consistent with the picture in sect. 3.1.1, and that the crack plane is slit-like, in the Irwin sense. An analogous model for plastic solids was developed about the same time by Dugdale (1960), in which Barenblatt's interatomic stresses are replaced by plastic yield stresses. As we shall see, the cohesion-zone approach leads to removal of the singularity. But, in so doing, it introduces a degree of arbitrariness into the mathematical location of the 'crack tip'.

3.3.1 Mechanics of the Barenblatt crack

Consider a continuum slit terminating at a mathematical line along C, fig. 3.3. We transform the nonlinear cohesive stresses $p_y(u)$ for *in-vacuo* separation (fig. 3.1) across the walls of the slit into a smoothly distributed function $p_y(X, 0) = p_y(X)$ over a zone CZ of length $X = X_z = \lambda \ll c \ll L$, with L a characteristic specimen dimension. Across CZ a balance is maintained between these closing stresses and the opening stresses transmitted through the linear material from the remote boundaries. The condition $\lambda \ll c \ll L$ is to meet the requirements of Barenblatt's two 'hypotheses' (Barenblatt 1962):

(i) 'The width of the edge [cohesion] region of a crack is small compared to the size of the whole crack'.

(ii) 'The form of the normal section [profile] of the crack surface in the edge region ... does not depend on the acting [remote] loads and is always the same for a given material under given conditions ...'.

By satisfying these hypotheses we manage to avert some of the complexities

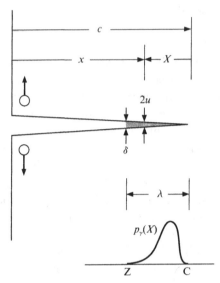

Fig. 3.3. Barenblatt cohesion-zone model. Equilibrium is determined as balance between externally applied loads and inner cohesive-stress function $p_\gamma(X)$ at CZ.

of a generally insoluble nonlinear problem, although we shall see later that the second of the two is somewhat superfluous.

A linear fracture mechanics analysis may then be applied to the system by evaluating both external *and* internal stress contributions as K-fields at the crack tip, noting that we can define stress-intensity factors for each contribution separately. The crux of the Barenblatt thesis is that the two stress-intensity factors represent singularities at C, but that these singularities are of opposite sign, and cancel at equilibrium. This cancellation removes the conceptual difficulties referred to earlier (sect. 3.1.1) in connection with rounded crack tip contours: the restraining effect of the nonlinear zone at CZ is to close the tip into a smooth cusp at C.

To place this description on a quantitative footing, we obtain first an expression for the stress-intensity factor K_0 associated with the cohesive stresses at the crack plane. This is readily done using the general formula (2.22a) for a straight crack in an infinite body. Substituting $\sigma_1(x,0) = -p_\gamma(X)$ over $0 < X \leqslant \lambda$, $X = c-x$, we obtain, in the small-zone limit $\lambda \ll c$,

$$K_0 = -(2/\pi)^{1/2} \int_0^\lambda p_\gamma(X)\,\mathrm{d}X/X^{1/2}. \qquad (3.6)$$

Now the external and internal contributions are superposable, giving a *net* stress-intensity factor

$$\mathcal{K} = K_A + K_0. \tag{3.7}$$

The Barenblatt requirement for equilibrium is that $\mathcal{K} = 0$, i.e. $K_A = -K_0$. Noting that the second term on the right of (3.7) is negative (reflecting its association with *closure* stresses), it is convenient to define an equivalent quantity of opposite sign, $T_0 = -K_0$, indicative of an equilibrium state,

$$T_0 = (2/\pi)^{1/2} \int_0^\lambda p_\gamma(X) \, dX / X^{1/2} \tag{3.8}$$

which we shall refer to simply as the *intrinsic toughness*.[1] The critical crack extension condition $K_A = K_C$ then reduces to

$$K_C = T_0. \tag{3.9}$$

The quantity T_0 in (3.8) is a characteristic material quantity, *independent of crack size c* or of any quantity relating to the form of applied loading or macroscopic specimen geometry, provided the condition $\lambda \ll c \ll L$ remains satisfied.[2] As a measure of intrinsic material resistance to crack propagation it may be used interchangeably with R_0 in (3.1), as is readily demonstrated from the K_C–G_C equivalence relation for mode I cracks in (2.32):

$$T_0 = (E'R_0)^{1/2} \tag{3.10}$$

recalling that $E' = E$ (plane stress) or $E/(1 - v^2)$ (plane strain).

Now let us turn our attention from the crack-tip stress-intensity factor to the crack-opening profile. The contribution from the internal cohesive stresses is (Barenblatt 1962)

$$u_\gamma(X) = -(2/\pi E') \int_0^\lambda p_\gamma(X') \ln |(X'^{1/2} + X^{1/2})/(X'^{1/2} - X^{1/2})| \, dX' \tag{3.11}$$

where X is a *field* point at which the displacement is to be evaluated and X'

[1] After Barenblatt's 'modulus of cohesion' for the quantity under the integral sign in this expression, i.e. differing from our intrinsic toughness term by $(\pi/2)^{1/2}$.
[2] As a check on the geometry independence, it is left as an exercise for the interested reader to repeat the derivation of (3.8) for *penny* cracks using the appropriate Green's function formula in (2.22b).

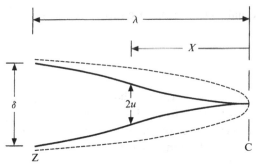

Fig. 3.4. Crack profile at cohesion zone CZ (solid curve, (3.13)) relative to Irwin parabola (dashed curve, $u \propto X^{1/2}$, (2.15)). Note how cohesive stresses close contour into Barenblatt cusp ($u \propto X^{3/2}$, (3.14)) at crack tip C.

is a *source* point for the stress function $p_y(X')$. The solution may be superposed onto $u_A(X)$ for the remotely loaded Irwin crack, which we obtain by inserting the equilibrium requirement $K_A = T_0$ of (3.8) into (2.15):

$$u_A(X) = (4/\pi E')X^{1/2} \int_0^\lambda p_y(X')\,\mathrm{d}X'/X'^{1/2}. \tag{3.12}$$

The singular, parabolic form of the crack contour, $u_A(X) \propto X^{1/2}$, remains evident in this expression. The *net* displacement field in the cohesion zone is then the sum of (3.12) and (3.11),

$$\begin{aligned} u(X) &= u_A(X) + u_y(X) \\ &= (4/\pi E') \int_0^\lambda p_y(X')[(X/X')^{1/2} \\ &\quad - \tfrac{1}{2}\ln|(X'^{1/2}+X^{1/2})/(X'^{1/2}-X^{1/2})|]\,\mathrm{d}X'. \end{aligned} \tag{3.13}$$

As with the intrinsic toughness (3.8), the near-tip profile (3.13) is independent of c, i.e. it has the quality of *invariance*.

As we shall see below, general analytical solutions to this displacement equation are unobtainable. However, there are special cases for which a solution *is* obtainable, at least in approximate form. One of special interest to us here relates to the shape of the crack contour very close to the slit edge C. If we argue that the closure stresses $p_y(X')$ do not exert their greatest influence in this region (justifiable on the grounds that these stresses must tail off near C, fig. 3.3), then the approximation $X \ll X'$ is appropriate. In this limit we can expand the logarithmic term in (3.13),

$$\ln\left[(X'^{1/2}+X^{1/2})/(X'^{1/2}-X^{1/2})\right] \simeq 2(X/X')^{1/2}+\tfrac{2}{3}(X/X')^{3/2}+\cdots$$

Terms in $X^{1/2}$ within the square bracket in (3.13) then cancel, and the displacement function reduces to

$$u(X) = (4/3\pi E')\, X^{3/2} \int_{\lambda}^{0} p_{\gamma}(X')\,\mathrm{d}X'/X'^{3/2}. \tag{3.14}$$

Thus the crack closes into the requisite cusp, $u(X) \propto X^{3/2}$, as depicted in fig. 3.4. The strain $\mathrm{d}u/\mathrm{d}X$ now tends smoothly to zero in the near-tip region $X \to 0$ (cf. infinity for rounded contours of stress-free crack walls in Irwin profile (2.15)), thereby justifying the retention of linear elastic fracture mechanics right down to the zone boundary.[3]

The above formulation for the equilibrium stress-intensity factor and the crack interface displacement field gives us the basis for a description of the crack-tip structure, within the limits of the continuum approximation. Unfortunately, we do not have a priori knowledge of the function $p_{\gamma}(X)$. We did remark in sect. 3.1.1 that (given sufficient fundamental information on the appropriate intersurface potential functions for the constituent material) it may be possible to predetermine the corresponding function $p_{\gamma}(u)$ (e.g. fig. 3.1). However, the critical T_0 and $u(X)$ relations (3.8) and (3.13) are coupled, via $p_{\gamma}(u)$, and must therefore be solved simultaneously. We are left with nonlinear integral equations for which general closed-form solutions are unavailable. As we shall see in subsequent chapters, this is a recurring difficulty in nonlinear fracture problems. On the other hand, if we are concerned *only* with global equilibrium conditions (i.e. the relation between applied loads and macroscopic crack dimensions), and if the two Barenblatt hypotheses remain valid, there is no need to invoke the $u(X)$ relations at all: we simply regard T_0 in (3.9) as a material constant, ultimately relatable to surface energies via (3.10).

There is one special case, that of a constant stress over the cohesion zone, which can be solved analytically.[4] Inserting $p_{\gamma}(X) = \bar{p}_{\gamma}$ (fig. 3.1) into (3.8) and integrating, we get

$$T_0 = (8/\pi)^{1/2}\bar{p}_{\gamma}\lambda^{1/2}. \tag{3.15}$$

---

[3] Note the inverse relation here in the integration limits, ensuring $u(X) > 0$. Note also that the condition $p_{\gamma}(X) \to 0$ at $X \to 0$ is now *essential* in order for the integral to remain finite.

[4] Note that constant $p_{\gamma}(X)$ violates the earlier requirement of zero stress at C. However, this is of minor significance if we confine our attention to displacements at Z, in which case we could avoid an infinity by excluding $p_{\gamma}(X)$ from as small a region about C as we please.

This relation may be inverted to give the Barenblatt zone length

$$\lambda = (\pi/8)\,(T_0/\bar{p}_\gamma)^2$$
$$= \pi E' R_0/8\bar{p}_\gamma^2 \tag{3.16}$$

where we have made use of (3.10) connecting T_0 to R_0. Making an analogous insertion $p_\gamma(X') = \bar{p}_\gamma$ into (3.13) gives us the displacement field

$$u(X) = (4\bar{p}_\gamma/\pi E')\,[(X/\lambda)^{1/2} - \tfrac{1}{2}(1 - X/\lambda)\ln|(\lambda^{1/2} + X^{1/2})/(\lambda^{1/2} - X^{1/2})|\,]. \tag{3.17}$$

At the edge Z of the cohesion zone at $2u = \delta$, $X = \lambda$, (3.17) reduces to

$$\delta = 8\bar{p}_\gamma\,\lambda/\pi E'$$
$$= T_0^2/E'\bar{p}_\gamma = R_0/\bar{p}_\gamma \tag{3.18}$$

in combination with (3.16).

Again, within the confines of Barenblatt's hypotheses, the cohesion-zone dimensions λ and δ in (3.16) and (3.18) are both *crack-size- and geometry-invariant*, as befits fundamental crack-tip quantities. Both also scale with the intrinsic cohesion energy, $R_0 = 2\gamma_B$. As a check on the analysis we may note the result $\bar{p}_\gamma\delta = T_0^2/E' = R_0$ from (3.18); the same result follows directly from the area under the curve in fig. 3.1, i.e. $R_0 = \bar{p}_\gamma\delta$. The solutions from the Barenblatt Green's function approach and the Griffith energy-balance approach are therefore *self-consistent*.

Let us evaluate λ and δ for typical brittle solids. Taking $R_0 = 2\gamma_B$ and $\bar{p}_\gamma \approx E/10$ (sect. 1.5) from table 3.1, we estimate $\lambda \approx 0.2$–1 nm and $\delta \approx 0.1$–0.4 nm. Even allowing for the approximations in our analysis, the estimate of just a few atomic dimensions for the Barenblatt zone length λ barely lends justification to our replacement of discrete interatomic forces by smoothly distributed cohesive stresses along the crack plane, at least for vacuum cracks. The atomic-scale crack-opening displacement δ highlights the intrinsic narrowness of the crack interface immediately behind the crack tip.

Finally, a subtle point. The nonzero cohesion-zone length is central to removal of the *K*-field singularity. It replaces the mathematical notion of an energy sink of line dimension where transformation of mechanical to surface energy is wholly concentrated. Exactly where now is the crack *tip* in fig. 3.3? Some might specify C, the point where the crack-opening displacement ostensibly goes to zero. Others might take it at Z, i.e. at the extreme range of the cohesive stresses. Still others might choose the point where the cohesion passes through a maximum. The price of eliminating the singularity is to introduce an element of arbitrariness into the crack location.

3.3.2 *Fundamental limitation of the continuum slit concept: the Elliot crack*

This last element of subjectivity forewarns of a fundamental limitation in the Barenblatt model. Barenblatt retains the Irwin representation of a crack as an infinitesimally narrow slit terminating at a line, C in fig. 3.3. That line is essentially defined by the displacement condition $u = 0$. But $u = 0$ necessarily corresponds to $p_y = 0$ in the interplanar stress–separation function (fig. 3.1). The notion of a well-defined terminus C at which the cohesive stresses vanish is not physically consistent with the intuitively obvious fact that crack-plane elements *ahead* of C must be elastically strained, i.e. must lie on the linear portion of the $p_y(u)$ curve. Effectively, C should be located at $x = \infty$.

This is another unfortunate manifestation of the continuum slit approximation. As pointed out in sect. 3.1, the true character of a brittle crack is more properly described by two separating lattice planes embedded in a discrete structure. Accordingly, provision should be made to incorporate an interplanar spacing into the description as a fundamental scaling dimension for the stress–separation function.

Such provision is made in a pseudo-lattice model developed by Elliot (1947), where the crack system is treated as two elastic half-spaces separated by b_0. The 'Elliot crack', depicted in fig. 3.5, retains the assumption that the interplanar forces may be represented by a smoothed-out cohesive-stress function. Thus $p_y(x)$ may be determined from the intrinsic separation function $p_y(y)$ in much the same way as in the Barenblatt analysis, except that now the zero-stress origin is taken at $2y = b_0$. The zero point in cohesive force along the crack plane is then located naturally at $x = \infty$, as physically required. Starting from the basic equations of elasticity theory, Elliot was able to obtain self-consistent (if numerical) solutions for a simple nonlinear force function $p_y(u)$ as a conventional boundary value problem.

The inclusion of a discrete lattice spacing as a critical element in Elliot's model takes us perceptibly closer to physical realism. Mathematically, however, there is a price to pay. The simplicity and generality of the slit-crack solutions are lost. Specifically, the power of the Irwin stress-intensity

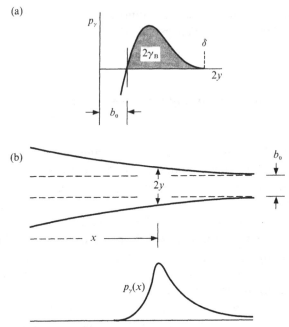

Fig. 3.5. Elliot crack. (a) Stress–separation function, with transformed coordinates $2y = 2u + b_0$ (cf. fig. 3.1). (b) Lattice-plane coordinates and stress distribution $p_y(x)$.

factor formalism, emphasised so strongly in chapter 2 and retained in the Barenblatt scheme, is sacrificed. For by replacing the singular slit edge with quasi-continuous bounding lattice planes we remove the very source of the K-field. Strictly speaking, a stress-intensity factor can not even be defined for the Elliot crack, although in principle one could specify an 'effective' K by appropriately matching asymptotic displacement solutions at $x \to \infty$ to the corresponding far-field solutions for the Barenblatt crack.

With these qualifying remarks we defer further enquiry into the structure of the crack tip to chapter 6, and revert to the slit-crack description.

3.4 Path-independent integrals about crack tip

At this point it is convenient to introduce an elegant and powerful mathematical device, proposed initially by Eshelby (see Thomson 1986) and developed later by Rice (1968a, b). Its appeal in the present context of nonlinear fracture mechanics is that it unifies the continuum description at all levels and links the thermodynamic and mechanistic viewpoints: and in a manner, moreover, that decouples the equilibrium condition for steady-state crack extension from the crack-profile equation. The full Rice treatment involves a rigorous formulation, but we concern ourselves here with just the basic principles.

The idea stems from energy-variation principles, and thereby retains close ties with the Irwin fracture mechanics. Consider the straight-crack system of length c in fig. 3.6. The mechanical energy may be written in the form

$$U_M = U_E + U_A = \int_A \mathcal{U} dA - \int_S \mathcal{T} \cdot \mathbf{u} \, ds \tag{3.19}$$

where dA is an element of cross-section A within a curve S linking the lower and upper crack surfaces, ds is an element of arc on this same curve, $\mathcal{U}$ is the strain energy density, $\mathcal{T}$ is the traction vector on S defined in relation to an outward normal unit vector and $\mathbf{u}$ is the corresponding displacement vector. For a crack system that *deforms reversibly* the mechanical energy variation conjugate to a virtual displacement through dc can be expressed (after much manipulation) as

$$-dU_M/dc = \int_S [\mathcal{U} \, dy - \mathcal{T} \cdot (\partial \mathbf{u}/\partial x) \, ds] \equiv J. \tag{3.20}$$

This defines the Rice line integral J. A feature of J is that it holds for *any* reversible deformation response, linear *or* nonlinear, however remote the integration contour S may be from the tip C.

For the *special case* where the applied loading response measured at the outer boundary A is linear elastic, as it effectively is in the small-scale zone approximation embodied in sects. 3.2 and 3.3, (3.20) is identically equal to the Irwin mechanical-energy-release rate,

$$J = J_A = G. \tag{3.21}$$

We emphasise 'special case': if the nonlinear zone becomes appreciably large, the essentially linear quantity G in (3.21) is not well-defined, and can no longer be identified interchangeably with J. Elaboration on this point is left to the end of the section.

The J-integral has several interesting properties, all of which arise from a path-independence characteristic. For any full circuit enclosing a region of elastic material free of body or surface forces or singularities it can be

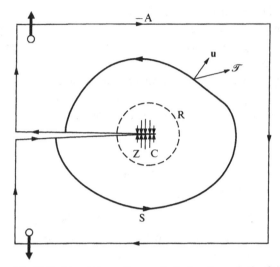

Fig. 3.6. Line integration path S about crack tip C in plane static system, unit thickness. *J*-integral around closed circuit S-A (arrows) is zero. Special *J*-integrals are taken around polar circuit R and cohesion zone CZ.

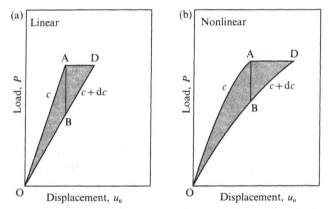

Fig. 3.7. Load vs displacement functions for cracked body with (a) linear and (b) nonlinear responses. Shaded area designates mechanical energy released through incremental crack extension. *J* may be identified with *G* in case (a) only.

shown, through an application of Green's theorem, that the line integral in (3.20) reduces to zero. Suppose we traverse the arrowed circuit in fig. 3.6. Then provided the region of crack surface traversed lies outside the cohesion zone CZ, $\mathscr{T} = 0$, $dy = 0$ there, and the appropriate contribution to the integral from this portion of the circuit is zero. This leaves $J_S + J_{-A} = 0$, with J_S and J_{-A} evaluated around the inner path S and outer path $-A$ in fig. 3.6, giving $J_S = -J_{-A} = J_A$. If now we adopt the convention that all paths be traversed in an anti-clockwise sense about C, starting at the lower crack surface and ending at the upper surface, we see that J is necessarily independent of S. We are therefore free to choose our path as we please. It is this flexibility that marks the power of the method.

Let us illustrate by considering just two special paths for S, selected to provide a comparison of the fracture energetics on different spatial scales. First, reduce S to a circular contour R of radius r about C, with r small compared to the characteristic dimensions of the crack system but large compared to the characteristic cohesion-zone dimensions (i.e. $\lambda \ll r \ll c$). A system of polar coordinates is then appropriate: writing $\mathrm{d}s = r\mathrm{d}\theta$, $\mathrm{d}y = r\cos\theta \mathrm{d}\theta$, (3.20) becomes

$$J = J_R = r \int_{-\pi}^{\pi} [\mathscr{U}(r,\theta)\cos\theta - \mathscr{T}(r,\theta)\cdot\partial\mathbf{u}(r,\theta)/\partial x]\,\mathrm{d}\theta. \qquad (3.22)$$

In this region the K-field of sect. 2.3.2 determines the stress–strain conditions at R. Using the mode I solutions (2.11) to specify the relevant components of $\mathscr{U}$, $\mathscr{T}$ and $\partial\mathbf{u}/\partial x$ in (3.22), integration gives

$$J_R = K^2/E'. \qquad (3.23)$$

Together, (3.21) and (3.23) confirm the essential equivalence between the linear G and K quantities demonstrated earlier in (2.18).

For our second special path, shrink S within R to coincidence with the zone boundary CZ in fig. 3.6 where cohesive closure stresses act, taking the contour along the bottom crack wall to the crack tip and back along the top wall to the end of the zone at $X = X_Z$, $2u = 2u_Z$. Everywhere along this contour $\mathrm{d}y = 0$, $\mathrm{d}s = \mathrm{d}x = -\mathrm{d}X$, $\mathscr{T} = p(X) = p(u)$. The integral in (3.20) then reduces to

$$J = J_Z = 2\int_0^{X_Z} p(X)(\partial u/\partial X)_{c,\lambda}\,\mathrm{d}X$$
$$= \int_0^{2u_Z} p(u)\,\mathrm{d}(2u). \qquad (3.24)$$

This is a perfectly general result. There is nothing to restrict it to small zone sizes (unless we seek to equate J_Z to G or K_A^2/E'), or specifically to intrinsic cohesive forces.

Before proceeding, some further comments on the general definability of J and G are in order. We mentioned that the integration contour for J may include nonlinear as well as linear material, provided the stress–strain characteristic along this contour remains reversible; but, that if it does so *and* the distance of S from the tip C is appreciably large, we can no longer identify J with G. The distinction may be usefully made on a load-point displacement, u_0–P, compliance diagram, as in fig. 3.7. The response for crack size c is represented by OA, for crack size $c + dc$ by OD. In (a) the u_0–P response is linear, in (b) it is nonlinear. Then the mechanical energy released through a crack extension dc is given by the shaded area in each of the plots; to first order this area does not depend on type of loading, e.g. OAB (fixed u) or OAD (fixed P). Whereas J in (3.20) is generally applicable to both (a) and (b) in fig. 3.7, we recall from chapter 2 that $G = G(u_0, P)$ requires the u_0–P response to be linear, i.e. as explicitly defined in (2.1) and implicit in the formulas for specific specimens in sect. 2.5. This has led some to regard J as a 'nonlinear equivalent' of G, and therefore a more universal fracture parameter.

There is one limitation that must be imposed even on J, however. If the crack system is unloaded the response *must* remain reversible, so that the load–displacement relation is single-valued. But irreversibility is manifest in any material with large-scale nonlinear zones containing dissipative elements. Actually, the Rice–Eshelby methodology may be extended to contours that intersect material with hysteretic stress–strain characteristics, although, strictly, not without some modification to the integral in (3.20).

In our treatment of shielding in sect. 3.6 we will retain the J-integral in its simplest form, but remain ever mindful that variants exist and that applications to dissipative systems require caution. The rigorous analysis of the line integral is truly the realm of the solid-mechanics theorist.

3.5 Equivalence of energy-balance and cohesion-zone approaches

The results in sects. 3.3 and 3.4 above call for special comment in connection with the relative merits of the energy-balance and cohesion-zone approaches to equilibrium fracture for *perfectly brittle* solids. Much has been written in the literature on this topic. There are those who argue that the Griffith energy-balance concept, with its essentially macroscopic view of the crack system, provides a necessary but not sufficient condition for extension. According to that school, the fundamental determinant is implied in the microscopic details of the crack-tip structure. Others maintain that the energy-balance concept is steeped in irrefutable thermodynamic principles, and thus not only meets the demands of sufficiency but moreover provides a sounder basis for defining equilibrium states. In fact, as we foreshadowed in sect. 3.3.1, both approaches are (within the bounds of continuum theory) fundamentally equivalent.

This equivalence can be demonstrated most strikingly for cracks with small Barenblatt zones by the *J*-integral results from the previous section. Comparing integration contours about the outer boundary, equation (3.21), and cohesion zone, (3.24) at $p = p_\gamma$, $2u_z = \delta$ (fig. 3.3), we identify

$$G_C = J_A$$
$$= J_Z = \int_0^\delta p_\gamma(u)\,\mathrm{d}(2u) \tag{3.25}$$

as the condition for equilibrium. But the integral quantity in (3.25) is just the intrinsic work of cohesion $R_0 = -G_0$ in (3.1). Thus we regain the Griffith equation

$$G_C = R_0 = W_{BB} = 2\gamma_B \tag{3.26}$$

directly from the cohesive-stress function $p_\gamma(u)$, but, contrary to the spirit of Barenblatt's second hypothesis, *without having to specify the exact form of this function*. In terms of configurational energy space, the system depends only on the difference between initial (unseparated) and final (fully separated) surface states: it is unnecessary to have any knowledge of the path *between* these states.

Herein lies the power of the path-independent integral. In demonstrating equivalence it avoids all the mathematical complexities of the cohesion-zone models, particularly those associated with the nonlinear integral equations (which, we recall, are generally insoluble). This avoidance is possible because we are dealing with a crack-tip configuration that is independent of all elements of macroscopic crack geometry. Under these special circumstances the upper limit in the cohesion-zone integral (3.25), the fundamental interatomic range parameter δ, is specifiable without any reference to a crack system at all (sect. 3.1.1). In other words, the equations for the crack driving force (3.8) and the crack profile (3.13) are mathematically *decoupled*.

In summary, there are several levels at which the nonlinear brittle fracture problem can be addressed, typified by the integrals J_A, J_R and J_Z

in sect. 3.4. The material parameters R_0 and T_0 associated with these integrals are demonstrably equivalent. Inevitably, there are many other fracture parameters that have been defined in the literature; the test of their validity is, as always, that the underlying theoretical basis must be consistent with the Griffith energy-balance concept.

3.6 Crack-tip shielding: the *R*-curve or *T*-curve

In sects. 3.3 and 3.5 we considered an ideally brittle crack system where the only forces acting were the applied opening loads at the outer boundaries and the cohesive closing stresses at the inner crack-tip walls. In that system we were able to establish the Griffith identity $G_C = R_0 = W_{BB} = 2\gamma_B$ at both the macroscopic and the microscopic levels. But as indicated in sect. 3.1.2 this identity is not always borne out by experimental results; the measured crack-resistance energy R is often substantially greater than the reversible cohesion component R_0. The Irwin–Orowan dissipative-zone model was put forward as a means of accommodating this discrepancy. However, Irwin–Orowan says nothing as to how the dissipative term in R might be predetermined: it is a purely phenomenological model. Our specific contention, after Thomson (see reviews by Weertman 1978, Thomson 1986), is that the dissipative processes influence the net toughness by *shielding* the crack tip from the remotely applied loads. These processes play no *direct* role in the underlying surface separation mechanism; in consequence, the intrinsic 'sharpness' of the brittle crack remains inviolate. The notion of shielding will be encountered on several occasions in the remaining chapters, so we develop the basic mechanics here.

Consider a brittle crack system as before, except that now we allow discrete 'source–sinks' (stress sources, energy sinks) to augment the closure from the intrinsic cohesion. These source–sinks, ultimately determined by the material *microstructure* (μ), are activated by the near field of the crack. Their activity is accordingly supposed to be restricted to an annular domain $r_0 < r < r_\mu$, as in fig. 3.8. The radius r_0 defines a crack-tip elastic 'enclave' $\lambda \ll r_0 \ll d$, with λ the Barenblatt cohesion-zone length as previously defined and d a characteristic spacing between sources. While acknowledging the discreteness of the source–sink elements as essential in establishing the existence of the enclave, we represent the influence of these same elements within the annular zone by homogeneous stress distributions so that the continuum basis of the fracture mechanics formalism

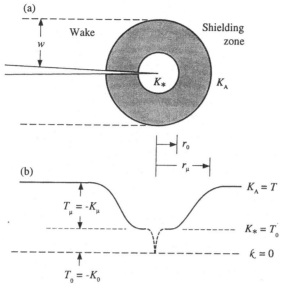

Fig. 3.8. Enclave-shielding model. (a) Outside shielding zone (shaded) one measures *global* K_A (or G_A), but the fundamental conditions for crack extension are decided by *enclave* K_* (or G_*). Moving crack trails wake zone, within which residual stresses may remain active. (b) Variation of effective K-field along radial coordinate ahead and behind crack tip, showing modifying effect of dissipative processes within zone $r_0 < r < r_\mu$. (Note zero K-field at crack tip, to meet Barenblatt equilibrium condition.)

may be retained. Note that a moving crack traces out a 'wake' zone of half-width $w = r_\mu$, within which the source–sinks may remain in a state of residual activity.

3.6.1 *Equilibrium relations*

To begin, we determine some equilibrium stress-intensity factor relations by generalising the formulation in sect. 3.3, linearly superposing a contribution from the shielding zone, K_μ, onto those from the outer applied and inner cohesion zones, K_A and K_0. A sketch of the modified K-field is included in fig. 3.8. The specific form of the K-relation now depends on the 'frame of reference' of our 'observer':

(i) *Crack-tip observer.* An observer stationed at the crack tip C perceives all contributions to the net stress-intensity factor k as deriving from 'external' mechanical forces on the crack. For such an observer the equilibrium requirement is that the net field should vanish, i.e.

$$\mathscr{k} = K_A + K_\mu + K_0 = 0, \quad \text{(cohesion zone)}. \tag{3.27a}$$

(ii) *Enclave observer.* An observer inside $r_0 \gg \lambda$ perceives the intrinsic cohesion term as a material resistance parameter, $T_0 = -K_0$, defined as in (3.8) so as to be positive for attractive forces. This second observer still regards the shielding term as part of an effective external mechanical field K_*, say. Then the equilibrium relation (3.27a) transforms to

$$\begin{aligned} K_* &= K_A + K_\mu \\ &= T_0 = (E'R_0)^{1/2}, \quad \text{(enclave)}. \end{aligned} \tag{3.27b}$$

We emphasise that the intrinsic toughness T_0 is, in the limit of small cohesion zones, independent of crack size.

(iii) *Global observer.* Our final observer, located outside r_μ, takes an alternative view of the shielding term, identifying it as an extrinsic part of the material resistance, $T_\mu = -K_\mu$, say. Again, the negative sign is a matter of convention, where we recognise that the shielding will usually (not always) act to close the crack. The corresponding form of the equilibrium relation determines the critical stress-intensity factor

$$\begin{aligned} K_A &= K_R \\ &= T_0 + T_\mu \equiv T, \quad \text{(global)} \end{aligned} \tag{3.27c}$$

where the subscript **R** is to indicate the presence of a shielding term in the resistance. We shall refer to T simply as the material toughness. In general, this quantity will be a function of crack size, $T = T(c)$, and perhaps other factors, depending on the history of the shielding zone, but nevertheless potentially amenable to first-principles determination for any specific shielding micromechanism (sect. 3.7).

Next let us turn to the analogous equilibrium relations for the energy-release rates. To obtain these relations we evaluate the J-integral along the three closed-loop contours shown in fig. 3.9. This loop configuration, if somewhat contrived and convoluted, conveniently allows us to identify the important contributing segments. The innermost loop is contained just within the elastic enclave and traverses the intrinsic cohesion zone along

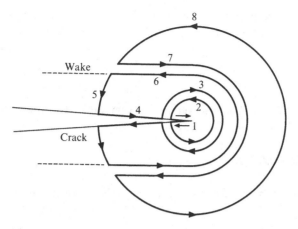

Fig. 3.9. *J*-integral construction for crack with shielding zone. Innermost contour 1–2 defines elastic enclave, within which Barenblatt cohesive forces are located. Intermediate contour 3–4–5–6 defines shielding zone. Outermost contour 7–8 senses the applied loads.

the crack surfaces, with a radius $r_2 \gg \lambda$. The intermediate loop runs around the inner and outer radii of the frontal shielding zone, but extends back a distance $r_5 \gg r_3$ behind the crack tip with segments that intersect the wake. The outermost loop lies entirely in elastic material outside the shielding zone, with a circular segment $r_8 \gg w$. We can write immediately

$$\left.\begin{array}{l} J_1 + J_2 = 0 \\ J_3 + J_4 + J_5 + J_6 = 0 \\ J_7 + J_8 = 0 \end{array}\right\}. \tag{3.28}$$

For the innermost and outermost loops we define (with due regard to the sign convention for path direction) 'bounding' relations (sects. 3.4, 3.5)[5]

$$J_1 = -J_{-1} = -J_Z = G_0$$
$$J_8 = J_A = G_A$$

in a linear system. For adjacent segments we similarly have (from the path-independence property) 'connecting' relations

$$J_2 = -J_3$$
$$J_6 = -J_7.$$

[5] The requirement $r_8 \gg w$ is so that we may ignore a small contribution to J_A from segment 5.

With these bounding and connecting relations inserted into (3.28), we may explore the equilibrium conditions from the perspective of the same three observers as in the derivation of the corresponding K relations in (3.27):

(i) *Crack-tip observer.* An observer at the very crack tip interprets all J contributions as mechanical-energy-release rates. If we write

$$G_\mu = J_\mu = J_4 + J_5,\qquad(3.29)$$

then the equilibrium requirement that the crack-extension force should vanish reduces to

$$\mathcal{G} = -\mathrm{d}U/\mathrm{d}c = G_A + G_\mu + G_0 = 0, \quad \text{(cohesion zone)}.\qquad(3.30\text{a})$$

The energy-release rate from the outer applied loading is negated by that in the combined shielding and cohesion zones; that is, energy is *absorbed* within these latter zones. Equation (3.29) tells us that G_μ is determinable solely from segments 4 and 5 in fig. 3.9. We explore the physical significance of these two segments separately in the next section.

(ii) *Enclave observer.* Our observer within the elastic enclave at $r_0 = r_2$ perceives the intrinsic cohesion term as a material resistance parameter, $R_0 = -G_0$. The shielding term is viewed by this observer as superposing onto the mechanical-energy-release rate from the applied loading:

$$\begin{aligned}G_* &= G_A + G_\mu \\ &= R_0 = T_0^2/E', \quad \text{(enclave)}.\end{aligned}\qquad(3.30\text{b})$$

In relation to its K counterpart in (3.27b), $G_* = R_0 = \text{const}$ represents the energy-balance condition within the screened K-field of the enclave.

(iii) *Global observer.* The observer at $r_\mu = r_8$ includes the shielding term in the internal material resistance, $R_\mu = -G_\mu$, and so, with satisfaction of the proviso $r_\mu \ll c$ for defining global G_A, writes the critical condition for extension

$$\begin{aligned}G_A &= G_R \\ &= R_0 + R_\mu \equiv R, \quad \text{(global)}.\end{aligned}\qquad(3.30\text{c})$$

Again, the subscript **R** is to indicate the presence of a shielding term in the crack-resistance energy R. Observe once more that R_μ is defined as a negative energy-release rate, appropriate to a dissipative process. Equation (3.30c) brings us back to the Irwin–Orowan formulation for crack extension in sect. 3.2. However, *now* the origin of the dissipative component is quite explicit; such that R might in principle be determined a priori once the shielding micromechanisms are specified. Like its counterpart T in (3.27c), the quantity R, although closely related to material properties, will generally be a function of crack size, $R = R(c)$, and other geometrical factors.

We reiterate that the three alternative forms of (3.27) and (3.30) are *equivalent*: it is simply a matter of the level at which we choose to view the system. The first, most fundamental form is a simple extension of the Barenblatt concept, and expresses the crack propagation condition as a net zero motive ($\ell = 0$ or $g = 0$). The second usefully relates the enclave K- and G-fields ($K_* = K_A + K_\mu$, $G_* = G_A + G_\mu$) to intrinsic material parameters ($K_* = T_0$, $G_* = R_0$). This condition is viewed in a region insensitive to details of the cohesion or shielding zones ($\lambda \ll r \ll d$). The third form, if least fundamental, is of most practical relevance, because it is via the screened far field (i.e. via the applied loading) that the experimentalist determines the critical conditions for propagation ($K_R = T_0 + T_\mu = T$, $G_R = R_0 + R_\mu = R$). If the first observer is the 'physicist', the third is the 'engineer'.

Finally in this section, let us examine the *inter-relationship* between the shielding terms $T_\mu = -K_\mu$ and $R_\mu = -G_\mu$. We may do this in the limit of long cracks, where the end conditions $G_A = K_A^2/E'$ and $G_* = K_*^2/E'$ remain valid. Inserting into (3.27c) and (3.30c), and eliminating G_A and K_A, we obtain

$$T_\mu = (E'R_\mu + K_*^2)^{1/2} - K_*, \quad (\lambda \ll d < r_\mu \ll c). \tag{3.31}$$

This equation has two limiting cases of interest, depending on the relative values of T_μ and K_*:

$$T_\mu = (E'R_\mu)^{1/2}, \quad (T_\mu \gg K_*, \text{'strong shielding'}) \tag{3.32a}$$

$$T_\mu = E'R_\mu/2K_*, \quad (T_\mu \ll K_*, \text{'weak shielding'}). \tag{3.32b}$$

Thus the functional dependence between T_μ and R_μ no longer has the same harmonic form as that between K_A and G_A, except in the strong-shielding limit. Cross terms are evident in the general case. These cross terms are manifest by the persistence of K_* in the weak-shielding limit.

3.6.2 Stability relations

The presence of a shielding zone can have a profound influence on crack stability (Mai & Lawn 1986). It can even transform an otherwise unstable configuration into a fully stable one. Such stabilisation is an important feature of resistance-curve or toughness-curve, '*R*-curve' or '*T*-curve', behaviour. Recall from sect. 2.7 that the instability requirement for a crack of area C is that $d^2U/dC^2 < 0$, i.e. $dg/dC > 0$. Writing $g = G_A - R$ from (3.30), or equivalently $k = K_A - T$ from (3.27), we derive

$$dG_A/dC > dR/dC, \quad dK_A/dC > dT/dC, \quad \text{(unstable)} \qquad (3.33a)$$

$$dG_A/dC < dR/dC, \quad dK_A/dC < dT/dC, \quad \text{(stable)} \qquad (3.33b)$$

in direct analogy to (2.33), except that now the right-hand side is generally nonzero. We remind ourselves yet again that, for large-scale nonlinear systems, G must be replaced by J.

Let us illustrate the principles embodied in (3.33) by the graphical constructions in fig. 3.10 for a hypothetical material of non-single-valued toughness, *R*-curve $R(c)$ in (a) and equivalent *T*-curve $T(c)$ in (b).[6] Consider a straight crack of length c extending from a sharp, traction-free starter notch of length c_0 under uniform tension σ_A. From (2.20) we have $K_A = \psi\sigma_A c^{1/2}$, thence $G_A = (\psi^2\sigma_A^2/E')c$, represented for successively increasing σ_A by the loading lines 1, 2, 3 in the plots. Suppose also that $R(c)$ and $T(c)$ are monotonically increasing functions of c. Then the condition for equilibrium is that the applied loading and material toughness curves should intersect ($G_A = R$, $K_A = T$). At loading line 1, the crack at $c = c_0$ is on the verge of extension, but stable ($dG_A/dc < dR/dc$, $dK_A/dc < dT/dc$). On progressing to line 2 the crack extends, but remains stable. Ultimately, at line 3, the crack reaches a critical point at $c = c_M$, and is thereafter unstable ($dG_A/dc > dR/dc$, $dK_A/dc > dT/dc$). This last is the so-called 'tangency condition' for failure, which defines the 'strength', $\sigma_A = \sigma_F = \sigma_M$.

---

[6] Some researchers, especially those with a solid mechanics background, prefer 'K_R-curve' $K_R(c)$ and 'G_R-curve' $G_R(c)$ from (3.27c) and (3.30c) as alternative terminology for these toughness functions.

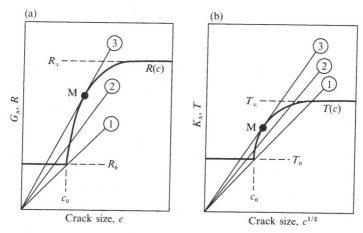

Fig. 3.10. (a) *R*-curve and (b) equivalent *T*-curve constructions for material with shielding. Curves $R(c)$ and $T(c)$ are toughness functions, lines 1, 2, 3 are applied loading $G_A(c)$ and $K_A(c)$ functions at increasing uniform applied stress σ_A. Crack undergoes precursor stable growth along curve from notch tip at $c = c_0$ to tangency point (M) at $c = c_M$, whence failure occurs.

What do these constructions tell us about the strength of our hypothetical material when failure occurs from 'Griffith flaws'? Suppose we replace the starter notch with an intrinsic flaw that experiences the totality of the *R*- or *T*-field 'history' during its formation; i.e. evolves in such a way that the shielding 'switches on' when the crack size c_f exceeds the characteristic dimension d between discrete source–sinks, corresponding to $c_0 = d$ in fig. 3.10. Then over a broad range of c the strength σ_F depends solely on c_M, and not at all on c_f (cf. sect. 2.7). If $c_f < d$ the crack will first pop-in and arrest on the *R*- or *T*-curve before eventually becoming unstable, unless c_f is *so* small that the loading line lies above configuration 3 in fig. 3.10, in which case the crack will propagate without precursor stable growth at $\sigma_A = \sigma_F > \sigma_M$. Similar unrestricted propagation will occur if $c_f > c_M$, at $\sigma_A = \sigma_F < \sigma_M$. Notwithstanding these lower and upper limits to c_f, the stabilisation of the crack-resistance curve imparts a certain quality of *flaw tolerance* to the strength characteristic.

3.7 Specific shielding configurations: bridged interfaces and frontal zones

The conclusion to be drawn from the preceding section is that the fracture mechanics of non-ideal, tough materials are most properly described by a resistance-curve or equivalent toughness-curve construction. Such a construction requires evaluation of the shielding component R_μ or T_μ. It is convenient at this point to consider two special types of shielding, corresponding to limiting steady-state configurations in which one or other of the segments J_4 and J_5 in fig. 3.9 vanishes in the expression (3.29) for J_μ $(= -R_\mu)$. These two limiting cases will form a basis for analyses of toughening processes in ceramic materials in chapter 7.

3.7.1 Bridged interfaces

For the first configuration, we take the shielding stresses $p_\mu(x,0) = p_\mu(x) = p_\mu(X)$ to be confined to the inner walls of the crack over a zone CZ of length $X_Z = c - x_Z \gg \lambda$, superimposed on (but, consistent with our notion of an elastic enclave, not overlapping) the Barenblatt cohesive tractions $p_y(X)$. We depict this distribution of 'bridging' tractions in fig. 3.11, along with the corresponding constitutive function $p_\mu(u)$.

This configuration can be analysed by invoking the J-integral. There is no contribution from segment 5 in fig. 3.9, i.e. $J_5 = 0$, so (3.29) reduces to $J_\mu = J_4 = -J_{-4} = -R_\mu$. Along segment 4 we have $dy = 0$, $ds = dx = -dX$, $\mathscr{T} = p_\mu(x) = p_\mu(X) = p_\mu(u)$, so that, in the limit $\lambda \ll X_Z$, we obtain

$$
\begin{aligned}
R_\mu &= J_{-4} \\
&= -2 \int_{c-x_Z}^{c} p_\mu(x)\,(\partial u/\partial x)\,dx \\
&= 2 \int_{0}^{X_Z} p_\mu(X)\,(\partial u/\partial X)\,dX \\
&= \int_{0}^{2u_Z} p_\mu(u)\,d(2u)
\end{aligned}
\tag{3.34}
$$

in the manner of J_Z in (3.24).

Equation (3.34) is striking for the simple way it highlights the fundamental source of the shielding contribution. In the *transient state*, corresponding to the rising portion of the R- or T-curve in fig. 3.10, the

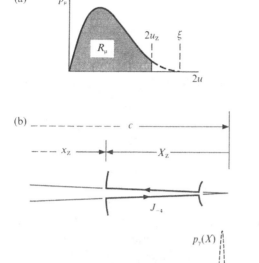

Fig. 3.11. Shielding by crack-interface bridging. (a) Stress–separation function $p_\mu(u)$, with range $2u = \xi$. (b) *J*-integral contours along crack surfaces with bridging tractions $p_\mu(X)$ over $\lambda \leqslant X \leqslant X_Z$, $\lambda \ll L$ (Barenblatt cohesive tractions $p_\gamma(X)$ over $0 \leqslant X \leqslant \lambda$ included as dashed curve). Increment in crack-resistance energy is given by area under curve between $0 \leqslant 2u \leqslant 2u_Z$ in (a). System achieves steady state at $2u_Z = \xi$, $X_Z = L$.

edge of the cohesion zone remains stationary while the crack advances, $x_Z = \mathrm{const}$, $(\partial u_Z / \partial c)_{x_Z} > 0$. Thus $R_\mu(c)$ is just the area under the $p_\mu(u)$ curve bounded by $2u_Z$ in fig. 3.11(a). In *steady state*, corresponding to the upper plateau $R_\infty = R_0 + R_\mu^\infty$ (or $T_\infty = T_0 + T_\mu^\infty$) in fig. 3.10, the zone edge translates with the advancing crack tip, $X_Z = \mathrm{const} = L$, $2u_Z = \mathrm{const} = \xi$, so (3.34) transforms to

$$R_\mu^\infty = \int_0^\xi p_\mu(u)\,\mathrm{d}(2u), \quad \text{(steady state)}, \tag{3.35}$$

i.e. the *total* area under the $p_\mu(u)$ curve.

Two features of the above formulation warrant discussion:

(i) *Decoupling of crack-resistance and displacement relations.* Given the underlying constitutive relation $p_\mu(u)$ for the source of the tractions, we may solve for $R_\mu(u_z)$ in (3.34) without any knowledge of the crack profile $u(X)$ over the traction zone. One could take an alternative route, by solving for $T_\mu(c) = -K_\mu(c)$ using the Green's function formalism of (2.22) (cf. Barenblatt relation (3.8)), but to do that would require *simultaneous* solution of the displacement equation $u(X) = u_A(X) + u_\mu(X) + u_\gamma(X)$. This would leave us with a nonlinear integral equation which (in the absence of simplistic approximations) is analytically insoluble. As it is, we still must solve for $u(X)$ at Z to convert $R_\mu(u_z)$ to the more conventional function $R_\mu(c)$; but at least the displacement equation is now mathematically decoupled from the toughness equation, and we need a solution *only* at Z. Moreover, at steady state, the upper limit can be specified without *any* solution of the displacement equation, because the critical range ξ at which the closure tractions cut off is determined independently by the shielding constitutive relation. Indeed, the entire constitutive relation $p_\mu(u)$ can, in principle, be specified without resort to a crack system at all, e.g. from a detailed consideration of the material micromechanics.

(ii) *Small-scale zone restrictions and non-uniqueness of R-curve.* There is nothing in the *J*-integral analysis that restricts $R_\mu(u_z)$ in (3.34) to small-scale zones ($L \ll c$). On the other hand, recall from sects. 3.4 and 3.6.1 that large zone sizes are manifest as nonlinear responses in the macroscopic compliance function, which excludes a proper definition of the Griffith–Irwin *G*. Experimentally, we are restricted in such cases to a *J* analysis (via compliance measurements), or *K* analysis (either in the manner indicated in (i) above, or via (3.31)). Most importantly, while $R_\mu(u_z)$ may be configuration-invariant, the same will not generally be true of $R_\mu(c)$. This is because the crack profile $u(X)$ inevitably depends on initial notch size (c_0 in fig. 3.10) and other characteristic dimensions of the crack system: the traction zone is not self-similar for different specimen configurations. Small-scale zones remain the only exception, e.g. 'semi-infinite' starting notches in 'infinite' specimens, such that Barenblatt's two hypotheses hold.

It is interesting to evaluate (3.35) in the approximation of an 'average shielding stress' $\bar{p}_\mu$, as we did for $\bar{p}_\gamma$ in sect. 3.3 for the Barenblatt zone (cf. $\bar{p}_\gamma$ in fig. 3.1):

$$R_\mu^\infty = \bar{p}_\mu \xi. \tag{3.36}$$

(a)

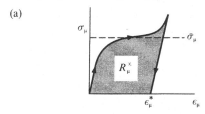

(b)

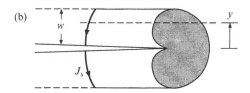

Fig. 3.12. Shielding by frontal-zone dilation. (a) Stress–strain function $\sigma_\mu(\varepsilon_\mu)$: $\bar{\sigma}_\mu$ is 'averaged' stress and ε_μ^* residual strain. (b) *J*-integral contours through wake. Distributions of dilation stresses $\sigma_\mu(x)$ and strains $\varepsilon_\mu(x)$ along plane $y = $ const indicated at bottom. Area under curve in (a) determines steady-state increment in crack-resistance energy.

For the effect of shielding to be comparable with the intrinsic cohesion energy (fig. 3.1), i.e. $R_\mu^\infty \approx R_0$, we would require $\bar{p}_\mu \xi \approx \bar{p}_\gamma \delta$. Using representative values $\delta \approx 1$ nm, $\bar{p}_\gamma \approx 10$ GPa (sect. 3.3), we see that if the interfacial tractions could be applied over a range $\xi \approx 1$ μm (typical of ceramic microstructural dimensions), the shielding stress need only be modest, $\bar{p}_\mu \approx 10$ MPa.

3.7.2 *Frontal zones*

For the second configuration, consider a dilation frontal zone around the crack tip, with persistent influence in a wake layer. Body stresses σ_μ and strains ε_μ within such a zone are defined by a constitutive relation $\sigma_\mu(\varepsilon_\mu)$, fig. 3.12(a). The stress–strain curve is hysteretic, corresponding to a residual dilation in the wake. An indication of $\sigma_\mu(x)$ and $\varepsilon_\mu(x)$ for volume elements on a plane $y = $ const intersecting the frontal zone is given in fig.

3.12(b). The zone width w defines a boundary outside of which volume elements remain on the elastic portion of the $\sigma_\mu(\varepsilon_\mu)$ curve.

Again, the configuration can be treated in terms of the J-integral. It is now segment 4 in fig. 3.9 that is traction-free, i.e. $J_4 = 0$, so (3.29) reduces to $J_\mu = J_5 = -R_\mu$. Along segment 5, back in the wake of the frontal zone where $\mathrm{d}x = 0$, $\varepsilon_\mu(x) = \mathrm{const}$, we have $\mathcal{U}(y) = \int \sigma_\mu(\varepsilon_\mu) \, \mathrm{d}\varepsilon_\mu = \mathrm{const}$ over $0 \leqslant y \leqslant w$, so the J-integral becomes

$$
\begin{aligned}
R_\mu &= -J_5 \\
&= -2 \int_w^0 \mathcal{U}(y) \, \mathrm{d}y \\
&= 2w \int_0^{\varepsilon_\mu} \sigma_\mu(\varepsilon_\mu) \, \mathrm{d}\varepsilon_\mu.
\end{aligned}
\tag{3.37}
$$

Thus as in the previous subsection, the energetics of shielding are governed by the area under a constitutive, $\sigma_\mu(\varepsilon_\mu)$, curve. In principle, given this curve, we need only specify the size w of the dilation zone to obtain a solution for R_μ. In reality, matters are somewhat more complicated. By allowing the system to unload *hysteretically* in stress–strain space (fig. 3.12(a)) we have exceeded the conditions for validity of the J-integral (sect. 3.4). For the scrupulous reader, a more rigorous treatment may be found elsewhere (Budiansky, Hutchinson & Lambropoulus 1983).

There are also subtle questions concerning how the zone develops. *Steady state* is straightforward. The wake is well-established, and the average stress $\bar{\sigma}_\mu$ and appropriate upper limit $\varepsilon_\mu = \varepsilon_\mu^*$ in the integral are independently determinable from the constitutive micromechanics of the dilation process. The corresponding limiting zone size $w = w_\mathrm{c}$ may be estimated (if only approximately) from solutions of the elastic near field (sects. 7.3, 7.4). Thus, in analogy to (3.36), the steady-state resistance $R_\infty = R_0 + R_\mu^\infty = \mathrm{const}$ (and corresponding T_∞) can be determined from

$$
R_\mu^\infty = 2\bar{\sigma}_\mu \varepsilon_\mu^* w_\mathrm{c}, \quad \text{(steady state).}
\tag{3.38}
$$

We note that whereas the shape of the frontal zone must be uniquely dependent on the constitutive relation, the steady-state solution (3.38) requires no specific information on this shape. For the *transient* state, matters are much more complex. In the embryonic stage of frontal-zone expansion from the initially stationary crack, (3.37) is not applicable because there is no established wake. In this initial stage the shielding contribution must be zero, because segment 5 (which we can move forward as close as we please to the boundary of the frontal zone in fig. 3.12(b)) traverses no material beyond the elastic range, i.e. it never extends into the irreversible region of the $\sigma_\mu(\varepsilon_\mu)$ curve in fig. 3.12(a), so the area under the unloaded stress–strain curve is zero. When the frontal zone is sufficiently well developed the crack begins to extend. That stage, which defines the *R*- or *T*-curve, now requires specification of the continually evolving shape of the dilation zone in the wake, and accordingly poses a considerable problem in theoretical fracture mechanics.

4

Unstable crack propagation: dynamic fracture

Thus far we have considered only static crack systems. Now, if an unbalanced force acts on any volume elements within a cracked body, that element will be accelerated, and will thereby acquire *kinetic energy*. The system is then a *dynamic* one, and the static equilibrium conditions of Griffith and Irwin–Orowan no longer apply. In certain instances, such as when stable cracks are made to grow slowly in controlled fracture surface energy tests, the kinetic energy component may be relatively insignificant in comparison with the system mechanical energy. The system may then be regarded as quasi-static, insofar as the static solutions describe the critical requirements for crack extension to sufficient accuracy.

There are two ways in which a crack system may become dynamic. The first arises when a crack reaches a point of instability in its length: the system acquires kinetic energy by virtue of the inertia of the material surrounding the rapidly separating crack walls. Such a dynamical state may be realised even under fixed loading conditions. A 'running' crack is typified by a rapid acceleration toward a *terminal velocity* governed by the speed of elastic waves. In most (but not all) test specimens the running crack divides the material into two or more fragments. The second type of dynamical state arises when the applied loading is subject to a rapid time variation, as in impact loading. In this case response may be limited by the characteristic duration of the loading pulse. Of the two states, the first is of greater interest to us here, for it takes us closer to the actual energy dissipation processes that operate in the vicinity of the crack tip.

A general approach to the dynamic fracture problem was outlined by Mott in 1948 in an extension of the Griffith concept. The approach is straightforward: one simply adds a kinetic energy term to the expression for the total system energy and seeks a configuration that maintains this total energy content constant. However, the evaluation of the kinetic

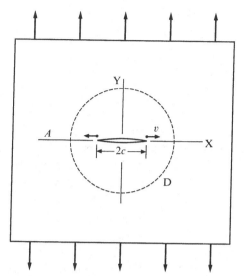

Fig. 4.1. Dynamic plane-crack system. Crack front propagates at velocity v. D denotes domain which receives stress-wave 'message' from crack front at instant t. Specimen has area A at crack plane, thickness unity.

energy itself, and the solution of the resulting energy equation for the motion of the crack, are generally not at all straightforward. For this reason we limit detailed analyses in the present chapter to comparatively simple case studies.

Thus it is that the element of *time* enters the fracture problem. We would emphasise that the dynamic processes envisaged here remain within the realm of the continuum approximation of matter. In chapter 5 we shall consider another kind of time-dependent crack propagation, due to thermal activation over discrete energy barriers at the atomic level.

4.1 Mott extension of the Griffith concept

We now reconsider the energy-balance formulation (sects. 1.2, 2.6, 3.2), dispensing with the restrictive assumption that the crack system be static. Following Mott's procedure (Mott 1948), we simply incorporate an inertial term U_K into the total system energy for the straight crack in fig. 1.4,

$$U = U_M + U_S + U_K. \tag{4.1}$$

Requiring that no energy cross the system boundary as the crack extends, i.e. $\mathcal{G} = -\mathrm{d}U/\mathrm{d}c = 0$ for all c ('pseudo equilibrium'), and invoking (2.8) and (2.29), we have

$$G - R_0 = \mathrm{d}U_\mathrm{K}/\mathrm{d}c. \tag{4.2}$$

Thus the kinetic term appears as a means for dissipating the excess energy resulting from an imbalance in the two opposing generalised forces.

To obtain a general expression for the kinetic energy U_K, consider a volume element of unit thickness at (x, y) in the plane-crack system of fig. 4.1. The mass of this element is $\rho\,\mathrm{d}x\,\mathrm{d}y$, where ρ is the density. Its velocity is $(\dot{u}_x^2 + \dot{u}_y^2)^{1/2}$, with $u_x = u_x(c, t)$ and $u_y = u_y(c, t)$ element displacements. The element velocity relates to the crack velocity through transformation relations of the type $\dot{u} = (\partial u/\partial c)_t v$, noting that $(\partial u/\partial t)_c = 0$ under conditions of fixed loading. The total kinetic energy is determined by integrating over all such elements within a domain D limited by the distance elastic waves carry 'information' from the tips during the growth interval t:

$$U_\mathrm{K} = \tfrac{1}{2}\rho v^2 \iint_\mathrm{D} [(\partial u_x/\partial c)^2 + (\partial u_y/\partial c)^2]\,\mathrm{d}x\,\mathrm{d}y. \tag{4.3}$$

The problem is thereby reduced to one of computing the displacements as a function of crack length from the equation of motion for the elastic solid. This is a formidable task, and certain simplifying assumptions must be made in obtaining solutions for even the simplest crack systems.

More formal expressions of the energy-balance concept in the dynamic problem may be given in relation to the crack-closure approach of sect. 2.4 or the path-independent integral of sect. 3.4 (Freund 1990).

4.2 Running crack in tensile specimen

In his 1948 paper Mott considered the special case of a crack in uniform tension, and derived an expression for the kinetic energy on dimensional grounds. His analysis assumed the following conditions: (i) the stresses and strains about the moving crack tip are adequately defined by the equations of static elasticity theory ('quasi-static approximation'); (ii) the domain D extends over the entire specimen (crack velocity small, cf. elastic wave velocity); (iii) the fracture surface energy remains invariant with crack velocity. The validity of these assumptions will be questioned later. While Mott's analysis lacks rigour, it is particularly instructive in the way it highlights some of the important features of a running crack without excessive mathematical complication.

One of the interesting facets of the dynamic crack problem is the manner in which the crack velocity varies as propagation proceeds. In particular, one would anticipate an upper limit to the crack velocity; for, as suggested above, the rate at which information concerning the local stress field can be communicated to the material immediately ahead of the crack tip is restricted by the velocity of elastic waves. It is in this context that the manner of loading discussed in sect. 2.2 becomes an important consideration: in 'dead-weight' loading the constant applied force might be expected to drive the (unstable) crack system rapidly toward a terminal velocity; in 'fixed-grips' loading the increase in compliance associated with crack extension in a finite specimen inevitably leads to a diminishing applied force, and any tendency toward a limiting velocity is correspondingly lessened.

4.2.1 Constant-force loading

Consider a homogeneous solid of single-valued toughness, cross-sectional area A at the crack plane, containing a straight internal crack of length $2c$, with $A \gg c^2$ (fig. 4.1). We treat first the case in which the applied tensile stress σ_A is raised incrementally beyond the Griffith instability level and is thereafter held constant as the crack extends from its initial length $2c_0$. This is the 'dead-weight' configuration of sect. 2.2. From the static solution of sect. 1.3, (4.1) may be written immediately as

$$U = -\pi c^2 \sigma_A^2 / E' + 2R_0 c + U_K(c), \quad (c \geqslant c_0) \tag{4.4}$$

with E' defined as in (1.8). Initially, the crack is at rest at $c = c_0$, $U_K = 0$, and therefore satisfies an equilibrium relation of the type (1.11). That relation allows us to evaluate the constant U and to eliminate R_0, yielding the following expression for the kinetic energy:

$$U_K(c) = (\pi c^2 \sigma_A^2 / E')(1 - c_0/c)^2. \tag{4.5}$$

Since, according to (4.2), the extrema in such functions indicate crack lengths where the Griffith condition $G = R_0$ obtains, it is apparent that the

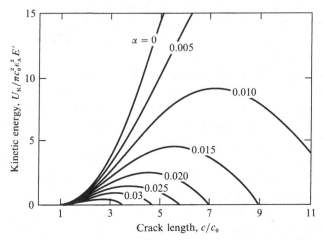

Fig. 4.2. Kinetic energy in (4.9) as function of crack length for system in fig. 4.1. Parameter α is a measure of initial crack length relative to specimen dimension. Case $\alpha = 0$ corresponds to dead-weight loading limit, cases $\alpha > 0$ to fixed-grips.

initial unstable configuration at $c = c_0$ represents the sole equilibrium state in constant-force loading. A plot of (4.5) is given as the curve at extreme left ($\alpha = 0$) in fig. 4.2.

To determine the kinetic energy in terms of *velocities* it is necessary to evaluate the integral in (4.3). To accomplish this Mott used an argument based on geometrical similarity: the spatial field of the crack must scale with the characteristic length c (small in comparison with the size of the domain D), such that the dimensions x, y, u_x and u_y in the integral of (4.3) are proportional to c. But the displacements must also scale with the strain level in the solid, such that the quantities $\partial u / \partial c$ are proportional to σ_A / E'. We may therefore write the kinetic energy

$$U_K = \tfrac{1}{2}(k' \rho c^2 \sigma_A^2 / E'^2) v^2 \qquad (4.6)$$

with k' a numerical constant.

4.2.2 Constant-displacement loading

Suppose now we load the tensile specimen to Griffith instability as before, but this time hold the outer boundaries at fixed displacement as the crack extends, i.e. in 'fixed grips'. A simplistic calculation may be used to obtain a solution for this configuration (Berry 1960). We have already determined the crack *formation* energy, $\pi c^2 \sigma_A^2 / E'$. The 'zero level' strain energy in the specimen *prior to* crack formation is $A\sigma_A^2 / 2E'$, so the *total* energy is $(A + 2\pi c^2)\sigma_A^2 / 2E'$, corresponding to an 'effective elastic modulus' $E/(1 + 2\pi c^2 / A)$. The applied stress on the crack plane therefore falls off with crack extension according to the compliance relation

$$\sigma_A(c) = \varepsilon_A E' / [1 + \alpha(c/c_0)^2] \tag{4.7}$$

where ε_A is the constant applied strain and the dimensionless quantity

$$\alpha = 2\pi c_0^2 / A \tag{4.8}$$

expresses the size of the crack relative to that of the specimen. Observe that in the limit $\alpha \to 0$ the modulus in (4.7) reduces to that for dead-weight loading, as expected.

We may now proceed as before, invoking the same initial conditions in (4.4), but using (4.7) to obtain an expression in ε_A:

$$U_K(c) = \pi c^2 \varepsilon_A^2 E' \{ 1 / [1 + \alpha(c/c_0)^2]^2 - (c_0/c)(2 - c_0/c)/(1 + \alpha)^2 \}. \tag{4.9}$$

This function is plotted in fig. 4.2 for selected values of α. The curves pass through a maximum, corresponding to an equilibrium configuration $G = R_0$ in static loading (recall (4.2)), and subsequently drop back to the abscissa, corresponding to a stationary system. Physically, this means that the crack, when perturbed from its initial (unstable) equilibrium length, will extend toward a second (stable) equilibrium length of lower potential energy, but will overshoot because of its inertia. If the system were to be perfectly reversible, it would then oscillate indefinitely between the intersection points for the appropriate curve on the abscissa of fig. 4.2. In practice, however, dissipative processes in the opening and closing half-cycles account for the excess kinetic energy, leaving a residual (unhealed) crack interface over the fully extended length. We note the limiting cases: $\alpha \to 0$, no arrest states, corresponding to dead-weight loading; and $\alpha \to \infty$, 'large' cracks, infinitesimal instability.

Again, we may combine Mott's kinetic energy relation (4.6) with (4.7) to obtain an expression in terms of velocity:

$$U_K = \tfrac{1}{2} k' \rho c^2 \varepsilon_A^2 v^2 / [1 + \alpha(c/c_0)^2]^2. \tag{4.10}$$

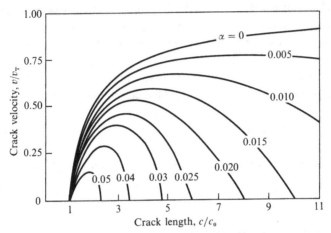

Fig. 4.3. Crack velocity as a function of crack length for system in fig. 4.1. Note asymptotic approach to terminal velocity v_T for $\alpha = 0$ (dead-weight limit), and arrest states $v = 0$ at $c > c_0$ for $\alpha > 0$ (fixed-grips).

4.2.3 Terminal velocity

Now consider the crack velocities attained in the crack system of fig. 4.1. Equating (4.5) and (4.6) in constant-force loading, (4.9) and (4.10) in constant-displacement loading, we obtain

$$v(c) = v_T f(c/c_0, \alpha), \quad (f \leqslant 1) \tag{4.11}$$

where v_T is the *terminal velocity*

$$v_T = (2\pi E'/k'\rho)^{1/2} = (2\pi/k)^{1/2} v_1, \tag{4.12}$$

with k another dimensionless constant and $v_1 = (E/\rho)^{1/2}$ the speed of longitudinal sound waves. The dimensionless f functions are given by

$$f(c/c_0, 0) = 1 - c_0/c, \quad \text{(dead-weight)} \tag{4.13a}$$

$$f(c/c_0, \alpha) = \{1 - (c_0/c)(2 - c_0/c)[(1 + \alpha c^2/c_0^2)/(1 + \alpha)]^2\}^{1/2},$$
$$\text{(fixed-grips).} \tag{4.13b}$$

Plots of (4.11) are given in fig. 4.3, again for selected values of α. In the most extreme case of constant-force loading in (4.13a) the crack approaches v_T asymptotically at large $c(c \gg c_0)$; in the other cases the maximum velocity diminishes as α increases.

4.3 Dynamical effects near terminal velocity

The above analysis indicates that a tensile running crack in an elastic solid is limited by the speed at which sonic waves can travel within the elastic medium, but does not provide us with absolute values: the constant k in (4.12) has yet to be determined. Roberts & Wells (1954) pointed out that if domain D in fig. 4.1 were to extend to the outermost reaches of a large-scale plate, as assumed in the Mott analysis of the previous section, the system inertia (and hence k' in (4.6) and (4.10)) would become too great for a reasonable terminal velocity to be attainable. Experimental determinations of crack velocities, using high-speed photography, ultrasonic techniques (see fig. 4.4), electrical grid techniques (e.g. measuring resistance increase as the crack severs conducting strips on specimen surface), etc., show that terminal velocities in many brittle solids can approach significant fractions of stress-wave velocities (Kerkhof 1957; Schardin 1959; Field 1971). To reconcile the theory with velocities of this magnitude it is necessary to consider the dynamic factors that set an upper bound to the scale of the domain D.

There are other phenomena associated with fast-running brittle cracks that require a dynamical interpretation, and that bear on the validity of the starting assumptions of the previous analysis. First, in many brittle materials, especially those homogeneous solids not constrained to a highly preferred cleavage plane, the accelerating crack is observed to bifurcate. This observation has no explanation in quasi-static fracture mechanics. Second, any plasticity processes operating within the crack-tip field, however limited in the more brittle covalent–ionic solids (see sect. 6.6), are largely suppressed at high velocities, because the underlying dislocation processes are themselves governed by dynamic factors (sect. 7.3.1). In addition, one might anticipate that other microstructural crack-tip shielding mechanisms (chapter 7) could be similarly influenced at high speeds.

Let us therefore take a closer look at some of the dynamical aspects of crack propagation. Our efforts will barely touch on a field of extreme theoretical complexity, and the reader is referred to specialist articles and texts (Erdogan 1968; Freund 1990) for greater rigour and depth. Our treatment below, while somewhat superficial, serves to illustrate some of the main features of the problem.

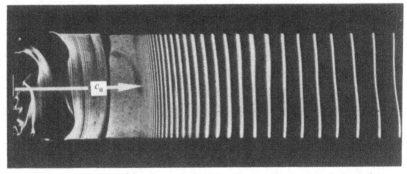

Fig. 4.4. Fracture surface of glass tensile specimen, showing crack approaching terminal velocity. A transducer fixed to the specimen generates ultrasonic waves transverse to the crack plane. Each wave slightly perturbs the crack-tip field, causing a slight undulation in the path and thereby producing a visible record of the crack-front position. Initial length c_0 indicated. Length of specimen 180 mm, frequency of oscillations 1 MHz. Reflected light. (After Kerkhof, F. & Richter, H. (1969) *Second Internat. Conference on Fracture, Brighton.* Chapman and Hall, London, paper 40.)

4.3.1 Estimates of terminal velocity

As already indicated, there is a limit to the size of the domain D in fig. 4.1 within which the material senses the field of the accelerating crack. Taking the boundary D to be a circle of radius r centred on the crack origin, Roberts & Wells estimated the cutoff as the distance $r \approx v_1 t$ travelled by the elastic waves in the time t for the crack to extend through $c \approx v_T t$ at terminal velocity. In combination with (4.12), this gives $r/c \approx v_1/v_T \approx (k/2\pi)^{1/2}$. A second condition for k was obtained by evaluating the integral in (4.3) numerically as a function of r/c. Simultaneous satisfaction of the two functional relationships between k and r/c uniquely determines the unknown constant in (4.12), giving $v_T \approx 0.38v_1$ for the terminal velocity. From typical values of v_1 we obtain $v_T \approx 1-5$ km s^{-1}.

This hand-waving calculation should be regarded as no more than a rough guide to the magnitude of the terminal velocity. Different treatments tend to predict different values of v_T/v_1. Indeed, there is no unanimous agreement as to which form of elastic wave velocity will control. There are many who argue, for instance, that the maximum velocity attainable by any moving surface discontinuity should identify with the velocity of Rayleigh surface waves ($0.58v_1$ for Poisson's ratio 0.25).

Moreover, the operation of any branching or shielding processes surely contributes to the toughness, and might therefore be expected to prevent crack velocities from reaching their theoretical limit. One may note, however, that of the terms in the energy equation (4.4) for constant-force loading, the surface energy term (proportional to c) will become insignificant in comparison with the mechanical energy term (proportional to c^2) as the crack expands, until the system ultimately tends to the steady-state limit $G \rightarrow dU_K/dc$ in (4.2). The contribution of shielding to the toughness therefore becomes irrelevant at high speeds. The same is *not* true of crack branching, which intensifies as the crack accelerates.

4.3.2 Crack branching

Crack branching marks various stages of kinetic energy dissipation. An illustrative example is shown in the glass tensile fracture of fig. 4.5. After initiating at a surface flaw the crack begins its acceleration on a relatively smooth surface ('mirror zone'). At some critical stage in the propagation the crack starts to bifurcate along its front, producing severe surface roughening ('hackle zone'), but without unduly retarding the rate of advance. Some workers distinguish an intervening transition region of fine-scale subsidiary fracturing ('mist zone'). When the crack plane is viewed side-on the onset of branching is seen to be quite abrupt, as in fig. 4.6. There we sketch observed crack paths in pre-cracked glass microscope slides fractured in tension: the sequence is one of increasing initial crack size c_0, thus of decreasing fracture stress σ_A. It is readily demonstrated that the stress-intensity factors (2.20) at prescribed branching points $c = c_B$,

$$K_B = \psi \sigma_A c_B^{1/2}, \quad (K_B > T_0), \tag{4.14}$$

are constant for a given solid. Systematic measurements of branching at mist, mirror and hackle regions on a wide range of materials, in different test geometries, confirm this result.

Several possible causes of crack branching have been advanced in the brittle fracture literature. We consider three below.

(i) *Dynamic crack-tip field distortion.* Theoretical treatments by Yoffe, Broberg and others show that the nature of the near field about the tip of a propagating crack changes with velocity (see Freund 1990). Most such treatments consider a crack running at constant speed in an infinite specimen, to simplify an otherwise intractable dynamic problem. The

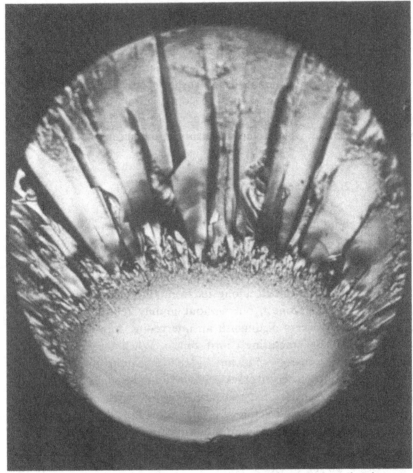

Fig. 4.5. Fracture surface of glass rod broken in tension, showing mirror, mist and hackle zones spreading outward from fracture origin (lower edge). Viewed in reflected light. Rod diameter 4.5 mm. (After Johnson, J. W. & Holloway, D. G. (1966) *Phil. Mag.* **14** 731.)

procedure involves finding steady-state solutions of the equations of motion for an elastic medium (the dynamic analogue of the static solutions for the crack-tip field in sect. 2.3). It is not easy to draw general conclusions from these available solutions, for the analyses are sensitive to the particular initial and (time-varying) boundary conditions imposed, but there does appear to be some general agreement as to the redistribution of the local crack-tip stresses at high velocities.

Fig. 4.7 shows the results of one calculation for the component $\sigma_{\theta\theta}$ at selected relative crack velocities. If we define a 'dynamic stress-intensity

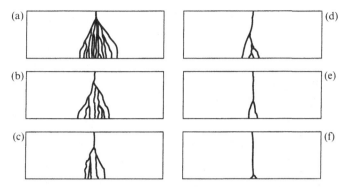

Fig. 4.6. Reproduced crack paths showing branching in glass slides broken in tension (tensile axis horizontal). Specimens pre-cracked at top edges, notch length increasing in sequence (a)–(f). (After Field, J. E. (1971) *Contemp. Phys.* **12** 1.)

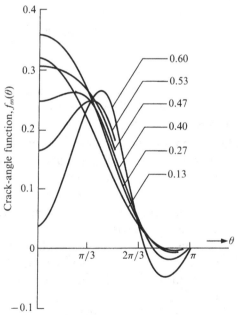

Fig. 4.7. Angular variation of dynamic crack-tip stress $\sigma_{\theta\theta}$, plotted in terms of function $f_{\theta\theta}(\theta)$ (see text), for several values of reduced crack velocity v/v_1 (indicated). (After Erdogan, F. (1968) *Fracture*, ed. H. Liebowitz, Academic Press, New York, Vol. 2, chapter 5.)

factor' $K' = K(v_1/v)^{1/2}$, then the angular function $f_{\theta\theta}(\theta) = \sigma_{\theta\theta}(2\pi r)^{1/2}/K'$ has the same form as in the static solution (2.14a). The pertinent feature in fig. 4.7 is the strong shift of the maximum local tension away from the existing crack plane at $\theta = 0$ onto an inclined plane at $\theta = 50$–$70°$ as the velocity approaches the predicted terminal velocity $v_T \approx 0.38v_l$. From our earlier discussion of crack paths in sect. 2.8 it can be readily demonstrated that the angular distribution of mechanical-energy-release rate must show a similar tendency. This leads to the so-called Yoffe hypothesis (Yoffe 1951): crack branching (in isotropic materials) is a natural consequence of the dynamical nature of the field. (In highly anisotropic crystals the reluctance of the crack to leave its cleavage plane would be expected to suppress branching until higher velocities are reached.) The Yoffe hypothesis appears to explain the qualitative features of branching in a wide range of brittle solids (Field 1971).

However, this explanation is open to some question on matters of quantitative detail. In practice, branching tends to occur over a crack velocity range $v/v_l \approx 0.2$–0.4, somewhat lower than one would predict from fig. 4.7. Also, the observed bifurcation angle in fig. 4.6 and other cases is considerably smaller than prediction. It is felt by some that these discrepancies are too great to be attributable to uncertainties in the dynamic field calculations of fig. 4.7.

(ii) *Secondary fractures.* The second possibility also relates to conditions in the primary crack-tip field, but proposes instead that secondary micro-fractures initiate ahead of, and link back up with, the advancing front. A potentially attractive feature of this proposal is the way it explains the mist-to-hackle transition. Immediately subsequent to initiation just ahead of the crack front the microfractures are overrun and absorbed by the primary front; this forms the mist. As the crack accelerates and the primary field intensifies, the microfractures form further ahead of the main crack and expand as separate entities; this produces the hackle.

Critics of this model question its generality. Secondary microfractures have been observed in certain brittle polymers where inhomogeneities for microfracture initiation are part of the microstructure, but not in homogeneous materials like glasses and monocrystals where internal defect sources are virtually non-existent. Moreover, the mechanics of initiation and interaction are complex (see sect. 7.3.2), and do not lend themselves to quantitative comparison with experiment.

(iii) *Stress-wave branching.* This possibility relates more to the geometry of the specimen than to the near field. The characteristic features are highly distinctive. Recall that a fast-running crack is accompanied by an outward-expanding domain of elastic disturbance (fig. 4.1). As we have already indicated, in a large specimen the disturbance may not extend to the outer boundaries. On the other hand, in a small specimen the stress waves not only may extend to the boundaries but may also suffer reflection there, and thereby interact with the advancing crack tip. In addition, secondary sources of stress waves may be activated at some point within the system: for example, from microstructural inhomogeneities (e.g. microcracks, phase-transforming particles – chapter 7), or from pulses in the loading system. Such interactions generate characteristic 'Wallner lines' (Field 1971), the naturally occurring analogue (and historical precursor) of the ultrasonic modulation technique illustrated in fig. 4.4. And, like the ultrasonically induced surface traces, the Wallner lines may be interpreted as a visual record of the crack growth.

If the interaction is strong, however, the ensuing crack deflection can be abrupt, leading in extreme cases to momentary arrest; in crystalline solids the front may deviate entirely onto an adjacent cleavage plane. This can be a highly effective mode of crack branching, mainly in small specimens within which a suitably intense wave pattern may be generated.

The study of dynamic fracture in brittle materials is of practical as well as academic interest. Much attention has been devoted by design engineers to acoustic emissions from running cracks, for potential use in non-destructive evaluation. The issue of stability, especially as it pertains to crack arrest, is also important. In many areas of materials technology the unstable crack is actually a useful instrument of surface preparation. With the fabrication of certain semiconductor devices, the cutting of gemstones, etc., there is a requirement for cleavage surfaces of relatively high perfection: conditions conducive to branching (e.g. dead-weight, impulsive loading) are then to be avoided. Conversely, in surface finishing processes, such as the grinding of ceramics and the comminution of minerals, these same conditions are to be encouraged.

4.4 Dynamical loading

In the preceding section passing mention was made of impulsive loading as a source of stress waves in crack systems. Such stress waves are generated when the applied forces displace the system boundaries at a rate comparable with that of sonic velocities (Kolsky 1953). The form of the

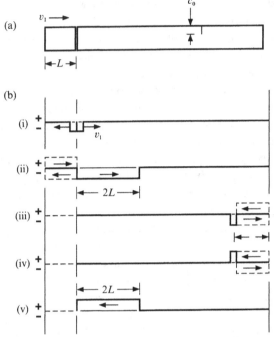

Fig. 4.8. Dynamic loading of impact bar containing crack. (a) Projectile (left) strikes pre-cracked test specimen (right). (b) Stress pulse generated within system, at stages (i)–(v) of impact.

ensuing pulse will depend on the time characteristics of these initial boundary conditions, but the subsequent propagation into the material depends only on the intrinsic wave velocities (as determined by an appropriate wave equation for the elastic medium). On traversing the crack system the propagating pulse will ultimately undergo reflection at an opposite boundary (and at the internal crack boundary itself, a fact again exploited in ultrasonic non-destructive evaluation). In all but the simplest specimen geometries this leads to an extremely complex stress pattern (both spatial and temporal) at the crack. Certain features of the attendant crack response have no parallel in the corresponding static loading state.

The basic principles involved here can be adequately demonstrated, with minimum geometrical complication, by means of the (idealised) impacting-rod system shown in fig. 4.8(a). Both projectile (length L, incident velocity v_i), and test specimen (length $> 2L$, initial velocity zero) are from the same original rod (cross-sectional area A, density ρ). The specimen contains a pre-existing transverse edge crack (effective length c_0). The dimensions of the specimen are such that the stress pulses propagate longitudinally as packets of plane waves (wave velocity v_1) without significant interference from the initial crack $(L^2 \gg A \gg c_0^2)$.

Consider first the stress pattern generated in the specimen by the impacting projectile, as indicated in fig. 4.8(b). At initial contact, $t = 0$, elastic disturbances propagate into both rods. After an interval $t < 2L/v_1$, at (i) in fig. 4.8(b), the front of each disturbance will have travelled a distance $v_1 t$ from the interface. During this interval a mass of material $\rho A v_1 t$ in each rod acquires a momentum $\frac{1}{2}\rho A v_1 v_1 t$, such that the total momentum of the loaded portions remains conserved throughout the impact (corresponding to a compressive displacement rate $\frac{1}{2}v_1$ relative to the centre of mass of each rod). The equation of motion giving the force required to supply the momentum in each rod is therefore

$$P = \mathrm{d}(\tfrac{1}{2}\rho A v_1 v_1 t)/\mathrm{d}t = \rho A v_1 v_1. \tag{4.15}$$

This gives the height of the stress pulse,

$$\begin{aligned} \sigma_P &= P/A \\ &= \tfrac{1}{2}\rho v_1 v_1 = \tfrac{1}{2}E v_1/v_1 \end{aligned} \tag{4.16}$$

with Young's modulus $E = \rho v_1^2$. It is clear that very large stresses can be generated when $v_1 \approx v_1$.

Eventually, the compressive wave in the projectile will reach the unloaded end, and reflect there as a tensile wave. The incident and reflected pulses cancel where overlap occurs, thereby maintaining a stress-free boundary. At $t = 2L/v_1$, (ii) in fig. 4.8(b), cancellation is complete, and the impacting interface 'unloads'. Meanwhile, a pulse of length $2L$ propagates along the specimen, with a similar reflection at the far end, configuration (iii). When the reflected pulse has travelled a distance L back along the rod the net stress becomes tensile, configuration (iv). A full tensile pulse of length $2L$ ultimately develops, configuration (v), and the reflection process repeats itself at alternate ends of the rod until internal friction dissipates the energy.

Now investigate the response of the pre-existing crack to the reflected tensile pulse. The crack will experience a uniform tension σ_P for a duration $2L/v_1$ (except when located within a distance L from the ends, where the duration drops off linearly to zero at the boundaries). If the strength of this field exceeds that required to maintain Griffith equilibrium under normal static loading the crack will run. Two possible results may be envisaged, depending on the duration of the pulse: (i) for short duration the fracture will be incomplete, and the crack will extend in a step-wise manner with each pass of the tensile pulse; (ii) for long duration the fracture will not only be complete, but any momentum 'trapped' in the separated portion of the rod will cause that portion to fly off ('spalling', or 'scabbing').

To illustrate the magnitude of the effect, consider an all-glass system containing a Griffith flaw of length $c_0 = c_f = 1$ μm. The critical stress needed to make the crack run may be estimated from Griffith's data (sect. 1.3) at $\sigma_P \approx 300$ MPa; inserting this value, along with $E = 70$ GPa, $v_1 = 5$ km s^{-1}, into (4.16) yields the corresponding critical impact velocity $v_1 \approx 50$ m s^{-1}. (This is the velocity a projectile would acquire in free-fall through 125 m.) Assuming that the crack rapidly accelerates to its steady-state limit, the resulting extension in a single pass of the tensile pulse may be approximated as the product of the terminal velocity and the pulse duration, $v_T (2L/v_1) \approx L$ (with $v_T \approx \frac{1}{2} v_1$, sect. 4.3). Within the limits of our original assumption, $L^2 \gg A$, this extension should be sufficient to ensure fracture.

We have considered only the simplest possible projectile–specimen geometry here. Among the complications that typify the more general case are the following: the initial disturbance does not resemble a square pulse, propagates in three dimensions, and contains shear and dilatational components; the waves reflected at the boundaries generate an intricate pattern of disturbance within the system; the specimen contains a population of microcracks, each of which may respond to the stress waves independently of the others. Analysis of such factors involves complex exercises in geometrical wave construction. Notwithstanding these complications, one can in practice exert a certain degree of control in the dynamic loading of any potential crack system: with a careful appraisal of the boundary conditions (geometrical configuration, free or fixed, etc.) the impulse may be delivered at such locations as to 'focus' the reflected stress waves onto a pre-selected internal target region; further, by appropriately varying the pulse duration (≈ 1 ms in mechanical impact, ≈ 1 μs in chemical explosive, ≈ 1 ns in laser-induced impulse) the resulting crack extension may be regulated to some extent. Such principles are used in industrial processes, of which rock blasting is just one important example.

4.5 Fracto-emission

We close this chapter with a brief look at a truly remarkable phenomenon, 'fracto-emission'. Fracto-emission refers to the copious ejections of photons and particles from crack tips during fast fracture in brittle solids. Passing mention was made of one form of it, 'triboluminescence', in our summary of Obreimoff's experiment on mica in chapter 1. Triboluminescence has been known to man for at least two centuries, principally as a visible discharge during the crushing of minerals and other brittle agglomerates.

For the most part prior to the 1980s fracto-emission was regarded as little more than a curiosity by the physics community. Since then the cause has been taken up in earnest by Dickinson (Dickinson, Donaldson & Park 1981; Dickinson 1990). Dickinson's group has reported emissions in semiconductors, inorganic crystals, glasses, minerals, oxide coatings on metals, composites and polymers. The variety of species detected is quite astonishing: phonons, electrons, negative and positive ions, neutral atoms and molecular species, photon radiation in the visible and radio bands. There has even been talk of neutron emission. All this points to highly excited states at the near-tip crack interface.

The customary experimental arrangement employs a fracture specimen with inherently spontaneous failure (sect. 4.2.1), e.g. flexure (sect. 2.5.3). Tests are almost invariably conducted in a high vacuum, with detectors located close to the newly exposed fracture surfaces. These detectors include electron and photon multipliers, mass spectrometers, radio-wave antennae, acoustic transducers, current monitors, and so forth. In most materials multiple emissions occur, with characteristic decay times ranging from nanoseconds to minutes. Ionic crystals tend to form charge domains on the newly separated surfaces, with consequent acceleration of ejected electrons to energies well in excess of 1 keV. The example shown in fig. 4.9 for a flexural specimen of mica indicates an eruption of electrons and positive ions at fracture, with protracted decay time. Crystals with the rocksalt structure appear to emit immediately *prior* to critical loading, indicative of precursor activity (sect. 9.2).

After a decade or more of accumulated observations, Dickinson has arrived at a general (albeit not universal) model:

(a) During fracture, charge separates on the pristine surfaces, which achieve a state of high excitation. This charge may be atomically localised or spread over surface domains behind the crack tip.

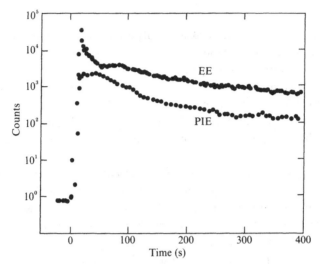

Fig. 4.9. Fracto-emission in mica, electrons (EE) and positive ions (PIE), as function of time after fracture. (After Dickinson, J. T., Donaldson, E. E. & Park, M. K. (1981) *J. Mater. Sci.* **16** 2897.)

(b) A micro-discharge ensues as the walls separate, causing electrons, photons and radio waves to be emitted. The discharge may be augmented by the dislodgement of neutral species into the crack-tip cohesion zone. Those species that enter the detectors without further encounter account for the emissions recorded during the actual separation process.

(c) An intense electric field is generated between oppositely charged surfaces at the narrowly separated interface behind the crack tip, notably in ionic solids, accelerating the emitted electrons to high energies.

(d) Some of the emitted species, especially accelerated electrons, collide with the crack walls, ejecting ions and atoms (electron-stimulated desorption). These account for the after-emissions.

Fracto-emission is now being used by physicists and chemists as a means of studying fundamental relaxations at highly excited surfaces. Materials scientists contemplate a role in non-destructive evaluation of brittle components, particularly in those materials that exhibit premature emissions. The most optimistic practitioners imagine fracto-emission as an earthquake predictor, monitoring precursor activity at stations along geological fault lines. Whatever its ultimate uses, the phenomenon is a powerful testament to the disruptive forces of bond rupture in brittle fracture.

We have included fracto-emission in this chapter on dynamic fracture because it is generally associated with fast cracks. But is it a true dynamic process? Is excess kinetic energy an essential component of excitation, or is the excitation exclusively a manifestation of the underlying bond rupture? What role might external environmental species play? Systematic experiments on slowly extending stable cracks in controlled atmospheres have yet to be performed.

5

Chemical processes in crack propagation: kinetic fracture

In developing the Griffith–Irwin fracture mechanics of chapters 2 and 3 we presumed that equilibrium brittle fracture properties are governed by *in-vacuo* surface energies. In practice, most fractures take place in a chemically interactive environment. The effects of environment on crack propagation can be strongly detrimental. One of the most distinctive manifestations is a *rate-dependent* growth, even at sustained applied stresses well below the 'inert strength', with velocities sufficiently high as to be clearly measurable but too low as to be considered inertial. We use the term *kinetics* (typical velocity range $\approx$ m s^{-1} down to and below nm s^{-1}) to distinguish from true *dynamics* (velocity range $\approx$ m s^{-1} to km s^{-1}). Kinetic crack propagation (alternatively referred to as 'slow' or 'subcritical' crack growth) is notable for its extreme sensitivity to applied load, specifically to G and K. It tends also to depend on concentration of environmental species, temperature, and other extraneous variables.

How does kinetic behaviour reconcile with the Griffith concept? A crack growing at constant velocity at a specific driving force implies a condition of steady state, whereas Griffith deals explicitly with equilibrium states. Experimentally, it is found that velocity diminishes with decreasing G or K until, at some *threshold*, motion ceases. On unloading the system still further the crack closes up and, under favourable conditions, *heals*, even in the presence of environmental species. This adds a new dimension to the Griffith conception of equilibrium, that of a special, quiescent kinetic state where forward and backward fluctuations just balance. However, healing is generally imperfect, in that the load to *re*propagate the crack tends to be measurably less than that for the initial propagation; the opening–closing cycle exhibits *hysteresis*. Such hysteresis raises questions as to the reversibility concept that lies at the very heart of the Griffith thermo-dynamics.

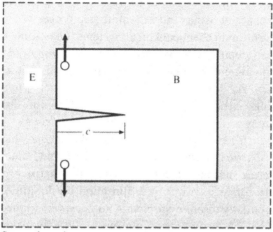

System boundary

Fig. 5.1. Cracked body (B) in contact with interactive fluid environment (E). Continuum description envisages fluid as having unrestricted access to adhesion zone.

In this chapter we seek to expand our fracture mechanics base to include *chemistry*. At first we restrict ourselves to ideally brittle solids. We consider a proposition by Orowan that adsorption can reduce the driving force for equilibrium propagation in brittle solids by lowering the surface (or interface) energy. This proposition can be incorporated naturally into our previous formalism via the Dupré work of adhesion. Then we explore an analysis by Rice that deals with kinetics by broadening the Griffith thesis to include the *second* law of thermodynamics. The Rice description treats crack velocity configurations as 'quasi-equilibrium, constrained states', such that stress-enhanced fluctuations drive the crack forward (*or* backward) at a steady rate. Quiescent states at zero velocity define threshold configurations. Generalisation of the treatment to materials with resistance- or toughness-curves is then implemented in the manner of chapter 3, i.e. via incorporation of a physically decoupled shielding term into G or K. We subsequently examine how the experimental evidence, conventionally presented on crack velocity vs mechanical-energy-release rate or stress-intensity factor (v–G or v–K) diagrams, supports these basic ideas. Generic activation-barrier models of kinetic crack growth are then considered. It is concluded that, at least for small departures from the quiescent states, the central theme of bond rupture as the limiting process of brittle fracture prevails in crack-tip chemistry.

While acknowledging bond rupture as the primary kinetic step in crack propagation there are other, subsidiary steps that make themselves felt in the crack velocity characteristics. Indeed, multi-step processes are the rule rather than the exception in chemical kinetic systems (Glasstone, Laidler & Eyring 1941). In particular, the active species must be *transported* along the interface of the moving crack before any adsorption interaction can take place in the cohesion zone. We shall see how such transport processes can become rate limiting, particularly at large deviations from threshold where the adsorption rate is so fast as to be virtually instantaneous.

Our approach in this chapter will be to consolidate the phenomeno-logical activation-barrier models into the continuum framework of Griffith–Irwin fracture mechanics, consistent with the sharp-crack theme. It will be necessary for us always to keep in mind the limitation of this approach to an essentially discrete problem – final answers to the question of chemistry must be obtained at the atomic level. Those final answers will be sought in the following chapter.

5.1 Orowan generalisation of the Griffith equilibrium concept: work of adhesion

One of the earliest theories of chemistry in brittle fracture was put forward by Orowan (1944). Orowan was intent on explaining a reduction by a factor of about three in the strength of glass specimens in air (specifically moist air) relative to that in a vacuum under sustained loading. He proposed that, on entering the crack interface and adsorbing onto the walls in the cohesion (adhesion) zone, the environmental molecules lower the surface energy of the solid. The magnitude of the observed strength reduction could be accounted for by reducing γ in the Griffith strength equation (1.11) by a factor of about ten. Orowan also speculated that this reduction would be limited by the rate at which the active molecules could diffuse to the critical adsorption region, consistent with a substantial time dependence in the experimental strength data for glass ('fatigue'). The appeal of the Orowan suggestion, apart from its simplicity, was its preservation of the Griffith energy-balance principle: now, however, it was not only G that one could vary to maintain a state of equilibrium (by adjusting the applied load), but also γ (by adjusting the environment).

Following the lead of sect. 2.6, let us develop the Orowan hypothesis in terms of the Dupré work of adhesion for the separation of two solid half-bodies (B) in a fluid environment (E). Thus for the environmentally

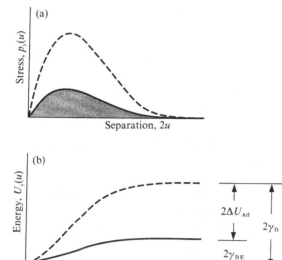

Fig. 5.2. Intersurface functions, (a) force and (b) adhesion energy, per unit area separation of *continuum* solid in fluid medium. Solid curves relate to separation in fluid, dashed curves to separation in a vacuum. Effect of environment is to 'screen' wall–wall attractions, and correspondingly lower work of adhesion from $W_{BB} = 2\gamma_B$ to $W_{BEB} = 2\gamma_{BE}$ (shaded area in (a)).

susceptible crack system in fig. 5.1 we have, in direct analogy with the earlier result (2.30a) for separation *in vacuo*,

$$R_0 = W_{BB} = 2\gamma_B, \quad \text{(vacuum)} \tag{5.1a}$$

$$R_E = W_{BEB} = 2\gamma_{BE}, \quad \text{(environment)} \tag{5.1b}$$

where γ_{BE} is the solid–fluid interface energy, i.e. the energy to form *one* unit area interface (B–E) relative to the initial cohesion state (B–B). The condition for equilibrium thus obtains simply by replacement of R_0 with R_E and γ_B with γ_{BE} in the Griffith balance relations (2.30)–(2.32).

It is instructive to interpret (5.1) in relation to a reversible surface–separation function for *adhesion*, in the same spirit as in sect. 3.1.1 for *cohesion*: i.e. acknowledging the nonlinear nature of the intersurface forces but nevertheless regarding both the fluid medium and the solid halves bounding the separation plane as essentially continuous. Accordingly, in fig. 5.2 we sketch the functions $p_\gamma(u)$, adhesion surface–surface force per unit area of separation plane (positive sign indicating attraction, consistent with previous convention – cf. fig. 3.1), and $U_\gamma(u)$, conjugate adhesion interaction energy (area under $p_\gamma(u)$ curve). The corresponding functions for separation *in vacuo* are included as reference baselines (cf. fig. 3.1). In the specific context of brittle fracture, the construction is consistent with the traditional continuum picture of a crack system in which the fluid has unrestricted access to the interaction zone, as is implicit in fig. 5.1 above. Then $R_E = W_{BEB} = 2\gamma_{BE}$ is defined as the work to translate the system along the solid curve in fig. 5.2(b) from initial (virginal closed) state at $2u = 0$ to final (adsorbed open) state at $2u = \infty$ (or, strictly, beyond some cutoff distance in the adhesion interaction).

Now we may arrive at the same final state by considering an alternative conservative path. First, let us (arbitrarily) set the zero level of energy $U_\gamma(u)$ in fig. 5.2(b) at the minimum, corresponding to the primary bonding 'ground state' in the virgin solid. Our path is taken in three steps: (i) begin with body in ground state (initial state, $U_i = 0$); (ii) separate halves (B–B) from $2u = 0$ along *dashed* curve to $2u = \infty$, in a vacuum, at cost of intrinsic surface energy of solid (intermediate state, $U_* = 2\gamma_B$); (iii) allow environmental species to enter the system and adsorb reversibly onto free surfaces, corresponding to transition from dashed to full curve, at gain in adsorption energy per unit area of separation plane (final state, $U_f = 2\gamma_B - 2\Delta U_{Ad}$). Since this path has the same end point as before we may write

$$R_E = W_{BEB} = U_f - U_i = U_f$$
$$= 2\gamma_B - 2\Delta U_{Ad}$$
$$= 2\gamma_{BE}. \tag{5.2}$$

Therefore, the requirement for the cohesion of a solid to be lowered by the environmental interaction is that $\Delta U_{Ad} > 0$, i.e. that adsorption occur. Note that there is nothing to preclude the inequality $\Delta U_{Ad} > \gamma_B$, corresponding to $\gamma_{BE} < 0$, in which case work would have to be done to *close* the B–B interface.

The path-independence quality attributed to the energy function in fig. 5.2(b) is reminiscent of the important result from the *J*-integral construction for *in-vacuo* fracture in sect. 3.5: we may derive the Griffith equilibrium condition '*directly* from the cohesive-stress function $p_\gamma(u)$, *but without having to specify the exact form of this function*'. Thus we see for environment-sensitive (as for environment-free) systems that, provided the path never deviates from equilibrium, it is only the end-point states that matter. The *initial* (closed) state corresponds to the virgin solid: its energetics are those of the primary bonding, and are consequently determined by solid-state chemistry, without having to specify the state of the environment. The *final* (open) state corresponds to the solid–fluid interface: in this case the energetics are those of adsorption chemistry, specifically the Gibbs adsorption equation (Adamson 1982)

$$d(\Delta U_{Ad})/d(\ln p_E) = kT\Gamma(p_E) \tag{5.3}$$

where k is Boltzmann's constant, T the absolute temperature, Γ the excess concentration of environmental species E at the B–E interface, and p_E the activity of that species (partial pressure for gases). The Gibbs equation opens the way to a determination of the role of chemical concentration in the fracture energetics without explicit reference to crack-tip geometry.

We reaffirm that the above construction for solid–solid separation in a fluid, based as it is on a reversible thermodynamic cycle, is in complete accord with the Griffith notion of brittle cracks. At $G > 2\gamma_{BE}$ the crack acquires a forward velocity; conversely, at $G < 2\gamma_{BE}$ the crack acquires a negative velocity (it heals). Beyond these inequalities, however, our description can make no specific statement about the non-equilibrium states. For instance, it can not explain how the velocities might achieve a steady velocity. Nor can it explain the commonly observed fact that brittle cracks generally do not close up and heal spontaneously on removal of an applied load. There is some vital component missing from our account.

Meanwhile, equations (5.1)–(5.3) may be used as a starting point for incorporating chemical terms into the Griffith–Irwin fracture mechanics:

(i) *Isotherms.* Evaluation of (5.3) for a given material–environment system may proceed if an appropriate adsorption isotherm for $\Gamma(p_E)$ is available. One of the most commonly cited is the Langmuir isotherm for gases (Adamson 1982). At equilibrated adsorption–desorption the fractional surface coverage is

$$\theta(p_E) = \Gamma(p_E)/\Gamma_m = p_E^\eta/(p_0^\eta + p_E^\eta) \tag{5.4}$$

where Γ_m is the excess at full coverage, η is the site occupancy of adsorbate E molecules per B surface, and p_0 is a characteristic pressure. The last term is an inverse measure of the adsorption activity: for sufficiently small p_0 $(p_0 \ll p_E)$ we may consider the coverage to be virtually complete $(\theta \to 1)$ at ordinary pressures ('strong, chemisorptive interactions'). Inserting (5.4) into (5.3) and integrating, and invoking (5.1) and (5.2), we obtain

$$R_E(p_E) = W_{BEB}(p_E)$$
$$= W_{BB} - (2\Gamma_m kT) \ln (p_E/p_0 \theta^{1/n}). \tag{5.5}$$

The work of adhesion therefore decreases with increasing pressure, in accordance with intuitive expectation.

In the event that the condensate wets the surface so that a capillary forms at the interface immediately behind the crack front, W_{BEB} will contain an additional, positive Laplace pressure term, equal at saturated atmosphere to the surface tension of the liquid in the meniscus.

(ii) *Interfaces.* The work of adhesion concept may be readily extended to unlike half-bodies A and B in environmental medium E (cf. (2.30b)), by replacing (5.1) with

$$R_E = W_{AEB} = \gamma_{AE} + \gamma_{BE} - \gamma_{AB}. \tag{5.6}$$

(iii) *Toughness parameters.* The equivalence between the Barenblatt toughness and crack-resistance energy expressed in (3.10) for vacuum cracks extends to chemically-exposed cracks,

$$T_E(p_E) = [E' R_E(p_E)]^{1/2}. \tag{5.7}$$

In the Barenblatt relation (3.8), one has only to modify the function $p_\gamma(X)$ to allow for the influence of the reactive environment on intersurface forces. We defer consideration of such modification to sect. 6.5. In accordance with our relation $R = R_E(p_E)$ in (5.5), we expect T_E to diminish with increasing chemical concentration.

5.2 Rice generalisation of the Griffith concept

The preceding section allows us to extend the Griffith description of equilibrium states to include the influence of chemistry, via the Dupré work of adhesion W. But Griffith theory is strictly bound by the first law of thermodynamics, and consequently has no provision to describe the response once the crack system is perturbed by more than an incremental amount from these equilibrium states. It can not deal with kinetics.

As pointed out by Rice (1978), this restriction can be removed by restating the Griffith–Irwin concept in the more general framework of *irreversible* thermodynamics. The essence of the Rice kinetic theorem can be summarised thus: for perturbations sufficiently small that the ensuing velocities remain negligible compared to the speed of sonic waves, the associated rate processes may be regarded as 'sequences of constrained equilibrium states', such that the crack system assumes the quasi-equilibrium configuration 'that would result if the internal variables of the system were frozen at their instantaneous values'. Then it is a requirement of the *second* law of thermodynamics that, for crack growth at any velocity, the entropy production rate must not diminish with time. We have to allow for a dissipative element in the bond-rupture process. Ultimately, a proper understanding of this dissipative element must be sought in terms of steady-state fluctuations over atomically localised energy barriers. Macroscopic thermodynamics is nevertheless equipped to provide powerful, if limited, insights into the permissible steady-state velocities.

A loose derivation of Rice's theorem can be presented as follows. We again consider a system of a body B containing a crack in contact with an internal environment E (cf. fig. 5.1), except that now we allow for isothermal transfer of heat Q across the system boundary from an external thermal reservoir at fixed absolute temperature T, fig. 5.3. Then the first law of thermodynamics requires that the internal energy of the system during an incremental crack extension area dC increase by an amount equal to the corresponding heat input,

$$dU = \delta Q \tag{5.8}$$

(recalling from sect. 1.2 that we define our system boundary to include the work of applied loading, so that the net *mechanical* work done by the system during any such infinitesimal crack growth is zero). From the second law of thermodynamics the heat content relates to the system entropy S,

$$dS = \delta Q / T + \Lambda dt \tag{5.9}$$

where $\Lambda \geqslant 0$ is the time rate of entropy production, i.e. the entropy rate in excess of that required to maintain a *reversible* alteration of state. Now for an ideally brittle solid the internal energy of the system may be decomposed into mechanical ($dU_M = -G\,dC$) and surface ($dU_S = R_E\,dC$) terms, as in (2.31), plus a term to allow for a change in the state variable S,

$$dU = -(G - R_E)\,dC + T\,dS. \tag{5.10}$$

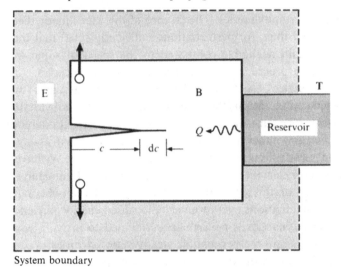

System boundary

Fig. 5.3. System of cracked body (B) with internal fluid environment (E) in contact with an external heat reservoir at absolute temperature T. Q is the heat input into the system during crack formation c.

Combining (5.8)–(5.10), we obtain the entropy production rate for irreversible crack growth,

$$\Lambda = \int (G - R_E) v \, ds / T \geqslant 0 \qquad (5.11a)$$

with $v = dc/dt$ the velocity normal to the crack at any position s along its front.

Assuming now the conditions at each point along the crack front to be governed by the local crack-extension force $\mathscr{g} = G - R_E$, the entropy rate line integral (5.11a) condenses to

$$\mathscr{g}v = (G - R_E)v \geqslant 0. \qquad (5.11b)$$

It is Rice's contention that this inequality constitutes a more general fracture criterion than $G_C = R_E$, since it admits kinetic states. If $\mathscr{g} > 0$, then $v \geqslant 0$, i.e. if $G > R_E$ the crack can only move forward; conversely if $\mathscr{g} < 0$, $v \leqslant 0$, i.e. if $G < R_E$ the crack can only move backward. (Observe, however, that the inequality does not exclude $v = 0$ at $G \neq R_E$, a point to which we shall return below.) The implication for ideally brittle solids is that there should exist a bounded kinetic region of mechanical-energy-release rates, within which we may define *quasi-equilibrium* states $G = R_E'$:

$$R'_{E} = R_{E} \pm \Delta R_{E}, \quad (R^{-}_{E} \leqslant R'_{E} \leqslant R^{+}_{E}) \tag{5.12}$$

with $\Delta R_{E} \geqslant 0$ a dissipative component and R^{+}_{E} and R^{-}_{E} limits for forward and backward propagation. (Outside this bounded region the system becomes dynamic.) This in turn implies the existence of a steady-state crack velocity function $v(G)$, with a well-defined threshold $v = 0$ at $G = R_{E} = W_{BEB} = 2\gamma_{BE}$. Note that there is nothing to preclude us extending the argument to vacuum cracks. However, without further elaboration we assert here (and justify in the next chapter) that the range $R^{-}_{E} \leqslant G \leqslant R^{+}_{E}$ is especially pronounced when the influence of chemistry is manifest. A simplistic velocity function containing the essential features of this description is represented by the 'virgin' curve at right in fig. 5.4.

The Griffith condition for crack propagation at $v = 0$ may now be seen as a very special case of (5.11b), one of true thermodynamic equilibrium at $g = 0$, $G = G_{C} = R_{E}$. At $G = R_{E}$ the entropy production rate Λ in (5.11a) is zero, affirming that the Griffith crack can, in principle, be extended and healed in a truly reversible fashion, even in an interactive environment. This theoretical affirmation of the principle of reversibility at the quiescent point further justifies the identification of R_{E} with reversible surface or interfacial energies in (5.1). As a corollary statement, the further the departure from Griffith equilibrium, as measured by the product gv, the greater the magnitude of the dissipative component ΔR_{E} in (5.12).

There is one further aspect of crack irreversibility, to which brief allusion was made in the introductory remarks to this chapter and toward the end of sect. 5.1, that may be accommodated within the Rice formalism. We refer to the common experimental observation that crack healing, when it does occur (see chapter 6), tends not to follow the negative branch of the 'virgin' v–G curve in fig. 5.4. Rather, G sometimes has to be lowered to some level below R_{E} before the crack can be made to retract, as represented by the 'healed' curve at left in the figure. On increasing G again the crack will generally repropagate along a continuous, positive branch of this second curve, as we shall verify experimentally in sect. 5.7 and theoretically in sect. 6.5. An indication was given above that such an indeterminate intervening region at $v = 0$, $G < R_{E}$, is not at all inconsistent with the inequality in (5.11). This may help to explain, at least in part, why in practice newly formed Griffith flaws tend not to self-annihilate on unloading.

The real power of the Rice thermodynamic theorem lies in the simple way it demonstrates that kinetics can exist *within the framework of the Griffith energy-balance concept*, without any need to identify specific

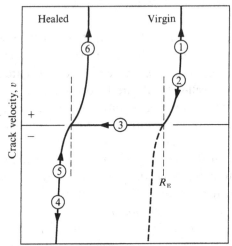

Fig. 5.4. Schematic v–G function, for forward $(+)$ and backward $(-)$ crack motion: right curve, reversible extension–healing response through virgin crack interface in fluid environment; left curve, corresponding healing–repropagation response in same environment. Initial loading ① drives crack through virgin solid along positive branch of right curve. On unloading ② the crack reaches threshold $G = R_{\mathrm{E}} = 2\gamma_{\mathrm{BE}}$, and remains stationary ③ until G intersects left curve, whence retraction occurs hysteretically along negative healing branch ④, i.e. bypassing negative virgin branch (dashed). On reloading ⑤ the crack grows through the healed interface along positive branch of the healed curve ⑥.

irreversible mechanisms. Herein lies a fundamental rationale for the construction of v–G or v–K diagrams (sect. 5.4). On the other hand, the thermodynamics can say nothing as to the form the v–G function might take. The description also allows for an intrinsic hysteresis in the crack reversibility on unloading (and subsequently reloading) the crack system. Again, no predictions can be made as to the degree of this hysteresis. As with any physical process decided at the atomic level, the final details can be obtained only with a deeper understanding of the nature of the discrete energy barriers, with ultimate resort to the fundamentally more powerful methods of statistical mechanics. This aspect will be considered more closely in our treatment of atomic processes in chapter 6.

5.3 Crack-tip chemistry and shielding

Thus far in our treatment of chemistry in fracture we have supposed our solids to be ideally brittle, essentially free of energy-dissipative elements beyond the adhesion zone. As described in chapter 3, the more practical brittle solids deviate from this ideal; they are subject to subsidiary (but often substantial) energy losses within some *shielding* zone encompassing the crack front. We have seen how this shielding manifests itself in the equilibrium fracture characteristics, via the resistance-curve or toughness-curve (R-curve or T-curve). We now examine the corresponding influence on the kinetic characteristics.

Consider the shielded crack in fig. 5.5. The mechanical-energy-release rate or stress-intensity factor sensed at the crack-tip adhesion zone is the 'enclave' G_* or K_* rather than the 'global' G_A or K_A (sect. 3.6.1). The intimation from our earlier R-curve analysis in chapter 3 is that the baseline toughness R_E, even if minutely small compared with the energy consumed within the shielding zone, retains a controlling role in the fracture mechanics. Physically, this is because it is the fundamental separation process in the (structurally invariant) adhesion zone that ultimately determines the condition for crack extension. Moreover, as we shall confirm in chapter 7, the enclave field intensity may control the size of the shielding zone, in such a way that R_E appears as a multiplicative rather than additive term in R_μ. It is interesting that this picture of an energy-dissipation zone whose exclusive role in the kinetics of fracture is one of mechanically screening the crack tip without altering the underlying bond-rupture process was first proposed for ductile metals (Hart 1980).

We formalise the discussion as follows (Lawn 1983). First we write down basic, general relations for kinetic crack propagation from the perspective of an enclave observer:

$$\left.\begin{array}{l} v = v(G_*) \\ v = v(K_*) \end{array}\right\} \quad \text{(kinetic)} \tag{5.13}$$

with zero-velocity thresholds at

$$\left.\begin{array}{l} G_* = R_E \\ K_* = T_E \end{array}\right\} \quad \text{(equilibrium)}. \tag{5.14}$$

The shielding is then introduced specifically via an independent set of

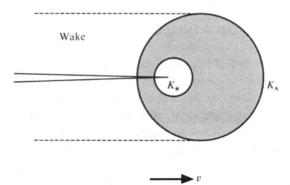

Fig. 5.5. Crack with shielding zone (shaded) translating at velocity v. Fundamental conditions for extension are determined uniquely by K_* (or G_*) within inner elastic enclave. Global toughness may be increased (true shielding), $K_A > K_*$ ($G_A > G_*$), or decreased (anti-shielding), $K_A < K_*$ ($G_A < G_*$), by activity of subsidiary energy source–sinks in shielding-wake zone.

relations that connect the mechanical conditions in the inner-enclave and outer-surround zones. These relations are expressible as a non-equilibrium form of (3.30) and (3.27)

$$\left.\begin{aligned} G_* &= G_A + G_\mu \\ K_* &= K_A + K_\mu \end{aligned}\right\} \tag{5.15}$$

where G_A and K_A pertain to a global observer and $-G_\mu = R_\mu$ and $-K_\mu = T_\mu$ denote shielding contributions to the material toughness.

Simplistically, the consequence of shielding may be viewed as a functional shift of the global v–G_A function along the abscissa of a v–G diagram, as indicated schematically in fig. 5.6. For a given material–environment system this shift will vary with crack 'history', because of the functional dependence $R_\mu = R_\mu(c)$. It all depends on how the crack system evolves in R-space: for initially 'short' cracks (near the lower portion of the R-curve) the shift will be relatively small; for 'long' cracks the shift can be substantial, and the entire shape of the velocity curve may change relative to that of the invariant enclave v–G_* curve. *Thus the global velocity function for materials with R-curves is not unique.* This strong conclusion is especially pertinent to those who would use traditional long-crack velocity data to predict lifetimes of materials with small flaws (chapter 10).

Bearing in mind our allusion above to the potential controlling role of enclave field intensity on the size of the shielding zone, we might also expect

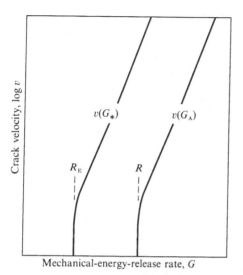

Fig. 5.6. Schematic representation of crack velocity functions: *without* shielding, intrinsic, invariant enclave v–G_* function; *with* shielding, global, history-dependent v–G_A function. Effect of shielding is thus manifest as a G_A shift (or K_A shift on v–K diagram).

the G-shift to scale in some sensitive way with R_E. This runs counter to the common assertion that the intrinsic surface energy term may be dropped from the fracture energetics (sect. 3.2), on the grounds that this term is negligibly small relative to the shielding contribution. All this is to focus even stronger attention on the fundamental crack-tip relations embodied in (5.13): those relations inevitably determine not only the G-scale, but also the environmental sensitivity, of the crack velocity response.

5.4 Crack velocity data

Let us now examine some experimental crack velocity data from the materials literature. Very few systematic studies of kinetic variables (e.g. applied stress, material composition or microstructure, temperature) in relation to a particular theoretical model have been documented. Perhaps the first reported evidence of 'slow' crack growth was made on mica by the Russian school, following Obreimoff's experiment (sect. 1.4). Much work followed on engineering metals (Johnson & Paris 1968). In the late 1960s a series of studies on silicate glasses in aqueous environments was initiated by glass scientists, notably by Wiederhorn (1967, 1970). Wiederhorn's

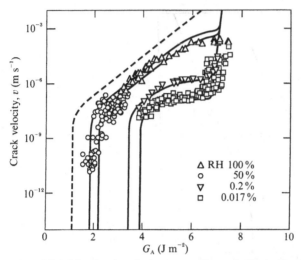

Fig. 5.7. Crack velocity curves for soda-lime silicate glass in air at different relative humidities RH (solid curves) and water (dashed curve). DCB (constant load) and indentation data at 25 °C. Curves are theoretical fits to data (sect. 5.6). (After Wan, K.-T., Lathabai, S. & Lawn, B. R. (1990) *European Ceram. Soc.* **6** 259. Including data from Wiederhorn, S. M. (1967) *J. Amer. Ceram. Soc.* **50** 407; Freiman, S. W., White, G. S. & Fuller, E. R. (1985) *J. Amer. Ceram. Soc.* **68** 108.)

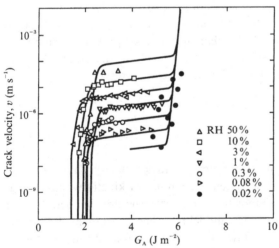

Fig. 5.8. Crack velocity curves for sapphire in moist air at different relative humidities. DCB data (constant load) at 25 °C. Solid curves are theoretical fits to data (sect. 5.6). (After Wiederhorn, S. M. (1969) *Mechanical and Thermal Properties of Ceramics*, ed. J. B. Wachtman Jr., N.B.S. Special Publication **303** 217.)

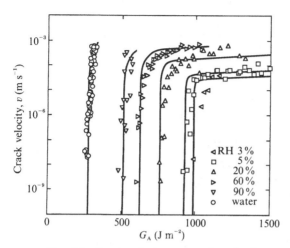

Fig. 5.9. Crack velocity curves for mica in water and moist air at specified relative humidities. DCB data (constant displacement), at 25 °C. Solid curves are theoretical fits to data (sect. 5.6). (After Wan, K-T., Aimard, N., Lathabai, S., Horn, R. G. & Lawn, B. R. (1990) *J. Mater. Res.* **5** 172.)

studies were to lay the groundwork for subsequent adoption of fracture mechanics techniques (sect. 2.5) in the construction of v–G_A or v–K_A curves for practical ceramics.

Accordingly, we examine below a selection of published crack velocity diagrams. Our selection is primarily to demonstrate salient features of the intrinsic velocity functions in 'model' brittle materials, transparent glasses and single crystals free of R-curve effects ($G = G_A = G_*$, $K = K_A = K_*$). Limited mention is made of tougher, polycrystalline ceramics and metals, and there mainly to bring out the importance of shielding. The diagrams are plotted with logarithmic ordinates to accommodate the typically wide range of velocities. At this stage we consider only crack *extension* through *virgin* material, leaving the greater question of crack healing and reversibility to sect. 5.7.

Consider first figs. 5.7–5.9 for soda-lime glass, sapphire, and mica in moist environments. These three examples exhibit common features:

(i) Generally, an extreme sensitivity of the velocity to applied load, suggesting exponential or power-law velocity functions with large co-efficients or exponents in G or K.

(ii) A tendency for the data to fall into three regions, implying transitions

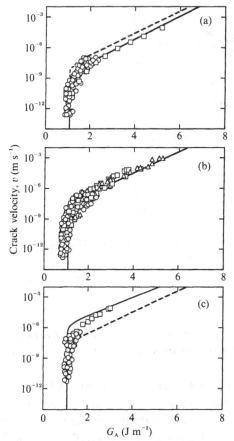

Fig. 5.10. Crack velocity curves for soda-lime silicate glass in water, at (a) 2–5 °C, (b) 25 °C and (c) 80–90 °C. Constant-load DCB data (□, △) and indentation (○) data. Solid curves are theoretical fits (dashed curves in (a) and (c) reproduce the fit at 25 °C as an indication of data shifts). (After Wan, K.-T., Lathabai, S. & Lawn, B. R. (1990) *European Ceram. Soc.* **6** 259; DCB data replotted from Wiederhorn, S. M. & Bolz, L. H. (1970) *J. Amer. Ceram. Soc.* **53** 543.)

in the rate-controlling process: *region I*, at low applied stresses, velocity highly dependent on G; *region II*, at intermediate stresses, velocity tending to a plateau (atmospheric environments); *region III*, at high stresses, velocity even more sensitive to G than in region I.

(iii) A dependence on chemical concentration or partial pressure of environmental species, notably water, in regions I and II. The most apparent manifestation of this dependence is a monotonic data shift at any given G to a higher velocity in these two regions. Region III is independent of environmental species, and reflects the response in a vacuum.

A steepening of the curves at the lower end of region I, especially for the mica in fig. 5.9, suggests the onset of a threshold, sometimes designated *region 0*. In practice, the existence of any such threshold is difficult to establish from forward velocity measurements alone, because of increasingly excessive time intervals needed to measure velocities as the cracks approach equilibrium; on what basis do we extrapolate the logarithmic data in figs. 5.7–5.9 to the abscissa? The true test for existence of a threshold is to unload the system so that the crack reverses. We shall affirm in sect. 5.7 that all our materials do in fact exhibit thresholds.

We now illustrate the role of two important variables in the kinetics, *temperature* and *chemical composition*, with data on glass in aqueous solutions. Fig. 5.10 shows data for soda-lime glass in water at three temperatures. The curve shifts correspond to velocity increases of about two orders of magnitude over a temperature range of ≈ 100 °C, typical of thermally activated processes. There is no indication that the threshold $G_A = G_* = R_E = 2\gamma_{BE} \approx 1$ J m^{-2} is significantly different for the three curves, consistent with the relative temperature insensitivity of intrinsic surface and interface energies.

The role of chemical composition is illustrated in the following three examples:

(i) Fig. 5.11, for several silicate glasses in liquid water. It is evident that the additive oxide components in the glass have a strong influence on the interaction with water. Interpretation of these results is not straightforward: there is no immediately obvious way of facilitating a simple data shift in fig. 5.11; the curves not only have different slopes, they also tend to cross one another.

(ii) Fig. 5.12, for fused silica in acidic and basic aqueous solutions. Again, there is clear evidence of a compositional influence, this time via the solution pH. In this case interpretation would appear to be slightly less complex: one has only to account for a change in slope.

(iii) Fig. 5.13, for soda-lime glass in long-chain alkane solutions. The data closely parallel the curve for water in region I, but then deviate onto a region II plateau in the manner of dilute atmospheres in fig. 5.7.

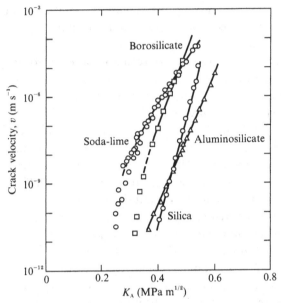

Fig. 5.11. Crack velocity curves for different silicate glasses in water. DCB data (constant load). (After Wiederhorn, S. M. & Bolz, L. H. (1970) *J. Amer. Ceram. Soc.* **53** 543.)

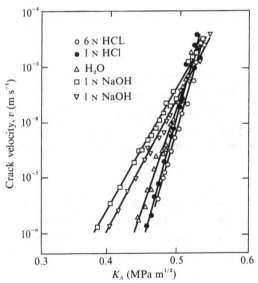

Fig. 5.12. Crack velocity curves for fused silica glass in acidic and basic aqueous solutions. DCB data (constant load). (After Wiederhorn, S. M. & Johnson, H. (1973) *J. Amer. Ceram. Soc.* **56** 192.)

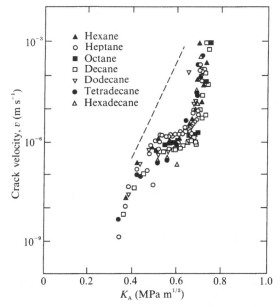

Fig. 5.13. Crack velocity curves for soda-lime glass in straight-chain
alkanes. Dashed curve for water included for comparison. DCB data
(constant moment). (After Freiman, S. W. (1975) *J. Amer. Ceram. Soc.*
58 339.)

The specific nature of the chemical interaction remains a contentious issue
in such complex systems, although the activity of the additive metal ions
within the open glass structure, or of H^+ and OH^- ions within the solution,
are key factors in the first two examples. In the third example, the presence
of region II leads one to conclude that it is trace amounts of water in the
alkanes, and not the alkanes themselves, that govern the kinetics.

At this point it is useful to summarise the principal features of the above
data for our model solids, to set the stage for the theoretical descriptions in
sect. 5.5. We do this schematically in fig. 5.14. Threshold, region 0, defines
a temperature-insensitive Griffith equilibrium state at $G_A = G_* = R_E =
W_{BEB} = 2\gamma_{BE}$. Region I depends strongly on extraneous variables, applied
stress, temperature and chemical concentration, indicative of a thermally-
activated process. Region II is insensitive to applied stress, suggestive of a
transport process in which the active environmental species are increasingly
unable to keep pace with the crack front as G increases. This intermediate
branch thereby 'connects' region I to region III, which identifies with the
velocity response in a vacuum.

Now consider polycrystalline ceramics. The literature data on materials

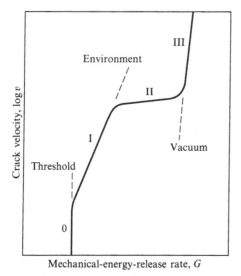

Fig. 5.14. Schematic summarising the different forward crack velocity regions observed in experimental velocity curves.

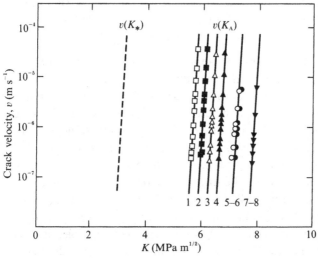

Fig. 5.15. Crack velocity curves for a high-density polycrystalline alumina (grain size 17 μm) in silicone oil. SENB data. Solid curves 1–8 represent v–K_A runs at increasing crack length on single specimen. Data shifts along K axis attributed to progression up T-curve. Dashed curve is deconvoluted v–K_* function obtained by separating out shielding component. (After Deuerler, F., Knehans, R. & Steinbrech, R. (1985) *Fortschrittsberichte der Deutschen Keramischen Gesellschaft* **1** 51.)

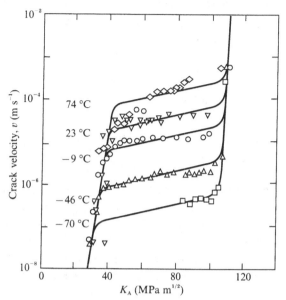

Fig. 5.16. Crack velocity curves for a Ti–5Al–2.5Sn alloy in hydrogen, at several temperatures. DCB data. (After Williams, D. P. (1973) *Int. J. Fract.* **9** 63.)

with 'microstructural inhomogeneity' (chapter 7) are conspicuous for their irreproducibility, not only in results from different material sources but also (even with specimens prepared from a common block of material) from different test geometries. Perhaps most telling, however, are inconsistencies in run-to-run data on single specimens. Such inconsistencies are attributable in large part to shielding. We demonstrate these in fig. 5.15 with some data on a monophase alumina ceramic. These data are from tests in silicone oil: the slow crack growth is presumably due to trace water in the oil. The measured (solid) curves shift further to the right with each successive run (cf. fig. 5.6), indicative of systematic progression up a T-curve. Such curve shifts reflect the history-dependence of global v–K_A functions, i.e. the functions determined by the experimentalist from the externally monitored applied loading. It is only the enclave v–K_* (dashed) curve, evaluated by subtracting out the shielding component $K_\mu(c)$ in (5.15), that is unique.

Finally, we show in fig. 5.16 some velocity data for a brittle metallic alloy in hydrogen, at several temperatures, to emphasise even further that the fundamental kinetic phenomena described here are not confined to ceramics. As in the case of glass in fig. 5.10, we see a strong temperature sensitivity. Well-defined v–K curves like this are evident in many metal systems, even some of those which ordinarily undergo ductile separation in the *absence* of environment (embrittlement).

5.5 Models of kinetic crack propagation

Consider now some of the theoretical quasi-continuum models used to describe crack velocity behaviour. Depicted in fig. 5.17 is a sequence of potential rate-limiting processes along the crack interface. These are, from mouth to tip: mass fluid flow (viscosity-limited); bulk diffusional flow (free molecular flow in gases, solute diffusion in solutions); constrained interfacial diffusion ('lattice-limited'); adsorptive interaction at crack tip. The crack geometry becomes ever more confining, and the dependence on G or K correspondingly stronger, as we proceed from one step to the next. Since it is the near-tip processes that determine the fundamental nature of the chemical interaction we give these our closest attention.

Use of the ensuing models to fit the experimental crack velocity data from the previous section will be deferred to sect. 5.6.

5.5.1 Crack-front reaction kinetics

We look first at a crack-tip interaction model promoted by Wiederhorn (1967, 1970), specifically to describe crack velocity curves in glass. Wiederhorn's model is actually a derivative of an earlier version by Charles & Hillig (1962), in which the crack is assumed to propagate by stress–corrosion at a rounded tip. In that it is based on reaction rate theory, the original Charles–Hillig approach has a certain universality: its phenomenological features extend to any stress-enhanced thermal activation process. Wiederhorn's model may therefore be discussed in relation to adsorption-enhanced rupture by environmental species at a sharp crack tip, consistent with our pervasive view of ideally brittle fracture. In this way, although we consider the role of extraneous variables such as stress, temperature and concentration in generic terms, the fundamental mechanism of bond rupture is implicit.

An idealisation of the model is presented in fig. 5.18(a). We suppose that the environmental species have unimpeded access to the active tip adsorption sites. The adjacent walls are assumed to be stress free, so that

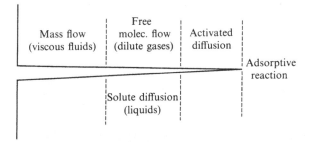

Fig. 5.17. Schematic of chemically interactive brittle crack showing potential rate-limiting steps. As fluid intrudes along the crack interface the restrictive influence of the confining crack walls is felt more strongly in the velocity function.

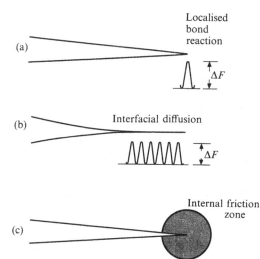

Fig. 5.18. Models for region I of v–G curve. 'Activated' processes: (a) concentrated reaction between active environmental species at crack front; (b) diffusion of active species within constrained (Barenblatt) adhesion zone. Barrier height ΔF is stress-biased for forward ($+$) and backward ($-$) activation. 'Shielding' process: (c) internal friction losses in immediate viscous zone around crack front.

the adsorption products there are passive: i.e. the critical bond-rupture interaction is localised entirely along the (singular) front. We assume also that the interactive species are sufficiently concentrated and chemisorptive that virtually all sites along this front are active ('strong-interaction approximation', $p_0 \ll p_E$, $\theta \to 1$ in (5.4)). Then the molecular 'reaction' rate between environmental fluid and solid surfaces may be written immediately from the classical theory of rate processes (Glasstone, Laidler & Eyring 1941):

$$K = v_0 [\exp(-\Delta F_+/kT) - \exp(-\Delta F_-/kT)] \qquad (5.16)$$

with v_0 a fundamental lattice frequency, kT the Boltzmann thermal energy, and ΔF_+ and ΔF_- molecular free energies of formation of a stress-activated adsorption complex from reactants and products respectively. Usually, the stress dependence is expressed via an arbitrary Taylor expansion in G or K. We choose instead to expand in $\mathcal{g} = G_* - R_E$, to maintain self-consistency with the Griffith balance (Pollett & Burns 1977):

$$F_\pm = \Delta F^\bullet \mp \alpha(G_* - R_E) + \dots \qquad (5.17)$$

with $\Delta F^\bullet$ a quiescent adsorption–desorption activation energy at $G_* = R_E$ and $\alpha = -(\partial F/\partial G_*)$ an activation area.[1]

The crack *velocity* is given by Ka_0 where a_0 is a characteristic atom spacing. Then (5.16) and (5.17) yield (cf. Pollett & Burns 1977)

$$v = 2v_0 a_0 \exp(-\Delta F^\bullet/kT) \sinh[\alpha(G_* - R_E)/kT]. \qquad (5.18)$$

The quiescent point $G_* = R_E$ represents a *threshold* configuration in v–G_* space. At $G_* > R_E$ the crack extends (adsorption-controlled), at $G_* < R_E$ it contracts (desorption-controlled). If we confine our attention to just forward velocities, $\Delta F_+ = \Delta F_- - \varepsilon$, $\varepsilon > kT$, the second exponential term in (5.16) may be neglected, so that (5.18) reduces to the simple form

$$v = v_0 \exp(\alpha G_*/kT), \quad (G_* > R_E) \qquad (5.19)$$

with pre-exponential term

$$v_0 = v_0 a_0 \exp(-\Delta F^\bullet/kT) \exp(-\alpha R_E/kT). \qquad (5.20)$$

The influence of concentration of active environmental species may now be introduced via $R_E = W_{BEB}(p_E)$. We see from (5.18) that a change in concentration amounts to a shift in the velocity curve along the G axis. For

[1] Strictly, the parameters $\Delta F^\bullet$ and α in region I are functions of surface coverage (θ); but, provided we remain in our assumed domain of 'strong interactions' $(p_0 \ll p_E)$, any such dependence may be disregarded. If on the other hand we were to enter the opposite domain of 'weak interactions' $(p_0 \gg p_E)$, these parameters would be expected to undergo rapid change, e.g. as reflected by the distinctive trend to higher slopes in region III of the velocity plots in sect. 5.4.

the special case of the Langmuir adsorption isotherm (5.5) for gases in the 'strong-interaction' limit ($p_0 \ll p_E$), the velocity in (5.19) and (5.20) assumes a simple power-law dependence on pressure, $v \propto p_E^\beta$, $\beta = 2\alpha\Gamma_m$ a dimensionless constant.

The above formalism accounts, at least qualitatively, for the dependence of region I crack velocity behaviour on the principal extraneous variables observed in the preceding section:

(i) *Applied load.* The strong dependence on G_* in (5.19) is consistent with steep region I behaviour in figs. 5.7–5.13.

(ii) *Temperature.* The comparable dependence on T in (5.19) and (5.20) accounts for the substantial velocity increases at higher temperatures in fig. 5.10.

(iii) *Concentration.* A modest (inverse) dependence on p_E in $R_E = W_{BEB}(p_E)$ explains the shift to higher velocities at increased humidities in figs. 5.7–5.9.

5.5.2 Transport-limited kinetics: activated interfacial diffusion

There is an alternative process that can lead us to the kind of stress-activated velocity function in (5.18): interfacial diffusion in the adhesion zone immediately behind the crack front where the intrusive species feel the influence of both crack walls simultaneously, as depicted in fig. 5.18(b). We may recall Orowan's speculation on this possibility from sect. 5.1. According to that view the crack-tip adsorption step is virtually instantaneous. The kinetics in region I are then attributable to solid-state diffusion of the ingressing molecules within an extended (Barenblatt-type) interface.

If we now assume that the energy barriers in fig. 5.18 are subject to bias from the applied loading, we may apply the phenomenological reaction rate formalism of the previous subsection to the interfacial transport process without any mathematical modification: the velocity dependence on applied load, temperature and chemical concentration is unchanged. We need only attach different physical meanings to the parameters in that formalism: ΔF^* and α relate to diffusion barriers rather than to adsorption barriers.

The implication of this mathematical correspondence is immediate: knowledge of the velocity function $v(G_*, T, p_E)$ alone is generally not sufficient to determine whether it is the adsorption or diffusion (or other) step that governs the region I kinetics.

5.5.3 Internal friction in intrinsic shielding zone

Yet another way that rate dependence might possibly enter the crack growth laws is by way of 'internal friction' in the material immediately surrounding the crack front, fig. 5.18(c). A continuum model based on this postulate has been described by Maugis (1985). It is based on the assertion that all materials are, to a greater or lesser extent, viscoelastic; and that it is the viscous component in the bulk constitutive law, not the adhesion or diffusion kinetics, that determines the velocity function.

A feature of the Maugis model is the continued preservation of the Griffith condition: the viscous loss process, although dominant in the energetics, remains subsidiary to the surface separation process. The underlying principle can perhaps be illustrated most simply by reference to a contrived experiment in which a film of viscous, wetting liquid is inserted between two narrowly separated but otherwise non-adhesive glass strips in a DCB configuration (Burns & Lawn 1968). For infinitely slow peeling of the glass strips, the mechanical-energy-release rate G is identical to the surface energy $W_{BEB} = 2\gamma_L$ of the liquid in the capillary (as is readily confirmed by independent surface tension measurements). For fast separations, however, there is energy dissipation as the liquid molecules migrate to or from the surface at the meniscus, the more so the greater the magnitude of $G - 2\gamma_L$, and the crack propagates forward (or backward) at a rate limited by the viscosity of the liquid. The viscous-loss and surface-separation processes operate simultaneously, but are physically decoupled.

This decoupling is reminiscent of the shielding description. The Maugis model may indeed be formulated as a shielding problem, by writing the condition for crack extension in terms of Rice's equation (5.12) for *quasi-equilibrium* states,

$$G_* = R'_E = R_E = \Delta R_E(v) \tag{5.21}$$

with $\Delta R_E(v)$ an internal viscous loss term, i.e. an intrinsic component of the constitutive stress–strain characteristic within the crack-tip enclave. The steady-state v–G_* function may then be deconvoluted from (5.21).

Maugis finds that an empirical function $\Delta R_E(v) = R_E(v/v_0)^{1/n}$, where v_0 and n are adjustable parameters, fits velocity data for viscoelastic solids. The proportionality between ΔR_E and R_E is indicative of the multiplicative scaling in the shielding energy foreshadowed in sect. 5.3. This results in the power-law velocity function

$$v = v_0[(G_* - R_E)/R_E]^n \tag{5.22}$$

where v_0 is an adjustable constant (cf. v_0 in (5.19)). Temperature and chemical concentration enter the description via v_0 and R_E, much as in the reaction-rate models.

Maugis's model, while demonstrably applicable to polymeric materials, is not widely acknowledged by those who deal with ceramics, especially dense monocrystalline structures like sapphire and mica. But there are certain crystals, notably those of the rocksalt type, where a limited amount of crack-tip 'visco-plasticity' can account for velocity effects, as we shall certify later (sect. 7.3).

5.5.4 Transport-limited kinetics: free molecular flow of 'dilute' gases

We alluded earlier (fig. 5.17) to fluid transport processes (additional to adhesion-zone diffusion) that might be instrumental in the crack kinetics, specifically in region II of the v–G curve. Such processes can be rate-controlling if the characteristic time to transport the active species along the crack interface to the interaction zone exceeds the interaction time itself. There are many possible transport mechanisms, for not only may the nature of the interacting species vary but so also may the transport medium. Thus, environmental molecules may gain access to the crack front by mass (viscosity-limited) flow in the wider interfacial regions near the crack mouth, by bulk diffusion of solute molecules in an otherwise inert liquid, by surface diffusion at the crack walls, by ionic diffusion through an 'open' solid structure (e.g. silicate glass) *adjacent* to the tip or corrosion product *behind* the tip, and so on. As a general rule, the v–G dependence becomes weaker the more remote the rate-controlling process from the crack tip.

Here we confine our attention to one specific example, that of free molecular flow in gaseous environments, fig. 5.19. This is probably the most commonly encountered of extrinsic transport mechanisms: recall the

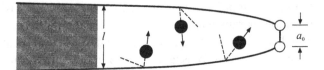

Fig. 5.19. Model for region II of v–G curve, gaseous environments. Shaded area at left is region of mass flow. Free molecular flow ensues when crack-opening displacement diminishes below intermolecular mean free path l. Attenuation in flow rate to crack tip at right is then governed by thermally diffuse scattering of molecules from walls. (After Lawn, B. R. (1974) *Mater. Sci. Eng.* **13** 277.)

distinctive region II plateaus in the curves for moist atmospheres, figs. 5.7–5.9. Moreover, the molecular flow rate may be derived relatively simply using the elementary kinetic theory of gases.

Consider a crack system containing an interactive gas. Suppose the underlying crack-tip interaction to be 'strong' ($p_0 \ll p_E$, sect. 5.1) and 'instantaneous', and the gas to be 'dense' (p_E const along the interface). Then the velocity is determined by the number of molecules M in the bulk gas impinging on each adsorption site along the crack front in unit time,

$$v = (a_0/\eta)\,dM/dt \qquad (5.23)$$

with a_0 the adsorption site spacing and η the site occupancy of adsorbate molecules. From kinetic theory, the impingement rate per site is

$$dM/dt = a_0^2 p_E/(2\pi m kT)^{1/2} \qquad (5.24)$$

where a_0^2 is the site cross-sectional area and m is the molecular mass. In combination with (5.23) we obtain the velocity

$$v = a_0^3 p_E/\eta(2\pi m kT)^{1/2}. \qquad (5.25)$$

Note that this velocity is independent of G or K.

Let us estimate the velocity for cracks in moist air. We take $\eta = 1$ (one molecule per crack-tip bond), $a_0 = 0.5$ nm (atom spacing), $m = 30 \times 10^{-27}$ kg (H_2O molecule), $T = 300$ K, $p_E = 3$ kPa (saturated water vapour pressure at room temperature), to obtain $v \approx 10$ mm s^{-1}. This value lies above the appropriate region II 'plateaus' in figs. 5.7–5.9.

In fact, we should expect (5.25) to overestimate region II velocities, because it makes no allowance for a pressure drop along the crack interface in steady-state flow. The mean free path for intermolecular collisions at the pressure and temperature used in the above calculation is ≈ 1 μm, which must exceed the crack-opening displacement at some distance behind the crack front (typically, at $X \approx 100$ μm in (2.15) for the more brittle solids). As the gas molecules migrate along the interface a point will be reached where collisions with the confining crack walls become more frequent than with other gas molecules (Knudsen gas). The gas is then effectively 'dilute', and enters a zone of 'free molecular flow' within which diffuse molecule–wall scattering attenuates the transport rate. The pressure differential between crack mouth (source) and crack front (sink) then regulates the flow: at low v the net adsorption rate at the front is comparatively small, so the crack-tip pressure equilibrates with that at the mouth; at intermediate v the adsorption rate accelerates, and the supply of molecules to the tip depletes. Accordingly, one may regard the transition between regions I and II as a two-step process in which extrinsic (chemically-assisted) bond rupture and free molecular flow act sequentially. Ultimately, at high v, the depletion is virtually complete, i.e. the crack front approximates to 'vacuum' conditions. The attendant transition between regions II to III involves a competition, where (flow-limited) extrinsic and intrinsic bond rupture act simultaneously. It is in this spirit of *series* and *parallel* processes that branch II may be seen as 'connecting' branches I and III in fig. 5.14.

These aspects of free molecular flow may be incorporated into the velocity formalism by including a dimensionless, geometry-dependent Knudsen attenuation term κ in the region II relation (5.25) (Lawn 1974):

$$v_{II} = \kappa(G_*) a_0^3 p_E / \eta (2\pi m k T)^{1/2}. \tag{5.26}$$

Evaluation of κ for parabolic Irwin cracks using (2.15) gives

$$\kappa(G_*) = 64 G_* / 3\pi E a_0 \ln(l/a_0), \quad (\kappa \leqslant 1)$$

with E Young's modulus and l the mean free path for the bulk gas. Thus the region II velocity varies, if relatively slowly, with G_*.

At this point let us rewrite the activation relation (5.18) for forward velocities specifically for regions I and III, using (5.1) to identify R_E:

$$v_I = 2v_0 a_0 \exp(-\Delta F_I^*/kT) \sinh[\alpha_I (G_* - W_{BEB})/kT],$$
$$(W_{BB} \geqslant G_* \geqslant W_{BEB}) \tag{5.27a}$$

$$v_{III} = 2v_0 a_0 \exp(-\Delta F_{III}^*/kT) \sinh[\alpha_{III} (G_* - W_{BB})/kT],$$
$$(G_* \geqslant W_{BB}). \tag{5.27b}$$

(a)

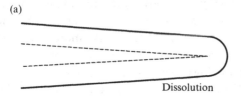

Dissolution

(b)

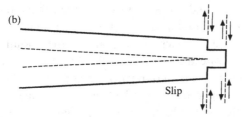

Slip

Fig. 5.20. Schematic depicting crack-tip blunting by (a) dissolutive corrosion, (b) plastic slip.

Then, simplistically, for the series I–II transition we have

$$v/v_{\mathrm{I}} + v/v_{\mathrm{II}} = 1 \qquad (5.28a)$$

so that the *slower* of the two steps is rate determining. Similarly, for the parallel II–III transition

$$v = v_{\mathrm{II}} + v_{\mathrm{III}} \qquad (5.28b)$$

so that the *faster* step is rate determining. Equations (5.26)–(5.28) may thereby be used to generate the entire, composite v–G curve in fig. 5.14.

A potential complication may be mentioned here. We have indicated (sect. 5.1) that in humid atmospheres water vapour can condense to form a capillary immediately behind the crack front. This capillary may act to shorten the zone of free molecular flow along the interface. In extreme cases where this meniscus extends back beyond the point at which the crack-opening displacement exceeds the mean free path of intermolecular collisions there will be no zone of free molecular flow, hence no region II. In this context we acknowledge the absence of plateaus at higher relative humidity for mica in fig. 5.9.

5.5.5 The blunt-crack hypothesis

If there is one pervasive theme that enables us to consolidate the preceding models into a universal description it is this: *the tips of propagating cracks are atomically sharp*. It is the sharp-crack, bond-rupture concept that embodies the very essence of brittleness. Yet there are crack-tip models that depart irrevocably from this ideal. Most notable are the *blunting* models, based on the postulate that extension necessarily involves a fundamental change in crack-tip structure. Such a change in structure determines not only the energetics but also the mechanism of extension.

Two classes of such models are sketched in fig. 5.20. Class (a), crack-tip rounding by stress-enhanced dissolution, was foreshadowed by the early Charles–Hillig theory (sect. 5.5.1). In class (b), an analogous rounding is attributed to shear-induced plasticity. The argument goes that as G or K decreases the propagating crack becomes increasingly blunt at the tip, and thereby slows down. Then the micromechanics of fracture are determined by extraneous material processes, in class (a) by corrosion chemistry, in class (b) by plastic flow.

The blunt-crack hypothesis is open to compelling physical objections:

(i) Acceptance of an ever-changing crack-tip structure is tantamount to rejection of the Griffith philosophy. Surface (interface) energies are no longer the fundamental determinants of equilibrium (threshold) states: additional, extrinsic parameters, e.g. crack-tip radius, must be specified. The basis for asserting the existence of unique v–G_* curves is thereby lost.

(ii) There is an *ad hoc* flavour to the modelling. Many variants exist, especially in the silicate glass literature.

(iii) The evidence for blunting is invariably circumstantial. Most commonly cited are correlations between crack velocity and rate dependence of corrosion or of hardness. Such correlations might well be expected of any stress-sensitive activation process.

In view of these and other, more specific, objections (chapters 6 and 8) we are led to suggest that 'atomic sharpness' is an innate quality of brittle cracks that is not easily negated. The sharp-crack 'laws' of crack propagation outlined in sects. 5.5.1–5.5.4 above, notwithstanding their phenomenological basis, possess a certain quality of invariance. At the same time, denial of the blunt-crack hypothesis is not to exclude the activity of corrosion or flow processes at the adjacent crack interface. Indeed, such processes *are* observed, e.g. in the form of interfacial corrosion products, even in the most brittle ceramics. Proponents of the sharp-crack concept would argue that the exclusive role of any such extraneous process is to shield the tip from the remote loading, and as such remains subsidiary to that of bond rupture: so that, whereas the global v–G_A function may become history dependent, the fundamental enclave v–G_* function *always remains uniquely defined*. In extreme dissolutive environments (e.g. as in the prolonged acid etching treatments used to strengthen glass surfaces) crack tips may be eliminated altogether, leaving remnant elongate cavities or pits ('notches'). In such instances the ensuing fracture falls more properly in the domain of crack (re-) initiation (chapter 9) than propagation.

5.6 Evaluation of crack velocity parameters

How well do specific theoretical crack velocity formulations fit the experimental data? This has been a topic of considerable debate in the literature for two decades. Our formulation for region I in sect. 5.5.1 indicates an exponential relationship between v and G. But we should remember that this relationship derives from a Taylor expansion in G in the activation energy equation, (5.17). Others argue that it is no less justifiable to expand in K. Still others insist that the appropriate v–G or v–K relation is a power law instead of an exponential (e.g. sect. 5.5.3). Scatter in the experimental data, coupled with the steepness of the curves in region I, make distinctions between the various possible relations impracticable.

Nevertheless, we have fitted the composite equations (5.26)–(5.28) for activated crack growth in gaseous environments to some of the data in figs. 5.7–5.10. Those fits, shown as the solid curves, have been made by adjusting the activation parameters collectively for each material data set, and independently adjusting R_E at each value of relative humidity in accordance with our simple notion of G-shifts. The resultant activation parameters listed in table 5.1 are typical of atomic-scale processes.

The reaction rate theory would therefore appear to account for the major features of the observed v–G characteristics, including transitions between regions I, II and III, thresholds, and curve shifts at different temperatures and concentrations. Again, however, it is well that the

Table 5.1. *Crack velocity activation area and energy parameters in* (5.28) *for solids represented in figs.* 5.7–5.10 (*water environment*).

Parameter		Soda-lime glass	Sapphire	Mica
a_0	(nm)	0.50	0.48	0.46
α_I	(nm^2 molec^{-1})	0.010	0.057	0.150
α_{III}	(nm^2 molec^{-1})	0.120	0.100	–
ΔF_I^*	(aJ molec^{-1})	0.105	0.106	0.071
ΔF_{III}^*	(aJ molec^{-1})	0.040	0.017	–

Values of a_0 are characteristic bond spacings.

phenomenological character of the analysis be re-iterated. The general stress, temperature and concentration dependence of region I crack velocity behaviour is consistent with *any* stress-enhanced thermal activation process. And there are anomalies in some velocity data that the activation relations, in their present simplistic form, have no provision to explain. Recall the tendency in figs. 5.11–5.13 for the curves to change their region I slope and even to cross one another as the solution or glass chemistry is altered.

Accordingly, we are not yet disposed to address the important question as to which environmental species might interact with which brittle solids.

5.7 Thresholds and hysteresis in crack healing–repropagation

One aspect of fracture chemistry that remains to be addressed is crack healing. Repeated reference has been made to the importance of healing in the brittle fracture description: theoretically, for its bearing on Griffith reversibility; practically, for its function in establishing the existence of crack velocity thresholds. Our continuum formalism, as implied in the simple interplanar functions of fig. 5.2 and the generic models of sect. 5.5, presumes that reversibility is total. For infinitesimal departures below the equilibrium condition $G_* = R_E = W_{BEB} = 2\gamma_{BE}$ the crack necessarily contracts at some nonzero rate. On the other hand, we pointed out that the more universal Rice description (sect. 5.2) imposes no such restriction on the reverse motion; so that there is an allowable hysteresis in the v–G curve, as depicted schematically in fig. 5.4. In this section we examine the experimental reality and foreshadow its repercussions on the continuum description of crack-tip chemistry.

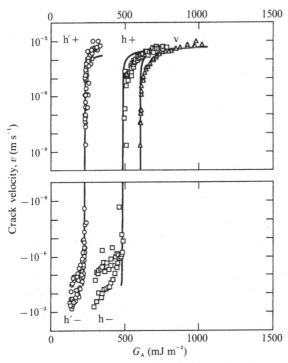

Fig. 5.21. Forward ($+$) and backward ($-$) crack velocity as function of G_A for mica. Data points for moist air (50% RH), in following sequence: loading through virgin interface (v); unloading–reloading through healed interface (h$-$, h$+$); unloading–reloading through healed–misoriented interface (h$'-$, h$'+$). Solid curves are theoretical fits (sect. 5.6). DCB data (constant displacement) at 25 °C. (After Wan, K.-T., Aimard, N., Lathabai, S., Horn, R. G. & Lawn, B. R. (1990) *J. Mater. Research* **5** 172.)

Of the few systematic experimental studies on the mechanics of crack healing, most have focussed on mica and silicate glass:

(i) *Mica* has long been considered the most amenable of all solids for healing studies, because of its atomically smooth cleavage (Bailey & Kay 1967). Fig. 5.21 shows some recent velocity data on this material for crack propagation through virgin (v) and healed (h) interfaces in moist air, including data for sheets recontacted and healed after complete separation and angular rotation (h$'$). The cracks do heal, so threshold behaviour is confirmed. But there is hysteresis in the loading–unloading–reloading cycle, as one traverses from curve v to h and from h to h$'$.

The threshold configurations at $v = 0$, $G = R_E$ measure the dependence

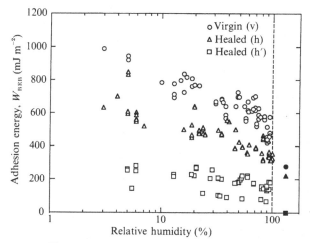

Fig. 5.22. Measured adhesion energy W_{BEB} for mica against partial pressure of water vapour (plotted as relative humidity RH $= p_E/p_S$, p_S pressure at saturation). Data from crack velocity thresholds for virgin interfaces, healed interfaces, and interfaces healed after angular misorientation of cleavage halves. Filled symbols at right are corresponding data for water. (After Wan, K.-T. & Lawn, B. R. *Acta Metall.* **38** 2073.)

of W_{BEB} on relative humidity, for healed as well as virgin interfaces. Results of such measurements are plotted in fig. 5.22. Note the general decline of W_{BEB} with relative humidity, in qualitative accordance with (5.5). The extrapolated values of W_{BEB} at 100% humidity are greater than those in water by about 150 mJ m^{-2}, indicating the presence of capillary formation in saturated atmospheres.

(ii) *Silicate glass*, because of its transparency, isotropy and homogeneity, is another favoured solid for healing studies. Velocity data for healed cracks in soda-lime glass in air are shown in fig. 5.23 (Stavrinidis & Holloway 1983). The hysteresis between propagation and repropagation branches is more pronounced than in mica. An important facet of these data is the history dependence of the healed interface: the hysteresis diminishes with increasing temperature and time between unloading and reloading. This is indicative of activated recovery in the interfacial energy. The development of corrosion products in overaged cracks in glass, ultimately leading to fusion of the interfaces, is one means of realising such recovery.

Qualitative observations of irreversible healing in other brittle solids,

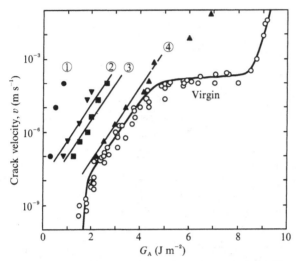

Fig. 5.23. Crack velocity data for soda-lime silicate glass in air (50–70 % RH). Curve at right denotes virgin crack; lines 1–4 denote healed interfaces; (1) aged 5 min, (2) 24 hr, (3) 30 days at 25 °C, and (4) heat treated to 120 °C. DT data. (After Stavrinidis, B. & Holloway, D. G. (1983) *Phys. Chem. Glasses* **24** 19.)

including sapphire (e.g. fig. 6.23), have been reported. One may conclude that, even for the best behaved brittle solids, loading–unloading hysteresis is the norm rather than the exception. There is the implication of metastability in the crack equilibrium states, an explanation of which awaits the atomic descriptions of the next chapter.

6

Atomic aspects of fracture

Until now we have approached crack propagation from the continuum viewpoint. Nonetheless, repeated allusions have been made in chapters 3 and 5 to the fundamental limitations of any such approach that disregards the atomic structure of solids. There we argued for the incorporation of a lattice-plane range parameter as a critical scaling dimension in the brittle crack description. We noted that the Barenblatt cohesion-zone model avoids reference to the atomic structure by resorting to the Irwin slit description of cracks; yet estimates of the critical crack-opening dimensions using this same model confirm that the intrinsic separation process indeed operates at the atomic level. The Elliot lattice half-space model of sect. 3.3.2 represents one attempt to incorporate an essential element of discreteness. The phenomenological kinetic models of sect. 5.5, with their presumption of energy barriers, represent another. However, those models are at the very least quasi-continuous. In brittle fracture, as in any thermodynamic process, the final answers must be sought at the atomic or molecular level.

On the other hand, while an atomistic approach provides greater physical insight into the crack problem, it inevitably involves greater mathematical complexity. Classically, solids may be represented as many-body assemblages of point masses (atoms) linked by springs (bonds). We will see that the mass–spring representation can lead us to a deeper understanding of brittle cracks. But even this representation is over-simplistic. In some cases, particularly when the crack interacts with environmental species, it is necessary to consider atoms as elastic spheres rather than point masses, to allow properly for molecular size effects. The solid-state mechanics equations that need to be solved in the general atomic modelling of brittle cracks are formidable, and several strategies have been adopted by researchers: devise 'naive' structural models that minimise computational detail yet preserve all the necessary elements of a discrete crack structure; revert to quasi-continuous descriptions, replacing individual atom–atom potential functions with an integrated, surface–surface function for atom planes; or, in the last resort, take the fully discrete problem to the computer. We shall encounter all of these strategies in the ensuing sections of this chapter.

Our treatment begins with one of the simplest of atomistic models, due to Orowan and Gilman, in which attention is focussed exclusively on an individual nonlinear crack-tip bond. Then we introduce the more sophisticated point-atom/spring-bond lattice-crack models of Thomson, with general capacity to prescribe a nonlinear short-range force function for the critical bond. These models lead us to the concept of 'lattice trapping'. An elaboration of the nonlinear crack-tip interaction function allows us to proceed naturally to an account of environmental chemistry, for both equilibrium and kinetic states. However, the short-range force models are not well-equipped to explain the observed hysteresis in crack loading–unloading–reloading experiments described in chapter 5. It is necessary to pay closer attention to the fine detail of actual (independently measured) interplanar force functions; in particular, to longer-range secondary energy minima that reflect the structural discreteness of intervening fluid species at the constrained crack interface. A modified Barenblatt–Elliot, 'surface forces' model, with due allowance for metastable interfacial states, is accordingly proposed. Then we consider crack-tip plasticity, and reaffirm its status as a subsidiary factor in the micromechanics of crack propagation in covalent–ionic solids. Finally, we examine critical experimental evidence, notably from transmission electron microscopy, that bears directly on the fundamental nature of the crack-tip structure. This evidence will be used to reinforce our assertion of atomic sharpness as an innate quality that defines the very essence of brittleness.

6.1 Cohesive strength model

We now consider how the sequential bond-rupture picture of brittle fracture may be expressed in terms of an interatomic cohesive-force function. In chapter 3 we observed that the energetics of continuum cracks could be formulated as the integral of a cohesion stress–separation function, as in (3.1). However, that formulation makes no specific comment on the form that the cohesion function should take, or on how such a

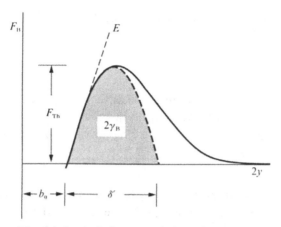

Fig. 6.1. Intrinsic interatomic force function. Dashed half-sine curve is Orowan–Gilman approximation. To break the bond (i.e. separate atoms from equilibrium spacing b_0 to infinity) the force transmitted to the crack tip by the applied load must exceed F_{Th}.

function may be adopted into an explicit criterion for crack extension at the level of the individual crack-tip bond. Orowan (1949) and Gilman (1960) were among the first to address these issues. Their approach, although semi-empirical, is useful in identifying the atomic parameters that control brittle fracture.

The *in-vacuo* force–separation function $F_B(y)$ for an individual crack-tip bond is depicted in fig. 6.1. Here we conform to the sign convention adopted in sect. 2.6, i.e. attractive cohesive force positive (i.e. $F_B(y) = +dU_B(y)/d(2y)$, where U_B is the conjugate interatomic potential function). Gilman approximated the intrinsic bond force–separation function in the attractive region to a half-sine curve

$$F_B(u) = F_{Th} \sin(2\pi u/\delta'), \quad (0 \leqslant 2u \leqslant \delta') \tag{6.1}$$

where F_{Th} is the theoretical limiting force for bond rupture, $2u = 2y - b_0$ is the displacement of the atom pair from its equilibrium interatomic separation b_0, and δ' is a 'range' parameter for the bond (analogous to δ in fig. 3.1). One may sensibly define an equivalent crack-tip cohesive-stress function

$$p_y(u) = F_B(u)/a_0^2 \tag{6.2}$$

where a_0^2 is the interfacial area of the bond. Equations (6.1) and (6.2) combine to give

$$p_y(u) = p_{Th} \sin(2\pi u/\delta'), \quad (0 \leqslant 2u \leqslant \delta') \qquad (6.3)$$

where $p_{Th} = F_{Th}/a_0^2$ defines the theoretical cohesive strength of the solid. For the crack to extend p_y must exceed p_{Th} at the crack tip.

An estimate of p_{Th} may be gained by matching the sine function in (6.3) to two familiar fracture parameters:

(i) *Young's modulus.* According to Hook's law, the slope of the stress–separation curve at equilibrium must give a measure of the extension elastic modulus. Thus

$$E = b_0[\mathrm{d}p_y/\mathrm{d}(2u)]_{u=0} = (\pi b_0/\delta')p_{Th}. \qquad (6.4)$$

(ii) *Surface energy.* The area under the curve must give a measure of the intrinsic surface energy for the body. Noting that the stress cuts off at $2u = \delta'$ we have

$$2\gamma_B = \int_0^{\delta'} p_y(u)\,\mathrm{d}(2u)$$
$$= (2\delta'/\pi)p_{Th}. \qquad (6.5)$$

Unfortunately, δ' depends strongly on the bond type, and is difficult to specify. We presume it to be of atomic dimensions, $\delta' \approx b_0$, so that

$$p_{Th} = E/\pi \qquad (6.6a)$$

$$2\gamma_B = 2Eb_0/\pi^2 \qquad (6.6b)$$

approximately. Alternatively, we may avoid any speculation as to the value of δ' by eliminating it from (6.4) and (6.5) to get

$$p_{Th} = (E\gamma_B/b_0)^{1/2}. \qquad (6.7)$$

Thus high intrinsic strength would appear to be favoured by large Young's modulus and surface energy, and by close packing of atomic planes.

With this estimate of the theoretical cohesive strength we return to the Inglis analysis (sect. 1.1) for the stress concentration at the crack tip. Inserting $\sigma_c = p_{Th}$ in (1.4), we obtain the critical applied stress $\sigma_A = \sigma_F$,

$$\sigma_F = \tfrac{1}{2} p_{Th}(\rho/c)^{1/2}$$
$$= [(\pi\rho/8b_0)(2E\gamma_B/\pi c)]^{1/2} \tag{6.8}$$

which is of the same form as the strength equation (1.11) obtained by Griffith. In fact, the two equations are identical if $\rho = 8b_0/\pi$.

It may seem fortuitous that the Orowan–Gilman and Griffith approaches should lead to near-identical results. After all, Griffith's treatment is based on global concepts while Orowan–Gilman is based on atomistic concepts. However, we may recall our earlier demonstration using the *J*-integral (sect. 3.5) that any soundly based crack-tip criterion must, if it is to satisfy the requirements of thermodynamic equilibrium, be *equivalent* to the Griffith condition. In this context it needs to be remembered that the stress equation (6.3) has been forced to match precisely those macroscopic parameters contained in the Griffith equations; the Griffith concept has been 'built in'. In considering the relative merits of the two approaches it might be argued that Orowan–Gilman is the more fundamental, in that it comes closer to a mechanistic description. If so, it is also subject to greater uncertainty, for implicit in that treatment are features that oversimplify the crack-tip structure, and that consequently cast doubt on the numerical constants in (6.8). Chief among these features is the assumption that the continuum concept of a crack-tip radius remains valid at the atomic level. Again, the model makes no attempt to distinguish between classes of bonding, and bond type is known to be an important factor in determining the intrinsic fracture resistance. In effect, Orowan–Gilman regards the solid as a simple array of parallel, uncoupled linear chains, subject only to longitudinal components of force; 'lattice constraint' effects are ignored. We address this last point in the next section.

The cohesive strength model, in its original form, was conceived strictly with *in-vacuo* bond rupture in mind. One might propose extending the model to include chemically-assisted crack propagation by inserting a half-sine function $p_\gamma(u)$ of reduced amplitude into (6.5), consistent with a reduction of surface energy from γ_B to γ_{BE}, as in fig. 5.2. That such an approach is inappropriate will become apparent in sect. 6.5 when we consider the more elaborate force functions that characterise crack-tip chemistry.

6.2 Lattice models and crack trapping: intrinsic bond rupture

The next level of discrete models for brittle cracks is that of a critical cohesive bond embedded in a point-atom lattice, as developed by Thomson and co-workers (Thomson, Hsieh & Rana 1971; Thomson 1973; Fuller & Thomson 1978). The main goal in this type of modelling is not so much a more realistic representation of true solid structures but, rather, a deeper physical insight into certain profound effects of material discreteness not evident from the simplistic Orowan–Gilman analysis. In particular, we shall explore the predicted existence of atomic-scale periodicity in the Griffith energetics as a natural consequence of the sequential bond-rupture process. This leads us to the concept of 'lattice trapping' (the fracture analogue of the Peierls resistance to dislocation motion), and thence to a fundamental description of crack kinetics.

6.2.1 Quasi-one-dimensional chain model

Consider the one-dimensional vacuum 'crack' structure in fig. 6.2. It consists of two semi-infinite chains of point atoms 'bonded' longitudinally by linear bendable elements and transversely ahead of the crack tip by linear stretchable elements, with zero-stress separations a_0 and b_0 and spring constants α and β, respectively. The only element assumed to be in the nonlinear force region is the nth stretchable bond B–B at the crack 'tip'. Opening forces P act at the free ends. Behind the crack tip the stretchable bonds are considered 'broken', corresponding to displacements beyond a 'cutoff' in the force–separation function. Propagation entails 'rupture' of the crack-tip bond, whence the force–separation function of the next bond along the chain assumes the nonlinear configuration of its predecessor. The aim of this model is to determine how the 'atomicity' of the structure makes itself felt in the global energetics of crack growth.

It might be argued by detractors that the unlikely looking quasi-one-dimensional assemblage of point atoms and springs in fig. 6.2 can have no possible relevance to any real solid. However, the structure does contain the salient features of a brittle crack: it has the quality of a sharp (Elliot-type) slit in a linear elastic matrix; it retains the atomistic features of the Orowan–Gilman nonlinear crack-tip description, with inclusion of lattice constraint on the bond rupture; and, as we shall see in sect. 6.4, it has

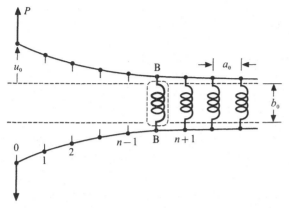

Fig. 6.2. Quasi-one-dimensional chain model of a sharp crack, with nonlinear crack-tip bond B–B embedded in a linear 'lattice'. (After Thomson, R. M., Hsieh, C. & Rana, V. (1971) *J. Appl. Phys.* **42** 3154.)

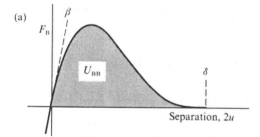

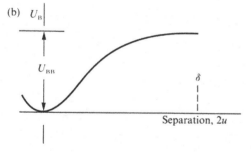

Fig. 6.3. Intrinsic (a) force and (b) potential energy functions, range δ, for bonds stretched across crack plane. In the force diagram the slope at equilibrium gives the stiffness β, and the area under the curve the bond-rupture energy U_{BB}.

provision to accommodate chemical interactions. Most importantly, it is amenable to analytical solution. There are in fact more sophisticated variants of this kind of lattice model in the literature. We concentrate on the quasi-one-dimensional chain because it allows us to make some very general conclusions concerning the role of discreteness in the crack resistance with minimum analytical complexity.

Formal analysis begins by seeking equilibrium configurations for the discrete structure. To set up the computation we first define some characteristic quantities in relation to the nonlinear stretchable bond, fig. 6.3. The cohesive force– and energy–displacement functions for the jth stretchable bond are connected in the usual way, i.e. $U_B(u_j)$ equal to the area under the $F_B (u_j)$ curve. In the region behind the crack tip ($j < n$) the bonds are stretched beyond the 'cutoff' displacement $2u_j = \delta$, so we may define a cohesion energy for the separated crack-tip bond

$$U_{BB} = \int_0^\delta F_B(u_j)\,d(2u_j). \tag{6.9}$$

Ahead of the tip ($j > n$) the bonds are stretched in the Hookean region. The stiffness in this region is determined self-consistently as

$$\beta = [dF_B(u_j)/d(2u_j)]_{u_j} = 0. \tag{6.10}$$

The total potential energy of the system may then be written as a function of the displacements u_j of atoms along the chain,

$$U = U_B(u_n) + nU_{BB} + 2\beta \sum_{j=n+1}^{\infty} u_j^2 + \alpha \sum_{j=1}^{\infty} (u_{j+1} - 2u_j + u_{j-1})^2 - 2Pu_0. \tag{6.11}$$

The first term on the right is the energy of the strained nonlinear crack-tip bond; the second term is the cohesion energy of all broken bonds behind the crack tip; the third is the strain energy of the stretchable elements ahead of the crack tip; the fourth is the strain energy of the bendable elements; and the last is the potential energy of the applied loading system.

There is one aspect of this description that calls for special comment. Since the transverse bonds ahead of the tip are elastically stretched they must lie part-way along the interatomic curves in fig. 6.3. A given bond at $j \gg n$ in fig. 6.2 therefore begins to contribute to the surface energy as soon as it senses the applied loading; its contribution increases until the crack 'tip' passes by and the bond is broken. The transformation from

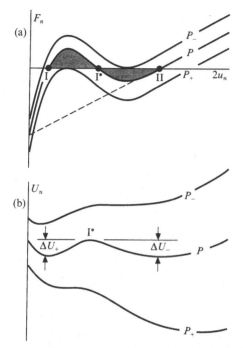

Fig. 6.4. Graphical construction showing lattice-modified bond-rupture functions, (a) force–separation and (b) energy–separation, for intrinsic chain model. Plotted for applied loads at lattice-trapping limits (P_+, P_-) and intermediate level $(P_- \leqslant P \leqslant P_+)$. Shaded area under curve between states I and I• in (a) denotes forward barrier ΔU_+, between I• and II backward barrier ΔU_-.

mechanical to surface energy is therefore not bond-localised, as presumed for singular cracks. This conclusion serves to reinforce earlier statements concerning the fundamental limitation of the slit models; statements, we recall, that led us to consider the Elliot crack in sect. 3.3.2.

Now consider the requirement that all j bonds in the crack system of fig. 6.2 should be in a state of mechanical equilibrium:

$$\partial U(u_j)/\partial(2u_j) = 0, \quad (0 \leqslant j \leqslant \infty). \tag{6.12}$$

An infinite set of fourth-order linear difference equations follows. Even without analytical solutions we can argue that if the tip bond at $j = n$ were to remain strictly Hookean, i.e. no cutoff, the crack could never advance or recede, and the system would be stable over an infinite range of applied loads. With this in mind we identify two critical mechanical states for the system *with* cutoff:

(i) *Crack extension.* If the system is loaded until the nonlinear bond at $j = n$ snaps, the crack will extend. This condition may be expressed as

$$2u_n = \delta, \quad P = P_+. \tag{6.13}$$

(ii) *Crack healing.* If the system is now gradually unloaded, a stage will be reached where the bond displacement at $n-1$ relaxes to within the cohesion range. At this point the previously broken bond will snap together and heal. The appropriate condition is

$$2u_{n-1} = \delta, \quad P = P_-. \tag{6.14}$$

Within the load range $P_- \leqslant P \leqslant P_+$ the crack is mechanically 'trapped' by the lattice.

Complete analytical solutions to (6.12) have been obtained by Fuller & Thomson (1978). They proceed by first obtaining separate sets of solutions for the equilibrium displacements u_j of all 'cracked' $(j < n)$ and 'uncracked' $(j > n)$ bonds. These two sets are coupled by the energy equation for the nonlinear bond. Substitution back into (6.11) gives us the 'lattice-modified' energy–separation function for the crack-tip bond

$$U_n(u_n) = U_B(u_n) + [nU_{BB} + \beta(\kappa - 1)u_n^2 - 2P(1 + n/\kappa)u_n$$
$$- (P^2/6\alpha)n(2n^2 + 3n\kappa + 1)] \tag{6.15}$$

where $\kappa = \{[1 + (1 + 8\alpha/\beta)^{1/2}]/2\}^{1/2}$ is a composite elastic constant. The corresponding crack-tip force function, $F_n(u_n) = +\partial U_n(u_n)/\partial(2u_n)$, is

$$F_n(u_n) = F_B(u_n) + [\beta(\kappa - 1)u_n - P(1 + n/\kappa)]. \tag{6.16}$$

For $F_n > 0$ the bond opens, for $F_n < 0$ it closes. The square brackets in (6.15) and (6.16) include contributions from the interacting lattice and applied load. Note, however, that these interaction terms are *mathematically decoupled* from the intrinsic $U_B(u_n)$ and $F_B(u_n)$ terms, a result that we shall turn to special advantage in our consideration of chemistry in sect. 6.4.

These results are represented graphically in fig. 6.4. Note that the interaction component in the force function of (6.16) is linear in u_n. In fig. 6.4(a) this interaction component (dashed inclined line) is the sum of two components: the 'applied load' term $-P(1 + n/\kappa)$, a driving force which effectively reduces the cohesion by displacing the entire $F_B(u_n)$

function downward along the F_n axis; the 'lattice-constraint' term $\beta(\kappa - 1) u_n$ associated with the elastic matrix, a restraining force which effectively increases the cohesion as the bond separates by biasing the $F_B(u_n)$ function upward. The net $F_n(u_n)$ function itself (solid curve) is constructed as the superposition of this interaction component onto the intrinsic cohesion function $F_B(u_n)$ (fig. 6.3(a)). The corresponding energy function $U_n(u_n)$ in (6.15), quadratic in u_n (with cross terms), is plotted in fig. 6.4(b).

We see that the composite functions in fig. 6.4 possess stationary values at I, II and I*, corresponding to intersection points at $F_n(u_n) = 0$ or to extrema in $U_n(u_n)$. These stationary values represent equilibrium states for the crack system. Those states at I and II are stable. At I the restoring force comes predominantly from the intrinsic cohesion (bond 'intact'), at II from the linear elastic matrix (bond 'broken'). The state at I* is unstable. At load P_+ (lower curve in fig. 6.4(b)), where solutions I and I* coalesce, the crack spontaneously *advances*; at P_- (upper curve), where II and I* coalesce, it spontaneously *retreats* (heals). Thus within the load range $P_- \leqslant P \leqslant P_+$ the crack is trapped in the lattice potential wells, i.e. is unable to move either forward or backward under the action of mechanical loads alone. We point out that this 'lattice trapping' persists even as the crack grows to macroscopic dimensions (i.e. large n in (6.15) and (6.16) above).

The lattice modelling above is more than just an esoteric mathematical exercise. As we shall see in subsequent subsections it allows us to make some powerful statements concerning the nature of equilibrium and kinetic crack states. Let us nevertheless acknowledge severe shortcomings in the way the assemblage in fig. 6.2 represents solid structures and associated force laws. First, we have imposed a nearest-neighbour force law with assumed nonlinearity at only a single crack-tip bond. This inevitably exaggerates the bond-snapping process. Then, the model is one-dimensional, and so excludes consideration of such features as 'kinks' along the crack front. Finally there is the issue of the point-atom and spring-bond representation: atom sizes are not taken into account. We will remain mindful of these factors as we seek to refine our atomistic models in the remainder of this chapter.

6.2.2 Lattice models and the Griffith condition

Consider now how the results of the discrete model in the preceding subsection may be reconciled with the thermodynamic Griffith condition for ideal brittle fracture. Let us construct a plane crack as a parallel array of identical (but uncoupled) double-chains of the kind in fig. 6.2, with out-of-plane periodicity a_0. Then in developing the connection between microscopic and macroscopic descriptions we may use the one-dimensional-chain structure to draw general conclusions about two-dimensional straight-front cracks.

We are led to suggest that the existence of energy barriers to crack motion may be accommodated within the Griffith framework by imposing a lattice-scale modulation on the intrinsic surface energy term. The rationale for such an oscillatory component is as follows. The requirement for a stationary value of the system energy $U(na_0)$ for a plane crack of length $c = na_0$ and unit width is that the net crack-extension force be zero, i.e. $g = -\mathrm{d}U(na_0)/a_0\,\mathrm{d}n = 0$. Suppose we try to advance the crack line through one atomic spacing, from $c = (n-\tfrac{1}{2})a_0$ to $c = (n+\tfrac{1}{2})a_0$ in fig. 6.2, by increasing the applied load. We saw in fig. 6.4 that there exists an energy barrier to any such discrete rupture event. Phenomenologically, we may incorporate this barrier into the energetics by writing the system energy as a *continuously* oscillating function of crack length c, $U(c) = U(na_0)$, with periodicity a_0. Realistically, the crack length can assume only *discrete* values (n an integer). In the parlance of reaction-rate theory, it is perhaps more appropriate to consider c as a 'reaction coordinate' linking the stable configurations at $(n-\tfrac{1}{2})a_0$ and $(n+\tfrac{1}{2})a_0$ via a saddle point in 'configurational energy space' (Glasstone, Laidler & Eyring 1941).

Now let us see how an oscillatory component in the lattice energetics in (6.16) might be partitioned into mechanical and surface components. The influence of any such oscillations will most certainly be felt globally, i.e. over the range $P_- \leqslant P \leqslant P_+$ in fig. 6.4. However, the attendant energy barriers are strictly a manifestation of the non-harmonicity in the function $U_B(u_n)$, confined in our models to the crack-tip bond. Following the reasoning used in sect. 3.2, we argue that the system mechanical energy U_M should be insensitive to any such detail in the crack-tip rupture micromechanics.[1] We conclude that although there will be a macroscopic trapping range in the (enclave) mechanical-energy-release rate, $G_* = -\mathrm{d}U_M/\mathrm{d}c$, the root of this trapping must lie in the effective surface energy.

Accordingly, let us impose an atomic-scale periodicity on the system

---

[1] For instance, one might reasonably expect the double-cantilever result (2.25) from continuum theory, $G \propto P^2c^2$, to be an adequate representation of the one-dimensional chain model within the mechanically stable trapping range. This is consistent with the result $Pn = \text{const}$ at $u_n = \text{const}$, $n \gg 1$, in (6.16).

surface energy U_s, via Rice's quasi-equilibrium crack-resistance term R'_0 (*in vacuo*) in (5.12). We first define the intrinsic surface energy for our lattice structure in terms of the cohesive bond energy in (6.9),

$$W_{BB} = 2\gamma_B = U_{BB}/a_0^2, \tag{6.17}$$

where a_0^2 is the interfacial bond area. Now one might imagine, in the extreme of a 'perfectly brittle' bond, that U_s should behave as the step function in fig. 6.5. That is, for the crack of length $c = na_0$, U_s should remain constant until the line of bonds at n snaps, at which point it should increase abruptly by an amount $2\gamma_B b_0$ per unit crack width. Since this step function may be decomposed into a linear term (continuum solution) plus an atomically periodic saw-tooth term (lattice modulation), it contains the necessary lattice-trapping characteristic. However, such a function clearly overestimates the effect, for it predicts alternate zero and infinite values of R'_0. Our simplistic description has omitted the contribution to the surface energy from the stretched elements *ahead* of the crack tip (sect. 6.2.1). This 'lattice contribution' tends to smooth out the discrete energy jumps, and the result is a curve somewhere between the linear and step functions in fig. 6.5. The exact form of this curve is complex for even the simplest lattice structures, requiring (among other things) a complete knowledge of the interatomic potential functions $U_B(u_n)$ (see sect. 6.3). An expedient empirical function for the quasi-equilibrium crack-resistance energy of (5.12) is

$$\begin{aligned} R'_0 &= R_0 \pm \Delta R_0 \\ &= 2\gamma_B + 2\Gamma_B \cos(2\pi c/a_0) \end{aligned} \tag{6.18}$$

where Γ_B is a modulating 'trapping' component. The fracture surface energy therefore oscillates within the trapping range $R_0^- \leqslant G_* \leqslant R_0^+$ where

$$\left. \begin{aligned} R_0^+ &= 2\gamma_B + 2\Gamma_B \\ R_0^- &= 2\gamma_B - 2\Gamma_B \end{aligned} \right\}. \tag{6.19}$$

Consider now how these results bear on the energy-balance condition, $\mathscr{g} = G_* - R'_0 = 0$. The function $R'_0(c)$ from (6.18) is plotted in fig. 6.6. Also plotted is $G_*(c)$ for an unstable crack system ($\mathrm{d}G_*/\mathrm{d}c > 0$), for a crack of sufficient length that $G_*(c)$ may effectively be taken as linear over the extension range of interest. The condition $G_* = R_0 = 2\gamma_B$ no longer provides an adequate fracture criterion, because the lattice constrains the

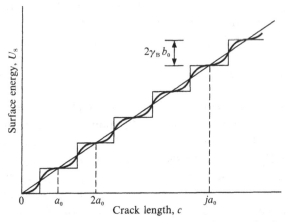

Fig. 6.5. Surface energy of straight lattice crack of unit width. Three representations: linear function (continuum limit); step function (extreme bond snapping); smooth periodic function ('realistic' discrete structure).

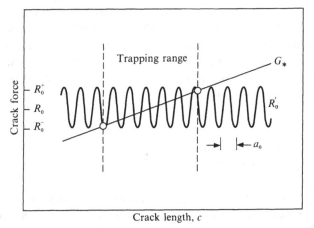

Fig. 6.6. Generalised crack force diagram, illustrating trapping of crack within range $R_0^- \leqslant G_* \leqslant R_0^+$. Within this range the crack satisfies the quasi-equilibrium requirement $\mathscr{g} = G_* - R_0' = 0$. Outside this range the crack is dynamic.

crack to subatomic displacements for all G_* within $R_0^- \leqslant G_* \leqslant R_0^+$. Within this trapping range the crack remains in a state of *mechanical* equilibrium. One must either raise or lower G_* to one of the limiting values in (6.19) before the crack may extend ($+$) or recede ($-$). The requirement for *forward* crack extension in the absence of thermal fluctuations is then

$$G_c = G_* = R_0^+ = 2\gamma_B + 2\Gamma_B.$$

6.2.3 Thermally activated crack propagation: kinetics and kinks

The impact of lattice discreteness is felt most strongly in the fracture mechanics when allowance is made for thermal fluctuations. Lattice trapping, as formalised by an oscillatory factor in the system energy function $U(c)$, has inbuilt provision for a quantitative description of crack kinetics. True Griffith *thermodynamic* equilibrium (as opposed to mechanical equilibrium described in the preceding subsection) obtains when the associated forward and backward energy barriers are equal, so that the net fluctuation rate is zero. Kinetic fracture may then be seen as an energy-dissipative process in which phonons (or other wave or particle fracto-emissions) are generated at ever-increasing rates as the systems departs further from the quiescent state.

The question arises as to how one may relate the barrier heights formally to the local mechanical-energy-release rate G_*. An analysis of the stress dependence of these barriers requires explicit information on the inter-atomic bond function $F_B(u_n)$ for any discrete model, and is usually intractable except for the simplest force laws and lattice structures (Fuller & Thomson 1978). Consequently, we shall seek an approximate solution for vacuum cracks by incorporating an oscillatory surface function of the kind described in the preceding subsection into the velocity formalism of chapter 5.

It is now time to acknowledge that the geometry of kinetically activated brittle cracks is essentially three-dimensional. The activation process can not involve a rigid line motion of the crack, as previously assumed, for that would require the cooperative opening (or closing) of every bond along the entire front. The thermal energy needed to surmount a potential barrier over an extended line increases in proportion to the total number of bonds along that line, and therefore becomes prohibitively large for cracks of macroscopic widths. The most favourable configuration is one in which the kinetic unit is atomically localised at an 'active' bond, e.g. the 'kink' site in fig. 6.7 (Lawn 1975). Longitudinal crack motion may then be interpreted in relation to the *lateral* motion of a population of such kinks. The kink action may be likened to that of a zipper along the crack front, with atomic-scale crack advance corresponding to the breaking of a whole line of bonds, crack retreat likewise to the remaking of this line of bonds. It is understood that the energy barriers to kink motion will inevitably be influenced by the configuration of nearest-neighbour bonds; also, where the barriers to kink nucleation (or annihilation) are relatively high, that

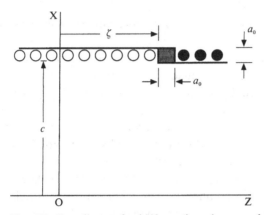

Fig. 6.7. Coordinates for kink motion along crack front of square lattice. Open circles denote ruptured bonds, filled circles unruptured bonds. Shading denotes active site.

some crack-front bonds will be inactive at any instant. However, we shall pass over such details here, remarking only that the picture of a vacuum crack with a full population of discretely active sites along its front is not unreasonable for those highly brittle solids with short-range covalent forces.

Accordingly, consider the idealised square-lattice kink structure in fig. 6.7. The coordinate ζ is used to locate an individual kink along the crack front. A successful activation results in a lateral jump of one lattice spacing a_0. We may now introduce the kinetic element by taking the rupture rate of active bond sites to be governed by the kind of Maxwell–Boltzmann statistics implied in the rate equations of sect. 5.5. The crack velocity therefore has the form

$$v = v_0 a_0 [\exp(-\Delta U_+/\mathbf{k}T) - \exp(-\Delta U_-/\mathbf{k}T)], \qquad (6.20)$$

with $v_0 = \mathbf{k}T/\mathbf{h}$ a natural lattice frequency, $\mathbf{k}$ and $\mathbf{h}$ Boltzmann and Planck constants respectively, T absolute temperature, and ΔU_+ and ΔU_- kink activation energies (in analogy to the forward and backward energies ΔF_+ and ΔF_- in (5.16)).

Now investigate the stress dependence of the activation energies for intrinsic kinks in this velocity equation. The total system energy may be partitioned into longitudinal and lateral components,

$$U(c, \zeta) = U(c)_\zeta + U(\zeta)_c. \qquad (6.21)$$

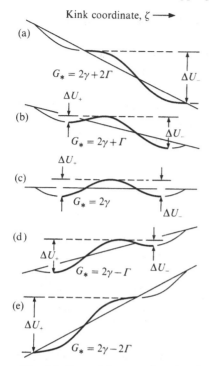

Kink coordinate, $\zeta \longrightarrow$

(a) $G_* = 2\gamma + 2\Gamma$ ΔU_+ ΔU_-

(b) ΔU_+ $G_* = 2\gamma + \Gamma$ ΔU_-

(c) ΔU_+ $G_* = 2\gamma$ ΔU_-

(d) ΔU_+ $G_* = 2\gamma - \Gamma$ ΔU_-

(e) ΔU_+ $G_* = 2\gamma - 2\Gamma$

Fig. 6.8. Potential energy function for kink motion. Solid curve represents activation barrier for motion through one atomic spacing. Straight line reflects bias of applied loading.

We are concerned with the energy variation as the kink translates through one lattice spacing. For an incremental variation in ζ at constant c

$$dU(c, \zeta)_c = dU(\zeta)_c = (-G_* + R_0') a_0 \, d\zeta. \tag{6.22}$$

Direct analogy with (6.18) yields

$$R_0' = 2\gamma_B + 2\Gamma_B \cos(2\pi\zeta/a_0). \tag{6.23}$$

Recall that γ_B and Γ_B are material quantities; and further, that G_* is insensitive to the crack-tip configuration, and is therefore independent of ζ. Inserting (6.23) into (6.22) and integrating with respect to ζ at constant c then gives

$$U(c, \zeta)_c = a_0[(-G_* + 2\gamma_B)\zeta + (\Gamma_B a_0/\pi) \sin(2\pi\zeta/a_0)]. \tag{6.24}$$

This equation provides us with the basis for determining kink-activated velocity functions. Accordingly, (6.24) is represented graphically for five values of G_* in fig. 6.8. Case (c) corresponds to true Griffith equilibrium ($G_* = 2\gamma_B$), cases (a) and (e) to extremes of the trapping range for crack advance and retreat ($G_* = 2\gamma_B \pm 2\Gamma_B$), and cases (b) and (d) to intermediate configurations ($G_* = 2\gamma_B \pm \Gamma_B$). The biasing effect of the applied loading is reflected in the relative barrier heights for expansion (ΔU_+) and contraction (ΔU_-) of the kink. Imposing the requirement $[\partial U(c, \zeta)/\partial \zeta]_c = 0$ for extrema allows us to determine these heights from (6.24). For small departures from the Griffith condition,

$$\Delta U_\pm(G_*) = \Delta U^\bullet[1 \mp \pi(G_* - 2\gamma_B)/4\Gamma_B \ldots], \quad (|G_* - 2\gamma_B| \ll 2\Gamma_B) \tag{6.25}$$

with $\Delta U^\bullet = \Delta U_+ = \Delta U_-$ the quiescent activation energy. For larger departures the activation energies contain higher terms in $G_* - 2\gamma_B$. Inserting (6.25) into (6.20), we obtain the vacuum crack velocity function

$$v = 2v_0 a_0 \exp(-\Delta U^\bullet/kT) \sinh[\alpha(G_* - 2\gamma_B)/kT \ldots],$$
$$(2\gamma_B - 2\Gamma_B < G_* < 2\gamma_B + 2\Gamma_B) \tag{6.26}$$

in the form of (5.18), with $\alpha = \pi \Delta U^\bullet/4\Gamma_B$. For forward extensions just above the quiescent point, (6.26) pertains to region III:

$$v_{III} = v_0 \exp(\alpha_{III} G_*/kT), \quad (2\gamma_B < G_* \ll 2\gamma_B + 2\Gamma_B) \tag{6.27}$$

with pre-exponential term

$$v_0 = v_0 a_0 \exp(-\Delta U^\bullet_{III}/kT) \exp(-2\alpha_{III}\gamma_B/kT) \tag{6.28}$$

in analogy to (5.19) and (5.20). Fig. 6.9 compares the velocity function (6.26) (higher-order terms over the entire trapping range included) with the approximation (6.27). The intrinsic nonlinearity of the general semi-logarithmic $v(G_*)$ (or equivalent $v(K_*)$) velocity function is apparent.

We note the following features of the above kink formulation:

(i) The quiescent activation energy and activation area can be determined explicitly in terms of the characteristic lattice dimension a_0 from (6.24):

$$\Delta U^\bullet = 2\Gamma_B a_0^2/\pi \tag{6.29a}$$

$$\alpha = \pi \Delta U^\bullet/4\Gamma_B = a_0^2/2. \tag{6.29b}$$

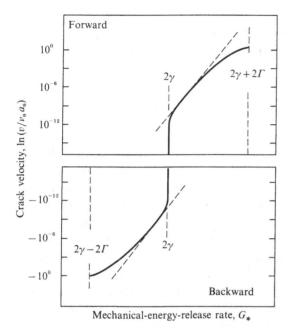

Fig. 6.9. Generic crack velocity curve for trapping model with sinusoidal kink energy function (6.24). Solid curve is exact function (6.26), inclined dashed lines are exponential approximations of the kind in (6.27). Plotted for $\Delta U^* = 25\,\mathbf{kT}$.

However, the numerical coefficients in (6.29) reflect the sinusoidal form of the potential function in (6.24). Generally, we must expect these coefficients to be sensitive to the specific interatomic force function $F_B(u)$. With this qualification, one may use (6.29) as a basis for prediction. Thus for vacuum fracture in soda-lime glass, say, the computed activation area $\alpha_{III} = 0.14$ nm^2 molec^{-1} (3.5 Si–O–Si bonds nm^{-2} of crack plane) compares with the fitted value $\alpha_{III} = 0.12$ nm^2 molec^{-1} in table 5.1.

(ii) The formulation is consistent with dynamic descriptions at the upper range of kinetic velocities. At the trapping limit in fig. 6.9 the velocity cuts off at a maximum $v_m = v_0 a_0$. Typically, $v_0 \approx 4$ THz, $a_0 \approx 0.5$ nm, giving $v_m \approx 2$ km s^{-1}, i.e. of order sonic velocities. Again, this estimate is by no means absolute: kink nucleation and entropy terms, which we have neglected, will appear as additional pre-exponential factors in the velocity equations.

(iii) The velocity relations are readily adapted to incorporate important extraneous factors: *shielding*, by substituting $G_* = G_A + G_\mu$ $(-G_\mu = R_\mu,$ sect. 5.3); *chemically assisted* bond rupture, by replacing γ_B with γ_{BE} and appropriately modifying the activation terms (sect. 6.4).

6.3 Computer-simulation models

The atomic models discussed thus far, while valuable for their insight into crack-tip rupture phenomena, do not lend themselves to quantitative descriptions of real solids. In principle, the progression from abstractions like the one-dimensional chain in fig. 6.2 to more realistic structures is straightforward: construct a lattice for a given solid, introduce a straight crack by negating attractive forces across a prescribed planar area, and allow the system to relax to equilibrium in accordance with some refined interatomic potential function. Unfortunately, two major obstacles present themselves: first, fundamental descriptions of the short-range interactions between atoms and molecules in the solids of most interest are still not readily available; second, the ensuing discrete lattice equations become totally intractable. One is generally forced to deal with the potential function in a semi-empirical manner, and to obtain solutions on a high-speed computer.

These difficulties are not confined to the crack problem, and computer-simulation techniques have been developed for a number of other lattice configurations (e.g. perfect crystal structures, to compute the theoretical cohesive strength; point defects; dislocations) in a wide range of crystal types. It is perhaps surprising that, despite the recent explosion of computer technology, some of the most significant crack simulation work belongs to the 1970s. Even with the great increases in computational speed and capacity the total number of atoms that can be handled in a nonlinear calculation remains limited, and one is usually forced to devise 'hybrid' atomic–continuum schemes for dealing with the problem.

Simulation studies of this latter kind have been made by Sinclair and co-workers (1972, 1975) on diamond-type solids and by Kanninen & Gehlen (1972), and others, on body-centred cubic iron. We focus here on the diamond structure, with specific attention to silicon, because of its immunity to crack-tip plasticity (sect. 6.6). The loaded crack system is separated into two domains: I, an atomistic 'core' at the tip; II, a constraining elastic continuum. Within the embedded core explicit consideration is given to the atomic structure. The anisotropic K-field solutions for a linear elastic slit are used to determine the displacement field for the outer domain, as well as to provide a 'starting configuration' for the core atom positions. One then allows the core to relax in accordance with a predetermined potential function.

The potential function demands careful attention, for it is through this function that the essential nonlinearity is introduced. It is desirable that its construction should conform to the following specifications:

(i) The potential energy of the crystal in its perfect, unstressed state should be a minimum at the equilibrium value of the lattice spacing.

(ii) For small strains the potential function should reduce to quadratic terms in displacements. The coefficients of these terms should match the linear elastic constants for the material in domain II.

(iii) For large strains the cohesive force should pass through a maximum and thereafter tend to a zero cutoff as separation increases.

(iv) The work done in separating atoms across the crack plane should match the surface energy.

For covalent structures like diamond it is also important to take into consideration the directional nature of the bond, by including second-nearest-neighbour (bond-bending) as well as the usual nearest-neighbour (bond-stretching) interactions.

An iterative relaxation procedure is then used to reassign atom co-ordinates for the starting configuration within the core so as to minimise the system energy. The usual procedure is to make incremental adjustments, and to cycle the program until some tolerance condition is satisfied. There are several ways in which these adjustments may be made; therein lies the skill of the numerical analyst. Part of this skill involves proper attention to boundary conditions to ensure self-consistency; it is important that the boundary between domains I and II is itself allowed to 'flex' during the adjustments, to avoid the artificial generation of 'back stresses' on the core.

The results of such a simulation for a (111) equilibrium crack in silicon (using $K_c = T_0$, table 3.1, to set the K-field boundary conditions) are shown in fig. 6.10. The configuration in fig. 6.10(a) is for a quasi-two-dimensional crystal, with crack front perpendicular to the symmetry plane $(01\bar{1})$. (With such symmetry, periodic lattice translations along the normal to the plane of the diagram infinitely repeat the configuration.) Cleavage-plane bonds

(a)

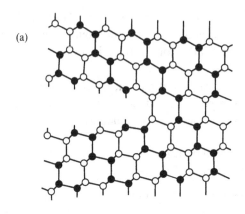

(b)

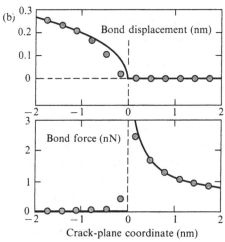

Fig. 6.10. (a) Computer-generated atomic crack configuration for silicon (111) crack (plane of diagram (01$\bar{1}$)). Two adjacent atom planes shown. (b) Corresponding atom-pair displacements and forces at crack plane, along with Irwin slit solutions (solid curves). (After Sinclair, J. E. & Lawn, B. R. (1972) *Proc. Roy. Soc. Lond.* A **329** 83.)

are seen to be effectively ruptured behind the tip and stretched within the force–separation cutoff ahead. Corresponding atom displacements and bond forces are plotted in fig. 6.10(b), together with continuum-slit K-field solutions for comparison. The discrete displacement and force distributions along the crack plane are manifestly non-singular.

What general conclusions may we draw from such computations in relation to specific crack-tip descriptors in preceding sections and chapters? The Irwin crack-tip radius from the K-field profile in fig. 6.10(b) is calculated at $\rho \approx 0.1$ atom spacings, confirming that the widely held notion of a smoothly rounded tip is of little physical significance. The alternative conception of the crack as a concentrated line defect, i.e. closely adjacent, stress-free half-planes held together at a terminal line (analogous to a dislocation line) by bonds close to their rupture point, would appear to be more reasonable. However, general extension of this conception to other brittle solids may not be justifiable: silicon lies at the covalent extreme of the bonding spectrum, and is therefore a most favourable candidate for a narrow crack-tip core. The core would be expected to be substantially wider in solids with a significant component of (non-directional, longer-range) ionic bonding (even more so in metallic solids). Even for the silicon crack in fig. 6.10 the first few bonds ahead of the nominal tip are actually strained beyond the usual Hookean limit (i.e. 1% or more). Accordingly, we are led to conclude that of the principal crack-tip cohesion models, Orowan–Gilman (sect. 6.1), Barenblatt (sect. 3.3.1) and Elliot (sect. 3.3.2), the second is closer to the reality than the first, but the third is the most realistic of all three.

In a later study Sinclair (1975) extended his computational algorithms to three-dimensional cracks in silicon, to estimate the energy barriers for kink motion through the trapping range $2\gamma_B - 2\Gamma_B < G_* < 2\gamma_B + 2\Gamma_B$. The barriers were found to be small, i.e. $\Gamma_B/\gamma_B < 0.05$. Since the directional covalent bond in silicon is again expected to provide the most favourable conditions for trapping, we might anticipate that intrinsic kinetic effects are unlikely to be pronounced for highly brittle materials in general. We shall come to an altogether different conclusion when we consider the effects of environmental chemistry in the next section.

6.4 Chemistry: concentrated crack-tip reactions

How may we incorporate the chemistry of chapter 5 into the atomic models? In this section we explore the above notion of bond rupture at a concentrated line along the crack front. This 'line defect' notion, built in to the one-dimensional chain model of sect. 6.2, contains the substance of much of present-day thinking on crack-tip chemistry.

We begin by modifying the lattice models to accommodate, in a somewhat generic way, a 'concerted chemical reaction' at the critical nonlinear crack-tip bond. Then we consider the implications of this modification in relation to our previous equilibrium and kinetic crack

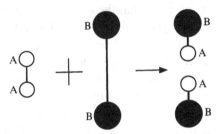

Fig. 6.11. Chemically induced bond rupture. Intrusive molecule A–A reacts dissociatively with crack-tip bond –B–B– to produce terminal bonds A–B–.

growth 'laws'. Finally, we examine a specific model that describes the concerted reaction in one brittle material, silica glass, from the standpoint of molecular orbital theory.

6.4.1 Chemically modified lattice model: incorporation of concerted reaction concept

We recall in the lattice-modulated energy– and force–separation functions of (6.15) and (6.16) that the terms $U_B(u_n)$ and $F_B(u_n)$ representing nonlinear crack-tip bonds are mathematically decoupled from the terms representing lattice constraint and applied load. This means that any chemical reaction at an individual crack-tip bond may be considered independently of the mechanical response of the remainder of the solid body: all that is required is the inclusion of an extrinsic component in the $U_B(u_n)$ or $F_B(u_n)$ functions. From the perspective of classical reaction-rate theory this component must ultimately be evaluated along a reaction coordinate in configurational-energy space. We confine ourselves here to graphical representations of the ensuing energy and force functions, using a simple chemisorption representation of the interaction (Fuller, Lawn & Thomson 1980).

Consider the generic interaction in fig. 6.11. An environmental molecule A–A reacts dissociatively with an initially intact crack-tip bond –B–B– to produce terminal groups A–B–:

$$(A–A) + (–B–B–) \rightarrow (–B–A \cdot A–B–). \tag{6.30}$$

Each such event ruptures the bond and expands the crack through an atomic area a_0^2, thereby exposing a new bond to the next incoming molecule. Decoupled from its otherwise constraining lattice matrix, the

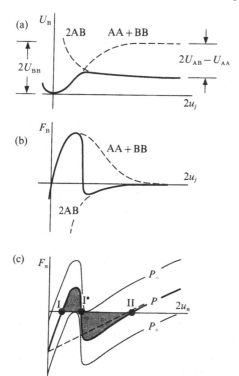

Fig. 6.12. Graphical construction incorporating dissociative chemical crack-tip bond interaction into lattice mechanics: (a) energy function for isolated 'diatomic-molecule' interaction, (b) equivalent force function. (c) Force function with lattice modification for loads at lattice-trapping limits (P_+, P_-) and intermediate level ($P_- \leqslant P \leqslant P_+$). Shaded area under curve between states I and I• denotes forward barrier ΔU_+, between I• and II backward barrier ΔU_-. (Cf. fig. 6.4 for intrinsic lattice.)

bond-rupture event (6.30) may be viewed in terms of a simple reaction between diatomic gas molecules, $AA + BB = 2AB$ (Slater 1939). Appropriate interaction curves are sketched in fig. 6.12 as a function of B–B atom-pair displacement (reaction coordinate for the crack-tip bond). Diagrams (a) and (b) represent the base energy and force functions for BB in isolation from the lattice (cf. fig. 6.3), (c) the corresponding force function with chain-lattice matrix constraint and applied load terms superposed (cf. fig. 6.4(a)):

(a) The potential energy $U_B(u_j)$ for the isolated diatomic-molecule system is plotted as a function of B–B displacement, with AA allowed at all separations to assume a position of minimum free energy,[2] At small separations the interaction between AA and BB is presumed to be negligible, so that the energy curve AA + BB for the unreacted state differs imperceptibly from that for an isolated bond (fig. 6.3(b)). At large separations, such that the bond B–B is effectively ruptured, the terminal configuration A–B is energetically favoured. The 2AB curve rises steeply as the bond separation diminishes because of polar and overlap repulsion between the like end groups. At any given separation the appropriate bond state, i.e. unreacted or reacted, will be determined by the lower of the two curves. Thus there is a critical separation at which reaction occurs, i.e. the crossover point ('rounded off' owing to resonance). This switch from bonding to anti-bonding states reduces the bond energy in (6.9).

(b) The equivalent force–separation function $F_B = +\partial U_B/\partial(2u_j)$ (cf. fig. 6.3(a)). Rupture is evident as the abrupt switch from attractive to repulsive force at crossover. (Note that the elastic stiffness β in (6.10) prior to rupture is unchanged.)

(c) A plot of the composite crack-tip force function $F_n(u_n)$ in (6.16), obtained as previously for the intrinsic lattice but using the chemically modified, adhesive force $F_B(u_n)$ in fig. 6.12(b). Equilibria at I and II now correspond to stable bond states along the reaction coordinate before (bond stretched) and after (bond broken) chemical interaction. Equilibrium at I* corresponds to the unstable complex (activated bond). From comparison with fig. 6.4(a) it is immediately evident that the chemistry significantly reduces the separation, and thus the activation area, for bond rupture. At the same time, recalling from (6.16) that the $F_n(u_n)$ curve intersects the ordinate at $-P(1 + n/\kappa)$, and that the crack is trapped in the range $P_- \leqslant P \leqslant P_+$, we see that the applied loads required to drive the crack either forward (P_+) or backward (P_-) over the modified barriers are reduced. Further, the trapping range $\Delta P = P_+ - P_-$ is increased, i.e. the 'snapping' characteristic is enhanced.[3]

It is well to remind ourselves that the modified lattice model is highly oversimplistic. There are questionable assumptions concerning the nature of the interaction, not to mention obvious limitations of the lattice

---

[2] The potential energy of the system depends on other coordinates as well, e.g. A–A bond length, AA–BB separation, BB orientation. In the crack problem it is the B–B coordinate that primarily determines the extension.
[3] Indeed, one should be able to convince oneself from the construction in Fig. 6.12(c) that even a crack system with zero intrinsic trapping may exhibit large extrinsic trapping.

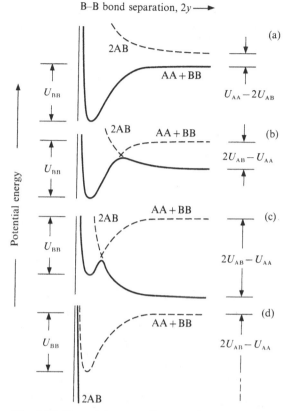

Fig. 6.13. Potential energy diagrams for dissociative chemical reaction (6.30). Sequence (a)–(d) represents increasing value of adsorption energy ΔU_{Ad} in (6.32), decreasing interface energy γ_{BE} in (6.31). Full curve is minimum energy composite of dashed curves, representing reactant (AA + BB) and product (2AB) states. To rupture bond chemically, B–B atoms must be separated over energy barrier.

structure itself. Nevertheless, relative to the continuum representation in fig. 5.2, the appearance of the crossover characteristic in fig. 6.12(a) and (b) takes us a step closer to a truly fundamental description of fracture chemistry.

6.4.2 Chemically modified lattice model and fracture mechanics

What are the implications of the above chemically modified lattice models concerning our basic fracture mechanics relations in sect. 6.2? We consider this question for both equilibrium and kinetic cracks:

(i) *Equilibrium cracks.* Begin by considering the end-point energy states for the lattice-isolated, diatomic-molecule system in fig. 6.12(a). The chemical interaction reduces the interface energy from intrinsic γ_B to extrinsic γ_{BE} in the manner of (5.1), so

$$
\begin{aligned}
W_{BEB} &= 2\gamma_{BE} \\
&= (U_{BB} + U_{AA} - 2U_{AB})/a_0^2.
\end{aligned}
\tag{6.31}
$$

The bond energies therefore determine the quiescent points $G_* = R_E = 2\gamma_{BE}$, i.e. thresholds, on v–G curves. The magnitude of the reduction is determined by the adsorption energy of (5.2) which, in conjunction with (6.17), is

$$
\begin{aligned}
2\Delta U_{Ad} &= 2\gamma_B - 2\gamma_{BE} \\
&= (2U_{AB} - U_{AA})/a_0^2.
\end{aligned}
\tag{6.32}
$$

We can now examine how the AA, BB and AB bond energies in (6.31) determine the nature of the quiescent state. Recall again that the constraint (square bracket) component of U_n in (6.15) simply load-biases the bonding energy U_B, without in any way affecting the relative disposition of the unreacted (AA + BB) and reacted (2AB) curves. We may therefore explore the influence of adsorption with minimum complication by focusing on the isolated-bond representation in fig. 6.12(a), noting that an increase in ΔU_{Ad} is equivalent to translation of the 2AB curve down the energy axis relative to the AA + BB curve. Fig. 6.13 accordingly replots the energy curves at progressively increasing ΔU_{Ad}:

(a) '*Spontaneous desorption*' state. $\Delta U_{Ad} < 0$, $\gamma_{BE} > \gamma_B > 0$. The system never leaves the unreacted curve, i.e. the reaction (6.30) is energetically unfavourable.

(b) '*Metastable adsorption*' state. $\Delta U_{Ad} > 0$, $\gamma_B > \gamma_{BE} > 0$. Reaction is activated reversibly, with the adsorption state metastable at zero applied stress. The forward barrier is substantial, corresponding to near-rupture of the primary bond before adsorption occurs. On unloading the extended crack the AA molecule desorbs relatively easily and the bond B–B heals.

(c) '*Stable adsorption*' state. $\Delta U_{Ad} > 0$, $\gamma_B > 0 > \gamma_{BE}$. Same as (b) but now the adsorption state is truly stable. The reaction occurs over a reduced barrier, at reduced B–B separation. The desorption barrier is now formidable.

(d) '*Spontaneous adsorption*' state. $\Delta U_{Ad} > 0, \gamma_B > 0 \gg \gamma_{BE}$. The system proceeds spontaneously and irreversibly to a strongly chemisorbed state, even in the absence of an applied stress.

It is self-evident from (6.31) and (6.32) that the strongest interactions, i.e. those lower in the sequence of fig. 6.13, are favoured by weak *co*hesion states (small U_{BB}, U_{AA}) and strong *ad*hesion states (large $2U_{AB}$).

(ii) *Kinetic cracks*. Suppose that the activated bond-rupture process for lattices with *ex*trinsic crack-tip interactions can be described by an oscillatory component in the kink translation potential function, as for lattices with *in*trinsic interactions. Then, in direct analogy to (6.27) and (6.28), we may write down crack velocity relations pertinent to region I:

$$v_I = v_0 \exp(\alpha_I G_* / kT), \quad (2\gamma_{BE} < G_* \ll 2\gamma_{BE} + 2\Gamma_{BE}) \tag{6.33}$$

with pre-exponential term

$$v_0 = v_0 a_0 \exp(-\Delta U_I^* / kT) \exp(-2\alpha_I \gamma_{BE} / kT). \tag{6.34}$$

It is therefore necessary only to reinterpret the barrier parameters α_I, ΔU_I^* and Γ_{BE} in these relations. It will be remembered from sect. 6.2 that first-principles analysis of these barriers is generally intractable, even for intrinsic bond rupture. There we circumvented the problem by imposing an empirical sinusoidal function for the crack resistance, leading to the evaluations in (6.29). Unfortunately, the relatively complex nature of the chemical interaction in fig. 6.13 raises doubts as to whether one is justified in attempting similar evaluations for extrinsic bond rupture. Nevertheless qualitative conclusions concerning the chemical influence may be reached by recalling our discussion of the modified chain model in sect. 6.4.1: the activation area is reduced ($\alpha_I < \alpha_{III}$); the quiescent activation barrier is reduced ($\Delta U_I^* < \Delta U_{III}^*$); the trapping range is increased ($\Gamma_{BE} > \Gamma_B$).

The role of chemistry on these parameters has become a topic of intense interest to those who concern themselves with slopes of v–G curves in silicate glasses. The experimental data evaluations for soda-lime glass in water in table 5.1 indicate a substantial reduction in activation area, $\alpha_I / \alpha_{III} \approx 0.1$: in light of our discussion of fig. 6.13 above one may conclude that the water–glass interaction must be strongly chemisorptive. Also, recall the conspicuous variability in the region I data for different glass compositions and aqueous solutions in figs. 5.11 and 5.12: the strength of the chemisorption is highly sensitive to small shifts in the reactant–product curves in fig. 6.13. Subtle changes in the chemical potential at the crack tip might be enough to effect those shifts.

6.4.3 Crack-tip reactions in glass

We have mentioned the reasons for fracture researchers to regard silicate glass in water as a model system for kinetic studies, not least its pronounced region I branch. Extrinsic bond rupture in this system is thought to be governed by the dissociative reaction

$$(H{-}O{-}H) + (-Si{-}O{-}Si{-}) \rightarrow (-Si{-}OH \cdot HO{-}Si{-}). \qquad (6.35)$$

An intrusive water molecule hydrolyses a siloxane bridging bond at the crack tip to form two terminal silanol groups.

While recognising the molecular discreteness of the interaction (6.35), most earlier studies of the glass–water fracture system relied on the kind of phenomenological crack-tip modelling described in chapter 5. That approach is restrictive: it can account in part for the role of certain intensive variables, such as chemical concentration and temperature, but contains no provision for using data from any one material–environment system to predict the response of another. Which chemical species are likely to promote fracture in glass or any other given material?

It is evident that answers to such questions must be sought at the molecular level. Consider the hydrolysis reaction in silica glass in fig. 6.14. Notwithstanding the point-mass, two-dimensional nature of this representation, the SiO_2 network structure is relatively 'open', so that intrusive water (or other) species might be expected to have direct access to individual crack-tip bonds. If this is so the concentrated-reaction description in sects. 6.4.1 and 6.4.2 above remains appropriate. In the light of this background, glass scientists have attempted to address the crack-tip dissociative interaction of (6.35) from a molecular orbital standpoint (Michalske & Freiman 1981; Michalske & Bunker 1987). Three distinct stages of bond strain are envisaged, as depicted in fig. 6.15:

(a) *Adsorption.* At small strains the water molecule attaches itself physically to the bridging Si–O–Si bond. The electron orbitals in the water are tetrahedrally coordinated (sp^3 hybrids) about the oxygen, two forming

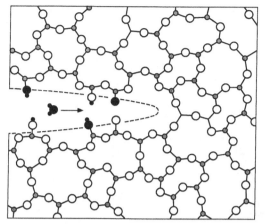

Fig. 6.14. Two-dimensional point-mass representation of water-induced bond rupture in silica glass, as per reaction in (6.35): silicon (shaded circles), oxygen (open circles), environmental water species (filled circles). Note open structure of –Si–O–Si– network. (Dashed curve is Irwin profile (2.15) for *in-vacuo* equilibrium slit-crack.)

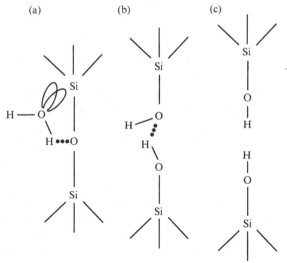

Fig. 6.15. Three-stage interaction between water molecule and strained crack-tip siloxane bond in glass: (a) adsorption, (b) reaction, (c) separation. (After Michalske, T. A. & Freiman, S. W. (1981) *Nature* **295** 511.)

bonds with the hydrogens (excess positive charge) and two forming lone pairs (excess negative). Since the bond also has some polar character (silicon excess positive), the water molecule aligns itself as shown.

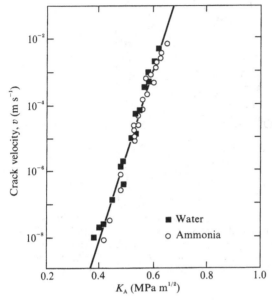

Fig. 6.16. Crack velocity data for silica glass in water and ammonia. (After Michalske, T. A. & Freiman, S. W. (1981) *Nature* **295** 511.)

(b) Reaction. At higher applied load, such that the bond is stretched beyond the crossover point on the energy curve of fig. 6.12(a), the water molecule donates an electron to one of the silicons and proton to the linking oxygen. Two new O–H bonds are thereby formed.

(c) Separation. After electron redistribution the polar terminal bonds mutually repel, thereby completing the bond-rupture process.

According to this prescription, the key ingredients for strong crack-tip interactions are the electron- and proton-donating capacity of the intrusive chemical species and the polarity of the bridging bond. With regard to the first of these, it has been demonstrated that ammonia, another molecule with lone-pair electrons (albeit a single lone pair, the other electrons bonding three hydrogens to a central nitrogen), is as effective as water in promoting crack growth. Indeed, the silica glass data in fig. 6.16 for water and ammonia environments overlap. On the other hand, molecules without *both* lone-pair electrons and hydrogens are effectively inert. With regard to bond polarity, it is observed that covalent structures like silicon seem to be immune to any form of chemically enhanced crack growth.

6.5 Chemistry: surface forces and metastable crack-interface states

The concentrated-reaction approach of the previous section, with its incorporation of an energy barrier into the crack-tip potential function, has the capacity to explain many important features of chemistry in fracture, down to *and including* threshold. But it is limited in one critical respect: it can not provide for a true account of crack velocity behaviour *below* threshold. According to the generic kink-model velocity function in fig. 6.9 incremental unloading below the quiescent point $G_* = R_E = W_{BEB} = 2\gamma_{BE}$ should cause the crack to retract immediately along the negative (desorption) branch of the v–G curve. Similarly, on reloading incrementally above threshold the crack should begin immediately to repropagate through the healed region, retracing the positive (adsorption, or now, readsorption) branch. We have made repeated reference (e.g. the Rice theoretical schematic of fig. 5.4 and the mica data in figs. 5.21 and 5.22) to departures from such continuity in the v–G response in unloading through the quiescent point: the reloading response is hysteretic. It is arguable that hysteresis could arise if the crack were to close too quickly, such that the chemisorbed species have insufficient time to desorb; but the unloaded system then remains on the 2AB curve in fig. 6.13, leaving the recontacting walls in mutual repulsion, thereby precluding healing. There are important elements missing from our description.

And now we make an assertion that will have far-reaching implications. The concentrated-reaction models are fundamentally limited insofar as the potential energy in fig. 6.13 possesses a single, primary minimum. *The true potential function must have at least one secondary minimum.* We must now extend our consideration to multiple *metastable* interfacial states.

From such considerations we are led to re-examine two other premises of the concentrated-reaction models. The first is the implicit assumption that the environment has unrestricted access to the crack tip. We have thus far treated atoms as point masses: no account has been taken of the *size* of constituent atoms in either the host substrate or the invasive molecules. We shall propose that the latter assume an ordered structure in the interfacial adhesion region; and, that it is this ordering which is responsible for metastability in the interfacial energy functions.

The second premise at issue is that the crack walls remain traction-free behind the frontal line of active bonds. Recall from the computer-simulation studies in sect. 6.3 that even for *intrinsic* fracture in structures with short-range covalent bonding the cohesion 'core' zone includes several lines of nonlinear bonds behind the (nominal) tip. We would expect the inclusion of long-range secondary minima in an *extrinsic* interaction function to extend the nonlinear zone still further. This takes us back toward the Barenblatt–Elliot crack description in which the discrete bond forces are replaced by a continuous distribution of adhesive 'surface forces'.

We outline below a model that incorporates these new elements (Lawn, Roach & Thomson 1987). The description appeals to the relative simplicity of continuum descriptions for cracks with well-defined cohesion zones, yet retains an essential discreteness by introducing lattice and molecular scaling dimensions into the surface force function.

6.5.1 Nature of surface forces

The potential for the existence of metastable states at interacting solid surfaces in a fluid medium has long been recognised by colloid scientists. Key to an understanding of colloids is the action of forces other than those involved in primary chemical bonding. Thus the stability of suspensions in aqueous solutions depends on the relative magnitudes of attractive van der Waals and repulsive double-layer forces. These ordinarily weak 'physical' forces, integrated over large surface areas, can act over unduly large distances ($\approx$ 1–100 nm) (Israelachvili 1985). The celebrated Derjaguin–Landau–Verwey–Overbeek (DLVO) theory used to calculate the interaction energy–separation function between spherical particles (Adamson 1982; Derjaguin, Churaev & Muller 1987) predicts a weak secondary minimum at a distance of order one particle diameter from the primary minimum, with an intervening energy barrier: if this secondary minimum is sufficiently deeper than kT the particles flocculate.

Over the last decade or so, Israelachvili (1985) and co-workers have developed an apparatus that allows intersurface forces to be measured *directly*. A sensitive displacement-control system is used to bring together opposing sheets of mica in a crossed-cylinder configuration. Mica is used because of its atomically smooth cleavage surfaces. The force is measured by deflection of a compliant spring and the separation by an optical interferometer. The early experiments by this school provided quantitative support for the DLVO theory. With gradual refinement in the instrumentation a striking new class of forces has been measured in various liquid media. Distinctive oscillatory characteristics are apparent in the force functions, with rapidly decaying amplitudes over separations up

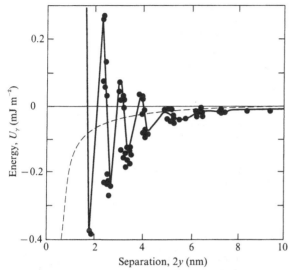

Fig. 6.17. Interplanar potential function $U_\gamma(y)$ for mica–mica surfaces in a non-polar organic liquid, OMCTS, measured in surface-forces apparatus. Periodicity of oscillations corresponds closely to diameter of the (approximately spherical) OMCTS molecule, ≈ 1 nm. (Energies measured relative to reference state at $2y = \infty$ in this experiment.) Dashed curve is continuum van der Waals prediction. (After Horn, R. G. & Israelachvili, J. N. (1981) *J. Chem. Phys.* **75** 1400.)

to ≈ 10 nm (Horn & Israelachvili 1981). The distance between secondary minima is approximately equal to the diameter of the intervening molecules. An example, for non-polar octamethylcyclotetrasiloxane (OMCTS) molecules ≈ 1 nm diameter, is shown in fig. 6.17. Such so-called 'structural' or 'steric' forces reflect a tendency for the fluid species to order in discrete layers as the solid surfaces approach mutual contact.

Let us consider in crystallographic detail how *water* molecules might intercalate an easy cleavage interface in *mica*. Separation occurs between adjacent oxygen layers in the aluminosilicate lamellar structure. Fig. 6.18(a) shows these bounding layers in the undistorted state, in elastic-sphere rather than in the usual point-atom representation. Coulombic attractions between potassium ions in interlayer interstitial sites and uncompensated aluminium ions in the adjacent sub-layers provide the interlamellar cohesion. There are other, smaller interlayer interstices that might be considered favourable ordering sites for water molecules. However, it is readily seen that accommodation of water in these smaller sites can be effected only by separating the oxygen layers to a normal strain

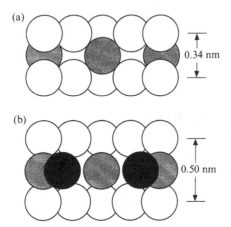

Fig. 6.18. Mica cleavage-plane interlayer structure, showing (a) undistorted lattice and (b) lattice strained to accommodate intercalating water molecules. Elastic-sphere representation, using ionic radii for oxygen (open circles), potassium (shaded) and water (filled). Intrinsic cohesion between cleavage planes due to Coulombic attractions between potassium ions and aluminium ions (not shown) in underlying lamellae lattice. Water molecules interact with structure to reduce Coulombic cohesion.

of some 50 %, as in fig. 6.18(b). Once the water molecules have penetrated the separating interface they interact with the mica structure to reduce the electrostatic cohesion. On closure, the surfaces experience an increasing overlap repulsion from the constrained molecules. Ultimately, the 'hydrolysed' interface becomes thermodynamically unstable and, *in an ideally reversible path*, expels the water, restoring the structure to its initial state. Herein lies the source of the secondary minima in fig. 6.17.

It is the extent to which path reversibility is enforcible that determines hysteresis in an opening–closing cycle. We have intimated that absorbed molecules may not always have sufficient time to expel during closure. Striking evidence for 'retarded expulsion' is obtained from mica–OMCTS–mica 'squeezing' experiments in the Israelachvili surface force apparatus (Chan & Horn 1985). On approaching contact, pronounced plateaus appear in the displacement–time response; the distance between successive plateaus is close to one molecular diameter. Even in prolonged compression the final molecular layers remain effectively immobile. The occluded mica–mica interface adopts a metastable 'extrinsic stacking fault' structure.

Our description of the structured solid–fluid–solid interface may be formalised by including secondary minima in the energy–separation

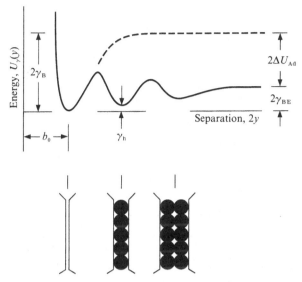

Fig. 6.19. Adhesion energy–separation function for (unit area) *quasi-discrete* solid in fluid medium. Position of primary minimum determined by equilibrium lattice spacing b_0 of the solid. Spacing between secondary minima (only two shown) determined by diameter of intervening fluid molecules. Cf. fig. 5.2(b). (After Lawn, B. R., Roach, D. H. & Thomson, R. M. (1987) *J. Mater. Sci.* **22** 4036.)

function of fig. 6.12(a), and re-expressing this function as a *quasi-continuous* version of fig. 5.2(b). An appropriate function is sketched in fig. 6.19, where we define adhesion energy $U_\gamma(y) = U_B(y)/a_0^2$ per unit area as a function of crack-wall separation $2y = 2u + b_0$. The dashed curve with primary minimum at ground state $2y = b_0$ represents the intrinsic, *in-vacuo* adhesion; the solid curve with subsidiary minima represents the chemically modified adhesion for integral numbers of 'ordered' occlusion layers between the walls, depicted at bottom in fig. 6.19. As with the construction for individual bond reactions in figs. 6.12(a) and 6.13, the intrusive molecules are allowed at all separations to adjust to a configuration of minimum free energy. Thus at increasing separation, as the constraining influence of the substrate surfaces diminishes, the interfacial phase tends increasingly to the properties of the bulk fluid, and the curve smoothes asymptotically to the continuum function of fig. 5.2.

Now take the system in fig. 6.19 through the kind of path-independent cycle as previously executed for the continuum surface–fluid–surface system in fig. 5.2 (sect. 5.1). The first three stages of the operation are unchanged: (i) begin with the surfaces in their virginal (v) ground state at

$2y = b_0$ ($^v U_i = 0$); (ii) separate the B–B surfaces along dashed curve to $2y = \infty$ in a vacuum ($U_* = 2\gamma_B$); (iii) admit the adsorbent molecules to the fully separated walls ($U_f = 2\gamma_B - 2\Delta U_{Ad}$). Now attempt to close the B–E–B interface along the solid curve *over the energy barriers* back to the virgin state. With infinitely slow closure these barriers may eventually be overcome ($^v U_i = 2\gamma_B - 2\Delta U_{Ad} - 2\gamma_{BE} = 0$). Then the work of adhesion W_{BEB} is identical to that in (5.1) and (5.2) for *thermodynamic* equilibrium:

$$
\begin{aligned}
^v W_{BEB} &= U_f - {}^v U_i = U_f \\
&= 2\gamma_B - 2\Delta U_{Ad} = 2\gamma_{BE} \\
&= {}^v W_{BB} - 2\Delta U_{Ad}, \quad \text{(virgin)}.
\end{aligned}
\tag{6.36a}
$$

However, if the closure is made too quickly the system is likely to become trapped in the first (or higher) subsidiary minimum, at an energy level higher than the ground state by the work of formation of the healed (h) interface with occluded species E ($^h U_i = \gamma_{hE}$). The work of adhesion is now that appropriate to *metastable* equilibrium:

$$
\begin{aligned}
^h W_{BEB} &= U_f - {}^h U_i \\
&= {}^v W_{BB} - 2\Delta U_{Ad} - \gamma_h = 2\gamma_{BE} - \gamma_h \\
&= {}^v W_{BEB} - \gamma_h, \quad \text{(healed)}.
\end{aligned}
\tag{6.36b}
$$

This last relation explicitly identifies the hysteresis between opening and re-opening half-cycles with the energy γ_h.

We reiterate our statement in sect. 5.2 that, for reversible paths, the W_{BEB} terms in (6.36) are determined *uniquely* by the end states U_f and U_i. Then as before, the final (open) solid–fluid interface state is governed by the energetics of the Gibbs adsorption equation. The initial (closed) state is governed by the energetics of solid state chemistry; or more strictly, for the metastable healed interface with occluded species, by the energetics of *defect* solid state chemistry.

6.5.2 Secondary interaction zone in brittle cracks

How do the subsidiary minima in fig. 6.19 influence our conception of the fundamental tip structure of a chemically interactive brittle crack? Specifically, how are the environmental molecules accommodated spatially at the near-tip crack interface? We demonstrate using the linear elastic displacement formulas for stress-free cracks in (2.11) to determine approximate elastic-sphere separations at the crack interface: i.e. profiles for ion centres initially on lattice planes at $X = \pm ma_0$ (m integer), $y = \pm\frac{1}{2}b_0$, rather than for continuous walls of slits at $\infty \leqslant X \leqslant 0$, $y = 0$. Such profiles ignore the very nonlinearity that we have emphasised as so central to the determination of the fundamental γ_B and γ_{BE} terms. Strictly, we should seek a self-consistent solution for $u(X)$, but the inclusion of oscillations in the $p_\gamma(y)$ function renders a generally intractable profile and toughness (T_E) analysis (sect. 3.3) even more formidable. Our elastic-sphere structure is in any event adequately equipped to accommodate all the qualitative aspects of the intercalation description in the preceding subsection.

Profiles for the oxygen cleavage-plane layers in mica (fig. 6.18) are shown in fig. 6.20. Cases (a)–(c) represent crack propagation through the virgin interface at increasing G_*, (d) *re*propagation at a healed, occluded interface. Consistent with the requirements of charge neutrality, the potassium ions are assumed to remain bound to the upper and lower surfaces in an alternating array. The leading water molecules are allowed to penetrate the virgin interface until the tight-fitting interstitial con-figuration of fig. 6.18(b) is attained.[4] It is immediately apparent, contrary to the concentrated-reaction modelling, that these molecules do not have unrestricted access to the Irwin crack-tip at C, nor can the crack walls be traction-free behind C. Note that the G_* values represented in fig. 6.20 embrace the experimental data range in the crack velocity plot for mica in figs. 5.9 and 5.21; in particular, case (b) corresponds closely to the threshold region for tests in water in fig. 5.9. Note also that the continuum notion of a crack-tip radius is again of little physical significance; for case (b), the Irwin continuum relation (2.15) yields $\rho \simeq 0.002$ nm.

A striking feature of fig. 6.20 is the clear delineation of the adhesion interface into two zones: an exclusive *primary* zone, within which the virgin bonds separate in 'virtual vacuum'; a broad *secondary* zone, defining the region of molecular penetration within which all extraneous chemical interactions are confined. Some consequential conclusions may be drawn from this picture:

(i) Because extraneous chemical interaction is confined to the secondary zone, the 'crack tip' is 'protected' from environmental attack by the steric

---

[4] Only a single layer of water molecules is included in fig. 6.20. Further back along the interface two, three, or more integral layers are accommodated until, ultimately, the water regains its fluid-like state.

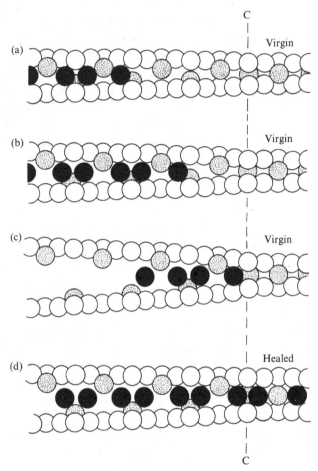

Fig. 6.20. Crack profiles for mica cleavage plane (fig. 6.17), computed from 'Irwin' elastic displacement field with origin (crack tip) at C–C. Loading at (a) $G_* = 100$ mJ m^{-2}, (b) $G_* = 200$ mJ m^{-2}, (c) $G_* = 800$ mJ m^{-2}, showing restricted penetration of water molecules along virgin interface. Region of penetration delineates *secondary adhesion zone*. Unloading at (d) $G_* = 100$ mJ m^{-2}, depicting entrapment of water molecules at healed interface. (After Lawn, B. R., Roach, D. H. & Thomson, R. M. (1987) *J. Mater. Sci.* **22** 4036.)

barriers. This is the root of the *structural invariance* of atomically sharp cracks alluded to in sect. 5.5.5. The primary and secondary zones are nevertheless *physically coupled*, via the embedding elastic lattice. It is this coupling that enables the adsorption energy in (6.36) to be 'recovered' by the crack system behind the tip.

(ii) The concentrated reaction of sect. 6.4 may be seen as a *limiting case* of the surface force model, in which the secondary zone contracts to a width of one atom dimension. Indications from fig. 6.20 are that for the mica–water system no such localisation occurs until $G_* \gg 1$ J m^{-2}.

(iii) Environmental molecules will experience significant steric impedance from the substrate surface structure as they attempt to migrate through an extended secondary zone, particularly at low (near-threshold) G_* levels (e.g. fig. 6.20(b)). This makes it likely that *activated interfacial diffusion* (sect. 5.5.2), and *not* bond–bond reaction chemistry, is the region-I rate-limiting mechanism in the mica–water system.

Can this concept of a well-defined secondary zone reasonably be extended to other material–environment systems? Of the spectrum of covalent–ionic solids, mica possesses one of the *least* densely packed cleavage atom-plane interlayers. In sapphire, a lattice strain $\approx 200\%$ is necessary to accommodate interfacial water molecules (cf. 50% in mica, fig. 6.18). We therefore suggest that the picture of a sterically constrained interface in fig. 6.20 has a broad generality. Silicate glass, with its unusually open network structure, is an extreme case where concentrated-reaction models may remain appropriate. Again, we have restricted our consideration of environmental species to water. Water is a relatively small molecule: other interactive molecules may be expected to encounter even stronger steric hindrance.

6.5.3 *Implications concerning fracture mechanics*

Consider now how we may adapt our earlier fracture mechanics formalisms to account for the hysteresis in crack propagation through virgin and healed interfaces:

(i) *Thermodynamic equilibrium.* We examine the condition $\mathscr{g} = G_* - W_{\mathrm{BEB}} = 0$ in (6.36) from the perspective of two enclave observers (cf. sect. 3.6). Observer **1**, located outside both primary and secondary zones, writes

$$G_* = {}^{\mathrm{v}}W_{\mathrm{BEB}} = {}^{\mathrm{v}}W_{\mathrm{BB}} - 2U_{\mathrm{Ad}}, \quad \text{(virgin)} \tag{6.37a}$$

$$G_* = {}^{\mathrm{h}}W_{\mathrm{BEB}} = {}^{\mathrm{v}}W_{\mathrm{BB}} - (2U_{\mathrm{Ad}} + \gamma_{\mathrm{h}}), \quad \text{(healed)}. \tag{6.37b}$$

In this frame of reference, the quantities $2U_{Ad}$ and γ_h are perceived as (negative) surface energy terms. The extraneous chemistry is regarded as an integral component of the intrinsic toughness. Observer **2**, located at the boundary between the primary and secondary adhesion zones, writes

$$G_* + 2U_{Ad} = {}^v W_{BB}, \quad \text{(virgin)} \tag{6.38a}$$

$$G_* + (2U_{Ad} + \gamma_h) = {}^v W_{BB}, \quad \text{(healed).} \tag{6.38b}$$

The quantities $2U_{Ad}$ and γ_h are now perceived as positive ('fictitious') mechanical-energy-release rates. This observer sees the crack 'tip' as structurally inert, characterised by a 'vacuum' surface energy ${}^v W_{BB} = 2\gamma_B$, and the intrusive chemical species in fig. 6.20 as delivering a pseudo-mechanical *anti-shielding* (opening) force, somewhat akin to a 'molecular wedge'.

(ii) *Kinetics.* The quasi-continuum crack velocity relation (5.18) is appropriate for near-threshold region I configurations:

$$v = 2v_0 a_0 \exp(-\Delta F_I^*/kT) \sinh[\alpha_1(G_* - {}^v W_{BEB})/kT], \quad \text{(virgin)} \tag{6.39a}$$

$$v = 2v_0 a_0 \exp(-\Delta F_I^*/kT) \sinh[\alpha_1(G_* - {}^h W_{BEB})/kT], \quad \text{(healed)} \tag{6.39b}$$

where α_1 and ΔF_I^* now define interfacial diffusion barriers. Equation (6.39) accounts for the Rice graphical construction in fig. 5.4. Thus, initial loading drives the crack through the virgin solid along the positive branch of (6.39a). On unloading, the crack becomes stationary at the threshold $G_* = {}^v W_{BEB}$, and is constrained by the entrapped species to remain so until $G_* = {}^h W_{BEB} = {}^v W_{BEB} - \gamma_h$, at which point it retracts along the negative branch of (6.39b). On reloading above $G_* = {}^h W_{BEB}$, the crack grows hysteretically through the healed interface along the positive branch of (6.39b).

The solid curves through the mica–air data in fig. 5.21 are fits in the manner of sect. 5.6 for virgin cracks, but with due regard to the hysteresis in (6.39). Accordingly, the G-shifts in the healing–repropagation branches correspond to fault energies $\gamma_h \simeq 100\,\text{mJ m}^{-2}$ (healed) and $\gamma_{h'} \simeq 300\,\text{mJ m}^{-2}$ (misoriented–healed) in (6.37) (typical of stacking fault energies in covalent–ionic crystals). The higher value for the misoriented interface is indication that the coherence of the solid–solid interface is no less crucial than the occlusion structure of the entrapped molecules in determining the fault configuration.

According to the above description, *the essence of all the important equilibrium and kinetic properties of brittle crack propagation, including hysteresis, is contained in the interplanar surface force function.* At present, our understanding of the nature of surface forces, particularly the oscillating components, is incomplete. It is interesting to reflect that brittle fracture is proving a useful adjunct to the Israelachvili apparatus in the study of these forces.

6.6. Crack-tip plasticity

We have assumed that the intrinsically brittle fracture is free of crack-tip plasticity. Lattice statics calculations provide some insight into this issue for specific material types. Thus computer simulations of diamond-type solids (Sinclair & Lawn 1972) reveal no evidence for shear instabilities that might be construed as indication of dislocation-generation processes. On the other hand, shear instabilities *are* observed in simulations of body-centred cubic iron (Kanninen & Gehlen 1972). But useful conclusions on crack-tip plasticity may be drawn without capitulation to the computer. We shall consider two quasi-continuous models to argue the case for the intrinsic brittleness of most covalent–ionic solids; and, indeed, of some metals.

6.6.1 Theoretical strength model

The first of the quasi-continuous models was developed by Kelly, Tyson & Cottrell (1967) in an attempt to derive a criterion for the onset of crack-tip flow. They proposed that an ideal (defect-free) solid may sustain an atomically sharp crack if the cohesive strength in tension is exceeded before the corresponding cohesive strength in shear within the near-field of the tip. If shear breakdown were to occur the crack would lose its atomic sharpness, i.e. it would be plastically blunted. No comment was made as to what form that blunting might take.

Their model builds on lattice statics calculations of theoretical cohesive strengths for monocrystalline and amorphous solids in homogeneous stress. They confirm Griffith's original estimate of the cohesive strength as one-tenth of an elastic modulus (sect. 1.5) as a useful general guide. However, detailed computations for individual materials reveal a distinct tendency toward a diminishing ratio of theoretical shear strength to tensile strength in proceeding from covalent to ionic to metallic bonding. This increased susceptibility to shear failure with less rigid bonding is reflected in the parallel tendency to increased fracture toughness in table 3.1.

It is necessary now to consider the theoretical cohesive strengths in relation to the stress distribution at the crack tip. To keep the analysis viable, Kelly–Tyson–Cottrell resorts to the continuum equations of linear, isotropic elasticity for an evaluation of this near-tip field. Here the approach is vulnerable; inadequacies of those equations as a basis for describing events in the tip region have been repeatedly emphasised. It is nevertheless not unreasonable to expect the self-similarity of stress components that characterise the field at any given angle to the crack plane to persist as r in (2.14a) decreases toward atomic dimensions. The $f_{ij}(\theta)$ plots of fig. 2.5 for mode I indicate that the tension favouring in-plane crack extension and shear favouring off-plane slip deformation are in the ratio of about two to one. Hence a simplistic requirement for crack-tip plasticity is that the theoretical strength in shear should be less than one-half that in tension.

For metallic materials the intrinsic shear strength is so low relative to the tensile strength that flow is almost inevitable: face-centred cubic metals (with their strong tendency to glide on close-packed planes) are predicted to be fully ductile. Conversely, for covalent materials the relative shear resistance is sufficiently high as to preclude the onset of flow: diamond-structure crystals, with their exceptionally rigid tetrahedrally coordinated covalent bonds, are predicted to be perfectly brittle. Classification of materials in the intermediate category, ionic solids and body-centred cubic metals, is less straightforward.

A number of additional factors makes prediction even more uncertain for most materials. Strictly, distinction should be made between plane strain and plane stress, since these two states lead to different maxima in shear stress. Then, the existence of a superposed tension (a manifestation of the biaxiality of the near-field) across the slip plane inevitably reduces the theoretical shear strength. Also, the influence of crystallographic anisotropy on the intensities of resolved stresses on primary cleavage and slip systems can be substantial. Not least, the issue as to whether shear breakdown is a sufficient condition for crack-tip plasticity remains to be addressed.

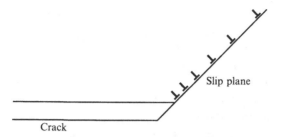

Fig. 6.21. Emission of dislocations from crack tip. To produce true blunting, these dislocations must have component of Burgers vector normal to crack plane, and slip plane must intersect crack front along its length.

6.6.2 Dislocation initiation model

Following the last point, suppose that in a given solid the cohesive shear strength *were* to be exceeded at the crack tip. What form might the ensuing deformation take, and how might this deformation influence the brittleness? These questions were addressed by Rice & Thomson (1974) in a somewhat more elaborate model of crack-tip instability.

Those authors pointed out that the shear deformation envisaged by Kelly *et al.* amounts to a *spontaneous* emission of dislocation loops from the crack tip. They suggested that such a picture must underestimate the resistance to breakdown. It is not sufficient to nucleate a dislocation loop; one must also propagate it out of the diminishing, $r^{-1/2}$, near field. In other words, there may exist an energy barrier to propagation, meaning that emission is an *activated* process. Brittle crystals are therefore those for which the energy barrier is large relative to k**T**. The barrier height can be calculated from a balance between three crack–dislocation interaction forces: (i) force on dislocation from stress field of crack (repulsive), (ii) line tension force from surface ledges formed by dislocation nucleation (attractive), (iii) image force of dislocation in free surface of crack (attractive). One then evaluates the mechanical stability of the configuration.

Rice & Thomson stipulated an additional requirement for crack-tip breakdown; the dislocations emitted must be of a 'blunting' type, fig. 6.21. For this to be so for a mode I crack the Burgers vector must have a component normal to the crack plane and the slip plane must intersect the crack front along its whole length. These are stringent restrictions on the crack geometry. Moreover, the surfaces and fronts of most real cracks are curved; the only way that curved cracks could attain the necessary crystallographic easy-slip orientations would be to degenerate into a configuration of low-index jogs and kinks.

With these considerations it is argued that crack-tip plasticity is probably less widespread than the model of Kelly *et al.* would indicate. Covalent and ionic solids are almost certainly truly brittle at room temperature. Even some metals, hexagonal close-packed to a greater extent and body-centred cubic to a lesser, may be stable against dislocation emission; face-centred cubic metals, on the other hand, remain clearly in the ductile category.

6.7 Fundamental atomic sharpness of brittle cracks: direct observations by transmission electron microscopy

Virtually all we say in this book is predicated on the atomic sharpness of crack tips. We have seen how naturally this elemental precept accounts for the innate brittleness of covalent–ionic solids and for the susceptibility of these solids to environmental chemistry. Yet, as indicated in sect. 5.5.5, the fundamentally sharp structure of brittle cracks remains an issue of some contention. There is the alternative conception that extending brittle cracks are more properly characterised by a continuum, changeable tip radius: i.e. that they are inherently blunt.

Substantiation of the blunt-crack hypothesis requires definitive evidence for one or other of the following features:

(i) In the corrosion-based models (fig. 5.20(a)), a smoothly rounded crack tip with root radius of order several atom diameters. (Recall the unphysically small, subatomic values estimated theoretically in sects. 6.3 and 6.5.)

(ii) In the crack-tip shear-deformation models (fig. 5.20(b)), a plastic zone of characteristic dimension (evaluated, for example, using a 'Dugdale yield stress' $\bar{p} = \sigma_C^Y \approx H/3$ in (3.16), H = hardness) somewhat in excess of

Fig. 6.22. Transmission electron micrographs of cracks in silicon: (a) at 25 °C, viewed down interface, width of field 1.0 μm; (b) same crack, after slight tilt; (c) another crack, formed at 500 °C, width of field 3.5 μm. CC denotes crack front. (After Lawn, B. R., Hockey, B. J. & Wiederhorn, S. M. (1980) *J. Mater. Sci.* **15** 1207.)

(a)

(b)

(c)

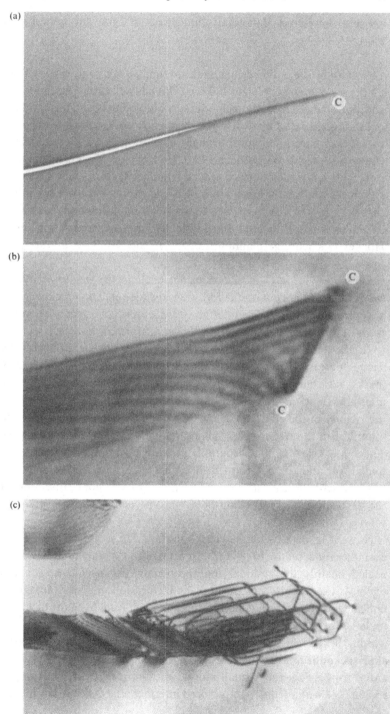

Fig. 6.22. For legend see facing page.

the corresponding Barenblatt dimension ≈ 1 nm for cohesive forces (sect. 3.3.1).

Estimates of these dimensions indicate that we are concerned with events on the scale of no more than 100 nm. This lies well below the limits of most conventional microscopy techniques available for observing cracks.

However, studies made using transmission electron microscopy (TEM) by Hockey and others have brought us close to the necessary resolution (Lawn, Hockey & Wiederhorn 1980). Hockey uses Vickers indentations to introduce controlled surface cracks (chapter 8) into single crystals. The specimens are then thinned from below to produce foils. Remnant crack segments in the foils are viewed in electron diffraction contrast. The examples cited below are typical of many hundred observations made by Hockey in silicon, germanium, silicon carbide, sapphire, and magnesium oxide.

Observe the cracks in silicon in fig. 6.22. In (a) we see a ribbon-like segment nearly normal to the foil, viewed edge-on. This crack was formed at room temperature. The interface remains partially open (due in part to the mouth-wedging action of the Vickers deformation zone – sect. 8.1.3). In (b) the same segment is viewed in tilted orientation, revealing interfacial fringe contrast. The enhanced band of contrast at the crack front C–C marks residual elastic strain associated with the imperfectly closed interface. In neither (a) nor (b) is there indication of dislocation activity in the vicinity of the crack tip, either in the region immediately surrounding the tip or in the wake region behind it. That any such activity should be detectable is clear from micrograph (c), showing a similar crack segment to that in (a) and (b) but formed at ≈ 500 °C, where silicon undergoes a brittle–ductile transition. Even in this last case it can not be discounted that the dislocations are of a non-blunting type activated from pre-existing sources (sect. 7.3.1).

Now let us investigate the crack-interface fringe contrast, with specific examples for sapphire in fig. 6.23. In (a) we show a crack that has partially healed against the residual mouth-opening displacement. The healed region is characterised by an in-plane dislocation network (at right), the unhealed region by broad moiré fringes degenerating from the dislocations (at left). In (b) the healing is apparent as the stacking-fault fringe patterns within the partial-dislocation network. Diagnostic diffraction contrast analysis confirms that the network patterns are *not* caused by slip. The moiré fringes indicate a small mutual rotation ($\approx 10^{-3}$ rad) of opposing crystal halves at an imperfectly closed interface; the dislocations indicate a

(a)

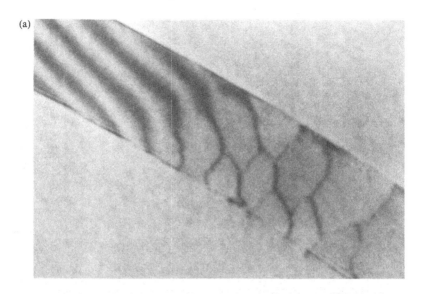

(b)

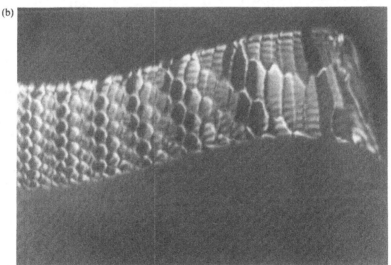

Fig. 6.23. Transmission electron micrographs of two crack segments in sapphire: (a) showing continuity between moiré fringe pattern and healing dislocation network (bright field) behind tip (located out of field of view at right), width of field 2.0 μm; (b) partial dislocation and 'stacking fault' contrast (dark field) at healed interface, width of field 6.0 μm. (After Hockey, B. J. (1983), in *Fracture Mechanics of Ceramics*, ed. R. C. Bradt, A. G. Evans, D. P. H. Hasselman, & F. F. Lange, Plenum, New York, Vol. 6, p. 637.)

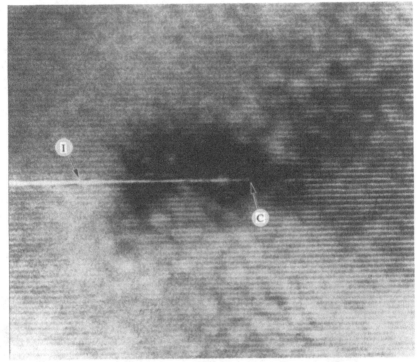

Fig. 6.24. High resolution transmission electron micrograph of crack-tip region in a magnesium Sialon lattice. The fringes correspond to the lattice spacing of the material. Open interface I indicates a state of residual elastic strain around crack tip C. (Courtesy D. R. Clarke.)

restored configuration in which surfaces have recontacted and bonds realigned within the constraints of the mismatch. Stacking fault patterns of the type shown in fig. 6.23(b) tend to degenerate with time, especially under conditions of beam heating in the electron microscope (Hockey 1983), suggestive of the metastable states associated with occluded (diffusive) water described in sect. 6.5.

It is the reproducibility of such observations as those in figs. 6.22 and 6.23 that leads us to discount the blunt-tip concept in brittle cracks. If propagation were to involve crack-tip *plasticity* we would certainly expect to observe some remnant glide dislocations or other slip elements about the crack tip, as in fig. 6.22(c). But such elements are conspicuously absent at room-temperature cracks in the crystals studied. (The ionic crystal magnesium oxide is an exception, and there the activity is attributed to pre-existing sources, sect. 7.3.1.) If the extension were to involve dissolution or any other kind of *chemical* rounding process it is unlikely that the interface would ever be able to close up elastically and heal against a residual mouth-opening displacement. Again, notwithstanding the constraints of spurious physical obstruction (steps, fracture debris), the incidence of near-tip healing is quite general (although to a greater degree in materials with a larger ionic/covalent ratio in the bonding, presumably because of a reduced tendency to structural rearrangement of broken bonds).

There is another kind of TEM observation that has the potential for a more quantitative vision of the true crack-tip structure. We refer to the technique of lattice imaging, in which the diffraction planes are revealed as a set of lattice fringes. The example in fig. 6.24 is for a crack in a magnesium Sialon crystal. Again, this crack is in a state of residual opening strain, as is evident from the dark contrast surrounding the tip. The continuity of lattice fringes gives no indication of a rounded tip. It should be acknowledged that interpretations of lattice images need to be made with extreme caution: indeed, all TEM imaging techniques at the level of near-atomic resolution are subject to artifacts and misinterpretations. Nevertheless, the sharp-crack picture is compelling. Stronger confirmation awaits the next generation of atomic resolution techniques.

7

Microstructure and toughness

We have thus far dealt with the resistance to crack propagation at opposite extremes of material representation, continuum solid and atomic lattice. It is now appropriate to investigate the problem at an intermediate level, that of the *microstructure*. By 'microstructure' we mean the compositional configuration of discrete structural 'defects': voids, inclusions, second-phase particles (volume defects); secondary crack surfaces, grain boundaries, stacking faults, twin or phase boundaries (surface defects); dislocations (line defects). It is principally at this intermediate level that significant improvements in the mechanical properties of traditional brittle polycrystalline ceramics (cf. table 3.1) may be realised. By tailoring the microstructure it is possible to introduce an interactive defect structure that acts as an effective restraint on crack propagation and thus enhances the material *toughness*.

In this chapter we examine some of these 'toughening' interactions. We identify two classes of restraint. The first involves purely *geometrical* processes, deflections along or across weak interfaces, etc. The responsible microstructural elements may be regarded as 'transitory obstacles', in the sense that their impeding influence lasts only for the duration of crack-front intersection. Because of their ephemeral nature such interactions are relatively ineffective as sources of toughening, accounting at very most for increases of a factor of four in crack-resistance energy R or, equivalently, a factor of two in toughness T.

The second class of restraint comprises *shielding* processes. The critical interactions occur away from the tip, within a 'frontal zone' ahead or at a 'bridged interface' behind. The most consequential are those with a large component of irreversibility, so that the restraining influence persists, via a pervasive state of residual closure strain, in the crack 'wake'. Because of their cumulative nature, shielding processes are potentially stronger sources of toughening (notwithstanding that deflection may on occasion assume an essential precursor role), and will accordingly receive the greater part of our attention. They are the underlying source of the phenomenological *R-curves* or *T-curves* introduced in sect. 3.6.

The groundwork for a description of the influence of microstructure on toughness has been set out in chapter 3. It is worthwhile restating the important elements of this groundwork, for they continue to be the source of much misunderstanding in the engineering materials literature. We stress again: *the validity of the Griffith–Irwin fracture mechanics is not lost at the microstructural level.* The only provisos are that the crack tip should retain its atomic sharpness, and that the characteristic spacing between discrete microstructural shielding elements should exceed the critical crack-tip dimensions (e.g. cohesion-zone length). Then the interactions may be considered to influence the intensity, but not the fundamental nature, of the crack-tip field. That is, the effective mechanical driving force on the crack may be determined in the usual way from the additive properties of K or G terms (depending on the fracture mode), and the conditions for extension of the (structurally invariant) crack tip thereby expressed in terms of the fundamental 'laws' defined in the previous chapter.

In the following sections we outline some of the major microstructural toughening mechanisms that operate in brittle materials. We start with geometrical considerations, intergranular vs transgranular fracture, deflection at second-phase inclusions, step formation, etc., highlighting the role of weak grain or interphase boundaries and internal residual stresses. Then we focus on the micromechanics of shielding. Frontal-zone processes, notably phase transformation in zirconia but also dislocation- and microcrack-cloud activity, are addressed. This is followed by a discussion of crack-interface bridging. Where possible we illustrate with monophase ceramic microstructures to bring out basic principles. At the same time, we underscore the potentially powerful role of controlled additive phases, culminating with a discussion on reinforced ceramic-matrix composites.

7.1 Geometrical crack-front perturbations

The intrinsic *in-vacuo* toughness of ideally homogeneous brittle materials with mechanical energy dissipation confined to the cohesion zone of a plane crack is determined by the Griffith condition,

$G_C = R = R_0 = W_{BB} = 2\gamma_B$ or, equivalently, by the Barenblatt condition, $K_C = T = T_0 = (E'R_0)^{1/2}$. 'Intersections' with otherwise inactive microstructural heterogeneities in the elastic field can cause a planar crack to deflect from its path and absorb additional energy, effectively increasing R or T. However, the crack retains no 'memory' of any such intersection once the intersection is complete.

The mechanics of energy absorption are rather subtle. Deflections 'toughen' the material by increasing the net resistance $G_C = R$. In evaluating R it is necessary to take due account of the mechanical energy release with respect to non-planar crack increments, with proper attention to mode II and III components on these increments. Such deflections can also lead to subsidiary energy dissipation, e.g. from local jump–arrest instabilities at the microstructural heterogeneities (with attendant acoustic emissions).

Another subtlety concerns the contribution of the intrinsic resistance R_0 to the net resistance R. Paradoxically, R may be enhanced by *reducing* R_0, via adjustments to the internal interface microstructure.[1] Weak interfaces provide preferred deflection paths, i.e. paths that maximise $\mathscr{g} = G - R_0$ (sect. 2.8). The key is to contrive the deflection geometry so that the immediate reduction in R_0 is more than offset by an ultimate increase in G_C.

7.1.1 Transgranular vs intergranular fracture

In a polycrystalline material the grains are crystallographically misoriented with respect to their neighbours. Depending on the relative values of $G - R_0$, a crack intersecting the boundary between two grains may either traverse the boundary and continue through the second grain with relatively minor adjustments to the plane of propagation (*transgranular* fracture), or deviate onto the grain boundary itself (*intergranular* fracture). Generally, we may expect a transition from transgranular to intergranular fracture as the grain boundaries become weaker and more misoriented.

We examine the two cases in turn below:

(i) *Transgranular fracture.* Consider an idealised configuration in which an initially plane cleavage crack orthogonally intersects and propagates

---

[1] As experimentalists, we have a certain degree of control over R_0. We have already seen in chapters 5 and 6 how $R_0 = W_{BB}$ may be lowered by extraneous environmental interaction from its intrinsic value, $2\gamma_B$, to $2\gamma_{BE}$, or even to the metastable quantity $2\gamma_{BE} - \gamma_{hE}$ in (6.36b).

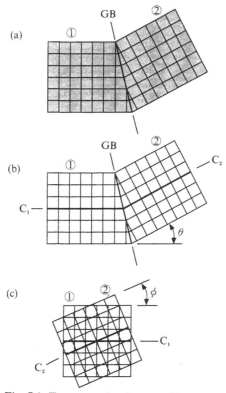

Fig. 7.1. Transgranular fracture, illustrated for simple cubic lattice: (a) grain misorientation about normal to plane, plan view of crack plane; (b) misorientation θ about line of crack front (tilt boundary), side view; (c) misorientation ϕ about direction of crack advance (twist boundary), end view. Note that only (b) and (c) affect the crack *plane*, although (a) does affect the crack *front*. GB denotes grain boundary, C_1 and C_2 the crack plane in (like) grains 1 and 2.

across a grain boundary between two otherwise perfect crystallites. The impedance will be determined by the type and degree of the cleavage-plane rotation. In general, a grain boundary has five degrees of freedom: three angular coordinates to specify the misorientation between the adjacent grains, and two more to specify the orientation of the boundary itself. Of these misorientation angles only two, the tilt θ and twist ϕ illustrated in fig. 7.1, have any bearing on the orientation of the cleavage plane. These configurations may be considered in relation to the earlier crack-path analysis of sect. 2.8. For crystals with pronounced cleavage tendencies, diagrams (a) and (b) of fig. 2.18 may be conveniently taken as representative of intersections with tilt and twist boundaries respectively. It

Cleavage planes ②

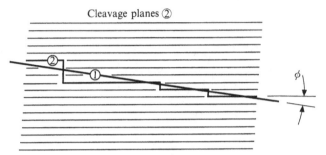

Fig. 7.2. Schematic of segmentation into partial fronts as crack traverses twist boundary from grain 1 to grain 2.

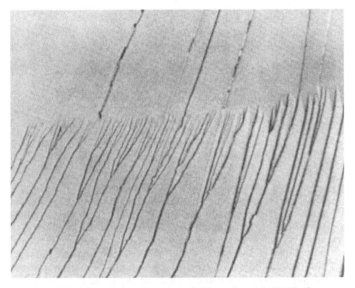

Fig. 7.3. 'River' pattern of steps on lithium fluoride (100) cleavage surface, formed upon traversal of a tilt–twist grain boundary ($\theta = 0.87°$, $\phi = 0.85°$). Crack advance from top to bottom. Viewed in reflected light. Width of field 500 μm. (After Gilman, J. J. (1958) *Trans. Met. Soc. A.I.M.E.* **212** 310.)

will be recalled from this earlier analysis that whereas a θ-rotation can be accommodated by a continuous adjustment of the advancing front, a ϕ-rotation can not. If in the latter case the crack is to remain on its preferred cleavage plane it must segment into partial fronts, linked by cleavage steps, as depicted schematically in fig. 7.2. A step-segmentation configuration of this kind is apparent in the micrograph of fig. 7.3. Such configurations are less likely to occur in crystals without a strong cleavage tendency: the crack

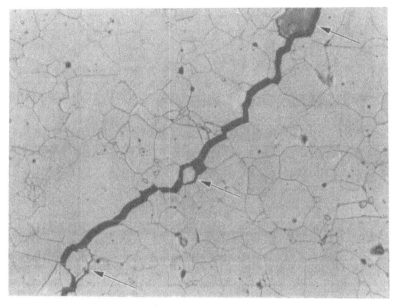

Fig. 7.4. Crack in alumina, mean grain size 25 μm, showing intergranular fracture. Note especially severe disturbances at some larger interface-adjacent grains (arrows). Surface thermally etched to reveal grain boundary structure, reflected light. Width of field 500 μm. (After Swanson, P. L., Fairbanks, C J., Lawn, B. R., Mai, Y-W. & Hockey, B. J. (1987) *J. Amer. Ceram. Soc.* **70** 279.)

then traverses the boundary without significant change in plane, i.e. the crack fails to 'recognise' the grain boundary.

With this picture we may gain an estimate of R in transgranular fracture. Suppose the crack in the first grain to grow stably at $G_C = R_0 = 2\gamma_B$ under pure mode I loading. It is evident from fig. 2.19 that the local value $G(\theta, \phi)$ must immediately decrease as the crack traverses the boundary and deflects into a state of mixed-mode loading. To maintain propagation the applied load must be increased until a new local deflection equilibrium state $G(\theta, \phi) = R_0$, corresponding to a global state $G_C = G(0) = R(\theta, \phi)$, obtains. Accordingly, the crack resistance $R(\theta, \phi)/R_0$ plots as the inverse of $G(\theta)/G(0)$ in fig. 2.19. For a pure tilt boundary of relatively large grain misorientation $\theta = 45°$, say, we have $R(\theta)/R_0 \approx 1.30$. We may similarly estimate $R(\phi)/R_0$ for twist boundaries, with additional toughening from the formation of cleavage steps (sect. 7.1.3).

(ii) *Intergranular fracture.* Grain boundaries are the most common

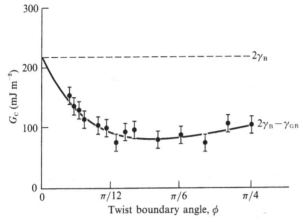

Fig. 7.5. Mechanical-energy-release rate for equilibrium crack propagation along (010) twist boundaries in potassium chloride, as function of misorientation. Dashed line represents monocrystal (100) cleavage value. DCB data. (After Class, W. H. & Machlin, E. S. (1966) *J. Amer. Ceram. Soc.* **49** 306.)

examples of weak interfaces in brittle materials. They are especially weak in ceramics because of the stringent directionality and charge requirements of covalent–ionic bonds. There is therefore an increasing tendency for cracks to propagate *around* instead of *through* grains as the angular misorientations between crystallites increase. Intergranular fracture is characterised by a tortuous path, fig. 7.4, with consequent increase in fracture surface area.

Consider a crack propagating along a planar grain boundary between two mutually misoriented crystallites. The work of adhesion is

$$R_0 = W_{BB} = 2\gamma_B - \gamma_{GB}, \quad \text{(grain boundary)} \tag{7.1}$$

where $\gamma_{GB} = \gamma_{GB}(\theta, \phi)$ is the *configurational* energy of the boundary relative to the virgin B–B state. For coherent, low angle boundaries ($\theta, \phi < 10°$) the interfacial configurations may be effectively treated as dislocation arrays (screw for tilt, edge for twist), strongly analogous to the dislocation networks at the healed crack interfaces in fig. 6.23. High-angle boundaries are more appropriately viewed as high-density layers of point defects. The lattice mismatch can reduce R_0 in (7.1) by more than one-half the single crystal value ($\gamma_{GB} > \gamma_B$), as demonstrated by the fracture data of fig. 7.5 for 'clean' boundaries in potassium chloride bicrystals. In practical ceramics, grain boundaries are highly susceptible to contamination by impurity or additive 'wetting' phases, especially during processing. Although perhaps only a few molecular layers thick these phases can exert a strong influence on the adhesion. Because of the geometrical constraints at the confined interface the molecular structure of the boundary phase may be highly 'ordered', in the manner discussed in relation to fig. 6.20. The work of adhesion is then determined by surface force functions pertinent to the species that define the immediate crack interface.

In polycrystalline materials, an intergranular path can be maintained only if the boundaries are sufficiently weak to compensate for the extra energy consumed in the attendant crack deflections. For a straight crack initially in pure mode I deflecting onto a grain boundary in mixed mode I + II, this condition is (sect. 2.8)

$$G(\theta, \phi)/G(0) > R_0(\text{GB})/R_0(\text{B})$$
$$= 1 - \gamma_{\text{GB}}/2\gamma_{\text{B}} \tag{7.2}$$

with $R_0(\text{GB})$ from (7.1) and $R_0(\text{B})$ for straight-ahead (transgranular) fracture through the bulk crystal. For a tilt deflection through $\theta = 90°$, say, we have $G(\theta)/G(0) \approx 0.25$ from fig. 2.19, so that $R_0(\text{GB})/R_0(\text{B}) < \frac{1}{4}$, corresponding to $\gamma_{\text{GB}} > \frac{3}{2}\gamma_{\text{B}}$. The 90° tilt represents an extreme case, so satisfaction of the less stringent condition $R_0(\text{GB})/R_0(\text{B}) < \frac{1}{2}$, $\gamma_{\text{GB}} > \gamma_{\text{B}}$, may be sufficient to realise intergranular fracture in most ordinary polycrystalline materials. Since intergranular fracture corresponds to a minimum energy path, any enhancement of the net resistance can arise only from a compensating increase in the ratio $\alpha = \text{actual/projected}$ fracture surface area, i.e. $R(\theta, \phi)/R_0(\text{B}) = \alpha R_0(\text{GB})/R_0(\text{B})$; and since $\alpha = 2-3$ at most, we see that the integrated resistance is unlikely to exceed unity by any substantial amount.

The reality of transgranular and intergranular fracture is much more complex than our idealised considerations here would indicate. First, our estimates of $R(\theta, \phi)$ are strictly valid only in the limit of infinitesimal extensions from an otherwise planar crack (fig. 2.18). Second, the crack does not generally deflect uniformly along its front as depicted in fig. 2.18. Rather, there is a *distribution* of local deflections (additional to the segmentations depicted in fig. 7.2), corresponding to intersections with grains of different orientation along the line. Thus while some intersections will deflect the crack away from the primary crack plane, others will simultaneously deflect it back again (with local instabilities and attendant phonon emissions). Third, we have assumed implicitly that granular rupture occurs exclusively *at* the primary crack front. Experimentally, this is not always the case. Grain fracture can occur *behind* the main front ('bridging' – see sect. 7.5). Finally, the presence of internal residual stresses, e.g. from differential thermal expansion mismatch in neighbouring non-cubic grains, will further perturb the crack propagation (although in the long term the competing compressive and tensile perturbations would be expected to average out). Because of the modest level of potential toughening, we omit detailed considerations of these additional factors here. On the other hand, the issue as to which of these two competing modes will dominate in a given polycrystalline material under given stress conditions is paramount to the effectiveness of certain *R*-curve mechanisms, as we shall see in various later stages of this chapter.

7.1.2 Fracture in two-phase materials

Much of the above discussion on polycrystalline ceramics carries over to two-phase ceramics (a basic form of ceramic 'composite' – see sect. 7.6). A crack may be forced to propagate either around or through any second-phase obstacle, with consequent effect on the toughness. This second phase may assume a variety of configurational forms, characterised by geometry (polyhedra, spheres, rods, plates, etc.), relative elastic stiffness (ranging from infinity – a rigid phase, to zero – a cavity), thermal expansion anisotropy mismatch, and so on.

Detailed analysis of the role of second-phase particles in the mechanics of deflection by Faber & Evans (1983) provides insight into the *geometrical* aspects of this kind of toughening. The Faber–Evans analysis stems from the same kind of incremental extension argument presented in the preceding subsection in relation to angular variations in the crack resistance, $R(\theta, \phi)$, with consideration of specific dispersions of particle geometries along the crack front. Those authors consider tilt and twist deflections of matrix-confined cracking around sphere-, rod- and disc-like particles, at variable interparticle spacing, fig. 7.6. The twist component is found to be more effective than the tilt in its capacity to induce crack-path deflections, and thence to amplify the integrated resistance. Of the different particle morphologies, rods induce the greatest levels of twist, corresponding to a maximum enhancement $R/R_0 \approx 4$ in crack resistance at the highest practical aspect ratios and packing densities, followed by discs then spheres.

This factor may be compromised if the deflected matrix crack physically

(a)

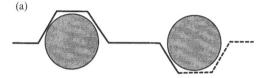

(b)

Fig. 7.6. Crack deflection geometries in two-phase solids: (a) tilt deflections around spheres, large mode II component (crack propagation left to right); (b) twist deflections around rods, large mode III component (crack propagation into plane of diagram – dashed line undisturbed crack plane). (After Faber, K. T. & Evans, A. G. (1983) *Acta Metall.* **31** 565, 577.)

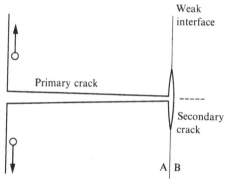

Fig. 7.7. Diversion of crack along extended weak interface in tensile specimen. Driving force for extension diminishes as secondary fracture grows further from primary plane. (After Cook, J. & Gordon, J. E. (1964) *Proc. Roy. Soc. Lond.* **A282** 508.)

intersects the interphase boundaries. Consider the case of a mode I primary crack intersecting an orthogonal interphase boundary in fig. 7.7. In analogy to (7.1), the work of adhesion for an interphase boundary is

$$R_0 = W_{AB} = \gamma_A + \gamma_B - \gamma_{IB}, \quad \text{(interphase boundary)} \quad (7.3)$$

where γ_{IB} is the formation energy of the interphase boundary. In general, γ_{IB} contains a *chemical* mismatch component (γ_{AB} for coherent boundary

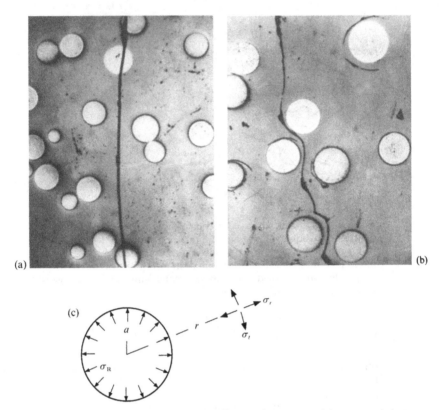

Fig. 7.8. Crack paths in two different glass compositions containing
10% volume fraction of thoria spheres: (a) $\Delta\alpha = +1.8 \times 10^{-6}$ K^{-1},
width of field 600 µm; (b) $\Delta\alpha = -5.1 \times 10^{-6}$ K^{-1}, width of field
1000 µm. Section views in reflected light. (c) Residual stress field of
spherical particle in infinite matrix. (After Davidge, R. W. & Green,
T. J. (1968) *J. Mater. Sci.* **3** 629.)

in (2.30b)) as well as a *configurational* mismatch component (cf. γ_{GB} in
(7.1)). The condition for deflection into the interface is similar to that for
intergranular fracture in (7.2), leading to $\gamma_{\text{IB}} > \gamma_{\text{A}} + \frac{1}{2}\gamma_{\text{B}}$. A more detailed
analysis of the configuration in fig. 7.7 allowing for modulus differences
between components A and B (Hutchinson 1990; Hutchinson & Suo
1991) reveals this to be a conservative bounding condition; i.e. the value of
γ_{IB} required to ensure deflection is generally reduced by elastic mismatch.

If the adhesion at an intersected boundary is too strong ($\gamma_{\text{IB}} \ll \gamma_{\text{A}} + \gamma_{\text{B}}$)
the crack may be forced to penetrate into the second phase. An advancing
front may then be *pinned*, causing local in-plane segments between

dispersed particles to *bow* into loops (much like the pinning and breakaway concept in dislocation-hardening theories for metals). Fracture mechanics analyses suggest that in special cases (e.g. ductile particles, spherical pores) pinning and bowing could overshadow out-of-plane deflection as a toughening mechanism. As we shall see (sect. 7.5), the most effective pinning is that which persists *after* the front has passed on ahead of the obstruction, i.e. 'interfacial bridging'.

Now let us explore the role of internal thermal expansion anisotropy stresses in the toughening processes. Consider a specific material system, that of a glass matrix with spherical inclusions, as illustrated in the micrographs of fig. 7.8(a), (b). Differential contraction on cooling from the glass melt gives rise to a residual outward pressure at the particle–matrix boundary, $r = a$ in fig. 7.8(c),

$$\sigma_R = \Delta\alpha\Delta T/[(1+v_M)/2E_M+(1-2v_P)/E_P] \tag{7.4}$$

where subscripts M and P refer to matrix and particle, $\Delta\alpha = \alpha_M - \alpha_P$ is the differential thermal expansion coefficient (controlled by the glass composition), ΔT is the temperature range of cooling, v and E are Poisson's ratio and Young's modulus. This outward pressure is manifested in the matrix as radial and tangential components of normal stress,

$$\left. \begin{array}{l} \sigma_r = -\sigma_R(a/r)^3 \\ \sigma_t = \tfrac{1}{2}\sigma_R(a/r)^3 \end{array} \right\} \quad (r \geqslant a). \tag{7.5}$$

When the matrix contracts more than the particle ($\Delta\alpha > 0$ in (7.4), $\sigma_r < 0$ and $\sigma_t > 0$ in (7.5)) the matrix is placed in 'hoop tension', and the crack is 'attracted' to the particle (fig. 7.8(a)). Conversely, when the contraction condition is reversed ($\Delta\alpha < 0$, $\sigma_r > 0$ and $\sigma_t < 0$) a state of 'radial tension' obtains, and the crack is 'repelled' (fig. 7.8(b)).

It is immediately apparent that internal stresses can be a source of deflection. The radial-tension field would appear to be particularly effective in this regard. On the other hand, this same radial-tension field maintains a continuous tensile crack path through the matrix, somewhat negating the increased crack resistance from deflection. With the hoop-tension field, deflection may be less important than pinning. This latter configuration enhances crack–particle intersections, in which case the bias to tensile stresses along the path in the matrix will be compensated by restraining compressive stresses in the particles.

Apart from their immediate role in deflection, second-phase inclusions make their presence felt in the toughness equation as adjuncts to shielding processes (sects. 7.3–7.5). They also serve as effective flaw centres (chapter 9).

7.1.3 Fracture surface steps

The typical brittle fracture surface contains a complex pattern of markings which reflect the crack history. Of these markings, the fracture *step* is the most prevalent. Fracture steps constitute a geometrical perturbation on the crack front, and are accordingly potential elements of restraint on propagation.

The incidence of fracture steps may be linked to any one of several possible origins: the primary crack may experience local disturbances and so segment into partial fronts on closely adjacent, non-coincident planes (e.g. mode III disturbances, fig. 2.22; twist-boundary misorientations, figs. 7.2, 7.3; deflections at obstacles, dislocations and other defects); a fast-running crack may bifurcate (e.g. figs. 4.5, 4.6); the primary crack may form by the coalescence of smaller cracks initiating from non-coplanar sources. Individual cracks can interact, and ultimately merge, with their neighbours, either to reinforce (steps of like sign) or annihilate (unlike sign). An example of this is seen in the way the many small steps created by intersection with the twist boundary in fig. 7.3 ultimately coalesce into a few large steps to form a characteristic 'river pattern'.

Step formation restrains the propagation of brittle cracks by absorbing extra energy in the vicinity of the connecting riser between adjacent crack planes. This additional expenditure is manifest as an effective pinning of the crack front along the step, rather like a trailing line obstacle. The link-up of partial fronts is in fact a convoluted process, occurring not only behind, but also transverse to, the main front. A consideration of the stress interaction between opposing, adjacent crack segments reveals a tendency to mutual attraction and subsequent link-up, not tip-to-tip as might be intuitively expected but rather tip-to-plane, fig. 7.9. The ensuing crack-plane overlap produces a connecting sliver of material that is subjected to an increasing bending moment as the primary crack front advances, until a state of local crack-path instability (sect. 2.8) is reached in one or other of the transverse base regions. At this stage the undercutting fracture deviates abruptly toward the adjacent crack plane. Because of the near symmetry of the configuration the slightest crack-front disturbance may be sufficient to deflect the base fracture to the opposite edge of the sliver, as

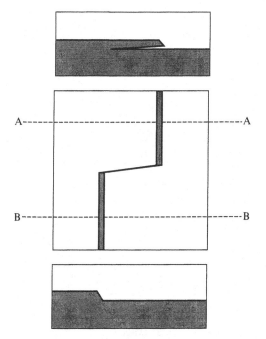

Fig. 7.9. Step-deflection mechanism. Centre diagram represents surface
view of lower fracture face. Upper and lower diagrams represent
section views through A–A and B–B respectively. Note undercutting at
A–A only. (Upper cleavage face would have complementary geometry,
except in the case of sliver detachment at both ends.)

depicted in fig. 7.9. Such discontinuities in step geometry are a characteristic feature of brittle fracture surfaces. Fig. 7.10 is an illustrative
example. In fortuitous instances base fractures may occur at both ends of
the sliver, accounting for the common observation of 'cleavage whiskers'
and other fracture surface 'debris'.

The increase in crack-resistance energy associated with step formation
will depend on the density and height of the steps. However, if total surface
area is any indicator (notwithstanding the possibility of additional
dissipation from instabilities in the undercutting process), this increase is
unlikely to amount to more than a factor of two in R/R_0.

We close this section by stressing yet again the transitory nature of
toughening by interactions that merely perturb the crack-front geometry.
And, as we have shown, the toughening effects are in any case modest, with
upper bounds at $R/R_0 \approx 4$, $T/T_0 \approx 2$. This is one reason that we have not
made a strong effort to support the above mechanisms with quantitative

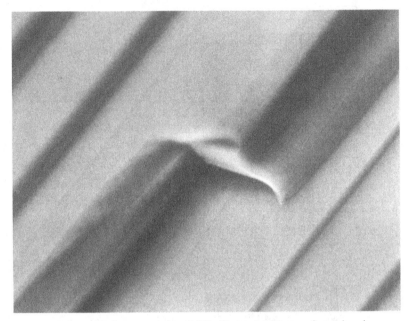

Fig. 7.10. Fracture steps on (111) silicon cleavage surface, showing abrupt discontinuity of kind depicted in fig. 7.9. Scanning electron micrograph. Width of field 85 μm. (After Swain, M. V., Lawn, B. R. & Burns, S. J. (1974) *J. Mater. Sci.* **9** 175.)

experimental data from the literature. Another reason is that many of the experimental observations used in the identification of particular mechanisms are subject to conflicting interpretation, based as they often are on post-mortem 'fractography'. While sometimes useful, fractography is limited and often misleading. Despite this element of uncertainty the micromechanics of deflection and internal-stress interactions considered above lay down an important groundwork for discussing some of the more effective, shielding processes in the following sections.

7.2 Toughening by crack-tip shielding: general considerations

Potentially the most effective means of compensating for the innate brittleness of ceramic materials is by crack-tip shielding. In this interlude we recapitulate the basic ingredients of shielding from chapter 3, and foreshadow some of the important material design issues in the context of microstructure. The requirements for shielding are (sect. 3.6):

(a) That there be a microstructural discreteness in the source–sink surround zone, such that there exists an elastic crack-tip enclave within which fundamental laws govern the equilibrium *in-vacuo* extension:

$$\left.\begin{array}{l} G_* = R_0 \\ K_* = T_0 \end{array}\right\} \text{ (enclave).} \tag{7.6}$$

(b) That the deformation of these elements contains a non-recoverable component, such that an ensuing closure term, $-G_\mu = R_\mu$ or $-K_\mu = T_\mu$, contributes to the equilibrium global resistance:

$$\left.\begin{array}{l} G_A = G_R = R_0 + R_\mu = R \\ K_A = K_R = T_0 + T_\mu = T \end{array}\right\} \text{ (global).} \tag{7.7}$$

The simplistic notion of a single-valued toughness (R_0 or T_0) is thereby lost: hence the resistance-curve (*R*-curve, G_R-curve) $R = R(c)$, or toughness-curve (*T*-curve, K_R-curve) $T = T(c)$. Neither the global nor crack-tip observer (our 'engineer' and 'physicist') of sect. 3.6 is now appropriate to a description of R_μ or T_μ. The micromechanics are determined by an intermediate observer (the 'materials scientist') within the shielding zone.

In the remaining sections of this chapter we analyse ways in which the microstructure can contribute to the *R*-curve. Examples are illustrated in fig. 7.11. In analysing such systems, special attention will be directed to the following elements of the shielding micromechanics:

(i) *Constitutive relations.* At the very heart of any formal fracture mechanics description is an appropriate relation for the integrated stress–strain response of the material over the shielding zone (figs. 3.11, 3.12). Such relations contain all the fundamental information on the controlling microstructural parameters.

(ii) *Frontal-zone and bridged-interface micromechanisms.* As indicated in sect. 3.7, we may distinguish two classes of shielding. Such a distinction allows us to make use of generic fracture mechanics formalisms for evaluating the *R*-curve. (Observe, however, that not all examples in fig. 7.11 fall unequivocally into one or other of the two categories.)

(iii) *Steady-state and transient solutions.* These two kinds of solution respectively determine the scale and shape of the *R*-curve (sect. 3.7). The shielding component R_μ in (3.34) and (3.37) requires explicit information

FRONTAL-WAKE   BRIDGED-INTERFACE

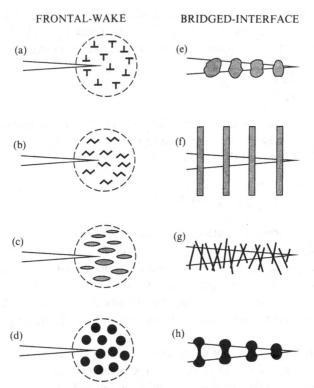

Fig. 7.11. Shielding mechanisms. *Frontal-wake*: (a) dislocation cloud, (b) microcrack cloud, (c) phase transformation, (d) ductile second phase. *Bridged-interface*: (e) grain interlock, (f) continuous-fibre reinforcement, (g) short-whisker reinforcement, (h) ductile second phase.

on the displacement field at the edge of the frontal or bridging zone. This displacement in turn depends on the stress intensities within the zone as well as without, in addition to crack size, e.g. $u_z = u_z(G_*, G_A, c)$. Calculation of $R_\mu(c)$ thereby entails numerical evaluation of a nonlinear integral equation (sects. 3.3, 3.7). To obtain a closed-form solution the analyst often resorts to simplified relations for the displacement field. In steady state the difficulty is circumvented because the zone-edge condition is identifiable with a limiting 'rupture' displacement (or strain), a material constant that may be specified without any reference to crack geometry.

(iv) *Zone size, and short vs long cracks.* Shielding zone sizes range from 1–1000 μm for typical structural ceramics to 1–1000 mm for composites and concrete (cf. cohesion zone ≈ 1 nm, sect. 3.3.1). Whereas large zones confer some desirable mechanical properties, they also pose difficulties in mathematical analysis. (Again, the definition of G_A and its quadratic relationship to K_A are contingent on a small shielding zone.) Traditional crack tests avoid the issue by measuring crack extensions in specimens with large starting notches. But strength is governed by flaws, whose dimensions may be comparable (or even smaller) than the zone size. Extrapolation from the macro- to the micro-scale is then open to question, not least because of non-uniqueness in $R_\mu(c)$ (sect. 3.7.1). This leads to a distinction between 'short' and 'long' cracks.

(v) *Weak interfaces and internal stresses.* Toughness may be achieved via weakness, by incorporating low energy interfaces and residual stresses into the microstructure. The intrinsic quantity R_0 may control the toughening increment R_μ: first, by determining the fracture path (e.g. transgranular vs intergranular); second, by determining the range of the shielding field. Then R_μ may be mathematically coupled to R_0, leading in some cases to a *multiplicative* rather than *additive* effect.

7.3 Frontal-zone shielding: dislocation and microcrack clouds

Frontal-zone dislocation activity and microcracking have their respective origins in metallurgical engineering and concrete mechanics. These concepts have more recently been extended to brittle ceramics, with somewhat limited success. They nonetheless serve as important precedents in the development of shielding micromechanics.

7.3.1 Dislocation clouds

It was acknowledged in chapter 3 that the tips of propagating cracks in metallic solids are usually encased by a plastic zone, and that the energy consumed within this zone can exceed that in bond rupture by several orders of magnitude. Analysis of the plastic zone has been the pre-occupation of solid mechanics practitioners for decades; so much so that it is still held by some that the same plasticity governs the toughness of ceramics, albeit on a lesser scale. Our consideration of the electron microscopy evidence in chapter 6 indicated that such is not the case: that the barriers to dislocation emission in covalent–ionic solids are insurmountably high at room temperature. At the same time, the possibility

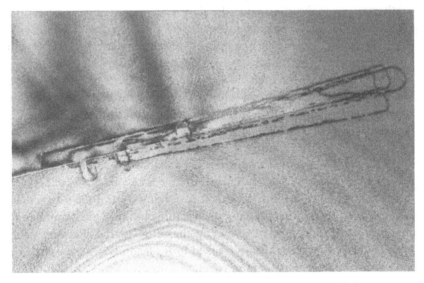

Fig. 7.12. Transmission electron micrograph of (100) crack in magnesium oxide foil with (001) surface, showing trailing dislocation cloud. Crack propagation left to right. Note dislocation segments trailing crack front. Width of field 5.0 μm. (Courtesy B. J. Hockey.) (Cf. silicon at high temperature, fig. 6.22(c).)

that pre-existent sources might be activated within the K-field, thereby developing a dislocation cloud which shields the crack tip, was not discounted. Active dislocation clouds have indeed been observed in ionic crystals with the rocksalt structure, where the Peierls lattice resistance is relatively low (Burns & Webb 1966). An example of such a cloud is shown in fig. 7.12.

The mechanics of toughening from crack–dislocation interactions are subtle. Start with the simple configuration of a single static screw dislocation i with Burgers vector, magnitude $+b$, located at (r, θ) and lying parallel to the front of an otherwise unstressed mode III crack. The dislocation imposes a shear stress on the crack tip, giving rise to a shielding stress-intensity factor (Majumdar & Burns 1981)

$$K_\mu^i = -[\mu b/(2\pi r_i)^{1/2}] \cos(\theta_i/2) \qquad (7.8)$$

with μ the shear modulus. This stress-intensity factor is equivalent to a generalised force, G_μ^i, on the crack; an equal and opposite, image force G_D^i, is exerted by the crack on the dislocation. Inserting $b \approx 2.5$ nm,

$\mu \approx 100$ GPa, $r \approx 1$ μm, $\theta = 0$, we see that a single dislocation contributes $K_\mu^i \approx -0.1$ MPa m$^{1/2}$. The total shielding $T_\mu = -K_\mu$ from a parallel array of N such static dislocations is determined from a superposition of K-fields:

$$T_\mu = -\sum_i^N K_\mu^i(r_i, \theta_i). \tag{7.9}$$

It might seem, therefore, that the density of dislocations would not have to be great for plasticity to make a significant contribution to the toughness.

However, if the field contains dislocations of opposite sign, e.g. $+b$ at (r, θ) and $-b$ at $(r, -\theta)$, the superposed K-fields of (7.8) cancel in (7.9). The mathematics are more complex if the crack is in mode I and if the dislocation array contains edge/screw mixtures, but the end result is the same: in a large, well-distributed array of immobile dislocations the sum total of the screening K-field tends to zero (Thomson 1986). A *static* dislocation population produces no effective toughening.

Suppose, on the other hand, the population is 'active' so that dislocation segments expand and multiply within the near field to form a *mobile* cloud. In extreme cases these segments may trail continuously behind the advancing front, as in the example of fig. 7.12. Then energy is consumed irreversibly as the dislocations are displaced against the 'lattice friction' stress (intrinsic Peierls or extrinsic viscous resistance), and residual strain is left in the wake. The toughness contribution from the array is no longer zero, and is governed by the dislocation mechanics.

These principles have been extensively explored by Burns & Webb (1966, 1970) in studies on single crystals of lithium fluoride. We illustrate here with minimum complication by referring to the idealised system in fig. 7.13, in which both slip and mode I fracture are confined to a single (basal) crystallographic plane (e.g. mica, zinc). Consider the evolution of two independent dislocation loops in response to the near-tip stress field. The lobes about the crack tip are loci of constant resolved shear stress on the slip plane, evaluated from (2.14a) at $K = K_A$ and $\sigma_{ij} = \sigma_{xy}$,

$$r(\theta) = (1/2\pi)[K_A f_{xy}(\theta)/\sigma_{xy}]^2. \tag{7.10}$$

Since σ_{xy} is the only nonzero shear component acting on the slip plane the crack will exert a glide force only on loops with components of Burgers vector parallel to OX. The sign of the Burgers vector then determines whether the dislocation loop expands or contracts. Thus the front and rear segments of the loop close to the approaching crack on the near slip plane

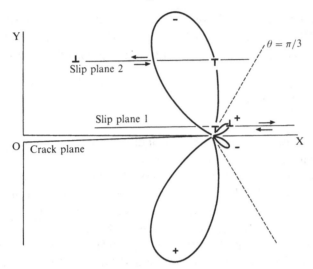

Fig. 7.13. Configuration of dislocation loop segments in crack-tip field, for ideal layer structure. Lobes represent locus of constant shear stress $\sigma_{\mathrm{C}}^{\mathrm{D}}$. Loops on slip planes intersecting lobes are swept up by crack. (After Burns, S. J. (1970) *Acta Metall.* **18** 969.)

1 in fig. 7.13 will tend to expand, toward the frontal region of zero stress at $\theta = 0$, $\pm \pi/3$. As the crack continues to advance the reverse action of the trailing negative lobe will restrain the rear segment of the loop from drifting behind the tip, and the entire dislocation will be carried along with the moving stress field. The loop of opposite sign on the same side of the crack plane will, on the other hand, tend to self-annihilate within the leading stress lobe: but if it survives, as it may on the more remote slip plane 2, it will again expand within the more intense trailing lobe. This time, however, there is no lobe of reverse sign to follow and only the front segment of the dislocation will be swept along with the crack.

Complex many-dislocation configurations are most easily handled by proceeding to the limit of infinitely dense populations within the shielding zone (but not within the inner crack-tip enclave). Then we may use the quasi-continuum analysis of sect. 3.7.2 for mode I cracks to determine the steady-state shielding contribution $R_{\mu}^{\infty} = -G_{\mu}^{\infty}$ to the resistance; for a fully developed wake (3.37) yields

$$R_{\mu}^{\infty} = 2 \int_{0}^{w_{\mathrm{C}}} \mathcal{U}(y) \, \mathrm{d}y = 2 \bar{U}_{\mathrm{D}} w_{\mathrm{C}} \tag{7.11}$$

where $\bar{U}_{\mathrm{D}}$ is the average strain energy density associated with the formation of remnant dislocation segments and w_C is the width of the wake layer. It is w_C that now controls the degree of toughening. For an equilibrium system the shear stress σ_{xy} to activate dislocation motion is given by the lattice friction stress σ_C^D, which therefore determines the size of the shielding lobes in fig. 7.13. In the approximation that K_A defines the outer field to the edge of the shielding zone (fig. 3.8), as is appropriate to 'weak shielding' (sect. 3.6), we insert $K_A = T_\infty$ at equilibrium, $\sigma_{xy} = \sigma_C^D$ and $r = w_C$ into (7.10) to get

$$w_C = \Omega_D (T_\infty / \sigma_C^D)^2 \qquad (7.12)$$

with $\Omega_D = (1/2\pi) f_{xy}^2(\theta_m) \sin\theta_m$ a zone-geometry coefficient, subscript m denoting the $r(\theta)$ tangency point furthermost from the crack plane ($\theta_m = \pm 111°$, $\Omega_D = 0.030$ for the ideal configuration in fig. 7.13). Equations (7.12) and (7.11), in conjunction with $R_\infty = R_0 + R_\mu^\infty$ and $T_\infty = (E' R_\infty)^{1/2}$ (sect. 3.6), yield an approximate steady-state resistance

$$R_\infty = R_0(1 + 2\Omega_D \bar{U}_D E'/\sigma_C^{D2}), \quad \text{(weak shielding).} \qquad (7.13)$$

Observe the multiplicative role of R_0 as foreshadowed earlier.

Thus far we have said nothing about rate and temperature, two variables that loom large in any material property involving dislocation activity. How might these variables be incorporated into the above description? Consider the response of the dislocation configuration in fig. 7.13 as we propagate the crack at velocity v_C. The motion of dislocations (as of cracks) in the domain $\sigma_{xy} = \sigma^D > \sigma_C^D$ is most generally governed by a kinetic 'law', $v_D = v_D(G_D, T)$. This law, in conjunction with the requirement $v_D = v_C$, then uniquely defines a steady-state glide force $G_D = G_D(\sigma^D)$. For monotonic $v_D(G_D)$ the shielding zone in fig. 7.13 contracts with increasing crack velocity, reducing the number of mobile dislocations as the system approaches a dynamic state. Analysis of the dislocation kinetics thereby constitutes a first step in the complex mechanics of brittle–ductile transitions.

It is acknowledged that significant dislocation activity in covalent–ionic solids at room temperature is the exception rather than the rule. Materials scientists concede that plasticity is not a practical route to toughening in ceramics, amounting at best to a factor of two in R in the softest, ionic crystals. It is of course of paramount import in metallics. In this context it is interesting to record the contention by several researchers (Thomson 1978; Weertman 1978; Hart 1980) that the validity of the dislocation-

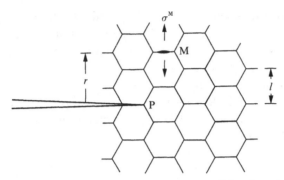

Fig. 7.14. Coordinates for evaluating condition for microcrack initiation in polycrystalline material. Intrinsic microcrack source (M) subjected to superposed opening stresses σ^M from field of primary crack (P) and thermal expansion mismatch stresses. It is assumed that the microstructure satisfies conditions of geometrical similarity, so that the flaw size scales with grain size l.

shielding/sharp-crack picture extends to metals in the brittle mode of fracture.

7.3.2 Microcrack clouds

Toughening by fracture-induced microcracking is strongly analogous in frontal-zone and *R*-curve phenomenology to that of dislocation-cloud activity, except that the discrete elements are stationary–dilational rather than mobile–deviatoric. In principle, microcracking can be triggered at incipient sources of weakness, e.g. grain and interphase sub-facet flaws, by the field of a primary crack. The primary-crack stresses may be augmented by internal residual tensile stresses, from differential thermal expansion or elastic mismatch. In relieving these tensile stresses the fully developed microcracks remain irreversibly open, typically over several grain dimensions, thereby imposing a dilatant closure field on the crack.

The issue of microcrack toughening raises two fundamental questions: *what are the conditions that a microcrack cloud should initiate in the field of a primary crack?; given that these conditions are met, what is the toughness increment?* For simplicity, we specifically consider monophase ceramics with intergranular fracture:

(i) *Cloud initiation.* We investigate the condition for microcrack initiation from the perspective of an observer within the enclave of an equilibrium primary crack (P) looking outward toward potential microcrack sources (M), fig. 7.14. The stress σ^M acting on an active source, assumed to be a grain boundary sub-facet at (r, θ), is the superposition of two components: the hydrostatic tensile near-tip stress of the primary crack, evaluated from (2.14a) at $K = K_*^P$ and $\sigma_{ij} = \bar{\sigma}_{ii} = \frac{1}{3}(\sigma_{rr} + \sigma_{\theta\theta} + \sigma_{zz})$,

$$\bar{\sigma}_{ii} = K_*^P \bar{f}_{ii}(\theta)/(2\pi r)^{1/2} \tag{7.14}$$

with $\bar{f}_{ii}(\theta) = \frac{2}{3}(1 + v)\cos(\theta/2)$; and (in non-cubic materials) the mean thermal expansion anisotropy stress

$$\sigma_R = E\Delta\alpha\Delta T/2(1 + v) \tag{7.15}$$

with $\Delta\alpha$ the difference in expansion coefficients along the principal crystal axes. Approximating the sources as uniformly stressed penny-shaped flaws of radius c_f, and recalling (2.20) and (2.21d), the critical condition is

$$K_*^M = 2\sigma_C^M(c_f/\pi)^{1/2}$$
$$= 2(\bar{\sigma}_{ii} + \sigma_R)(c_f/\pi)^{1/2} = T_0 \tag{7.16}$$

with T_0 the grain boundary toughness.

One of the microstructural parameters of greatest practical interest is grain size, l. There exists a limit l_C above which non-cubic ceramics tend to microcrack *spontaneously* during the original cooling cycle. This limit serves as a reference state for evaluating initiation conditions in the superposed primary-crack field. Note now that the intensity of the residual field in (7.15) is independent of grain size. The critical K-field condition for spontaneous microcracking must then be met by a scaling in the flaw radius, $c_f = \beta l$ for geometrically similar structures, β a scale-invariant quantity. Then we may determine the conditions for microcracking as follows:

(a) *Spontaneous (general) microcracking.* In the absence of a primary crack $(\bar{\sigma}_{ii} = 0)$ sources M initiate under the sole action of the internal tension $(+\sigma_R)$. Equation (7.16) is then satisfied at the critical grain size

$$l_C = (\pi/4\beta)(T_0/\sigma_R)^2. \tag{7.17}$$

Above l_C *general* microcracking occurs from active sources throughout the

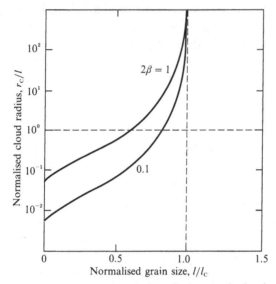

Fig. 7.15. Reduced plot of microcrack cloud radius around primary crack tip as function of grain size, for two values of penny-flaw diameter.

material. Note the limiting case $\sigma_R \to 0$, $l_c \to \infty$ in (7.17), confirming that spontaneous microcracking is not expected in cubic materials.

(b) *Activated* (*cloud*) *microcracking.* Now consider the activating influence of a primary crack in the grain-size domain $l < l_c$. Microcracking is confined to a cloud around the tip P, with a maximum radius determined by inserting $K_*^P = T_0$ in (7.14) and combining with (7.16):

$$r_C = 2\beta l \{\bar{f}_{ii}/\pi[1-(l/l_c)^{1/2}]\}^2, \quad (l \leqslant l_c). \tag{7.18}$$

Plots of r_C/l as a function of l/l_c are shown in fig. 7.15, at bounding values of crack diameter 2β, for $\bar{f}_{ii}(\theta) = 0.72$ ($\theta = 60°$, see below). There are two limits of interest: $l/l_c = 1$, the extreme of general microfracture discussed in (a) above; $l/l_c = 0$, ultra-fine-grain ($l \to 0$) or cubic ($l_c \to \infty$, $\sigma_R \to 0$ in (7.17)) material. Between these limits the cloud radius diminishes with diminishing grain size, more quickly at the smaller flaw size. The condition $r_C/l = 1$ indicates a point at which any microcrack sources simply coalesce into the primary crack, i.e. there is no detached cloud. Significant microcracking ($r_C/l > 10$, say) is therefore confined to a narrow 'window' in microstructural scale.

(ii) *Toughness increment.* Suppose conditions are such that a microcrack cloud does initiate. What then is the contribution to the toughening? We proceed as in the previous subsection for dislocation clouds, assuming a quasi-continuous density of microcracks within the dilational shielding zone (Faber & Evans 1983). The dilation zone is indicated in fig. 7.16 as the locus of constant hydrostatic stress $\bar{\sigma}_{ii}$ in (7.14). Included in fig. 7.16 is the variation $\sigma_\mu(x)$ and $\varepsilon_\mu(x)$ for a volume element in a plane at $y = \mathrm{const} < w_C$ intersecting the microcrack cloud, along with the constitutive dilational stress–strain relation for this element. The steady-state resistance increment $R_\mu^\infty = -G_\mu^\infty$ is determined directly from (3.38):

$$R_\mu^\infty = 2\sigma_C^M \varepsilon^M w_C: \tag{7.19}$$

σ_C^M is the critical stress for microcrack initiation defined in (7.16) and $\varepsilon^M \approx e^M V_f$ is the residual strain in the microcracked wake, with $e^M \approx \frac{1}{2}\Delta\alpha\Delta T$ unconstrained strain per source associated with release of internal stress σ_R in (7.15) and V_f volume fraction of sources (fraction of active grains).[2] The 'weak-shielding' zone size is (cf. (7.12))

$$w_C = \Omega_M (T_x/\sigma_C^M)^2 \tag{7.20}$$

with $\Omega_M = (1/2\pi)\bar{f}_{ii}^2(\theta_m)\sin\theta_m$ ($\theta_m = \pm 60°$ at furthermost tangency point, $\Omega_M = 0.062$). In analogy to (7.13), we obtain

$$R_x = R_0(1 + 2\Omega_M e^M V_f E'/\sigma_C^M), \quad \text{(weak shielding)} \tag{7.21}$$

approximately, so that R_0 is again multiplicative in the toughness relation.

There is a size effect implicit in this toughness description. Whereas $e^M (\approx \frac{1}{2}\Delta\alpha\Delta T)$ is independent of l, σ_C^M for microcracking in (7.16) for self-similar grain structures ($c_f = \beta l$) is an inverse (square root) function of l. (Here is yet another example in violation of the constant critical stress concept.) Thus (within the window of fig. 7.15) R_x in (7.21) should increase with the scale of the microstructure, up to the limit of spontaneous microcracking.

Accordingly, let us estimate an upper bound to the toughening for alumina. Taking $\sigma_C^M \approx \sigma_R \approx 250$ MPa (spontaneous limit, $\bar{\sigma}_{ii} \to 0$ in (7.16)), $e^M \approx 0.0005$ ($\Delta\alpha \approx 1 \times 10^{-6}\,°\mathrm{C}^{-1}$, $\Delta T \approx 1000\,°\mathrm{C}$), $V_f \approx 1$, $E' \approx 400$ GPa, we

---

[2] An elastic compliance term in R_μ^∞ should strictly be included in (7.13) (see fig. 7.16). However, this component, due to elastic 'softening' of the material within the shielding zone, is important only for very high microcrack densities, and we neglect it here.

(a)

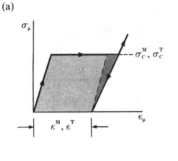

(b)

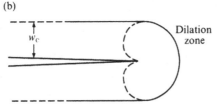

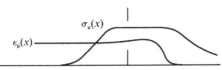

Fig. 7.16. Frontal-zone toughening. (a) Constitutive stress–strain relation $\sigma_\mu(\varepsilon_\mu)$ for microcracking (M) or transformation toughening (T) at volume element within dilation zone. (Dark shaded area to right of dashed line is secondary contribution to toughening associated with modulus reduction.) (b) Dilation zone, locus of constant hydrostatic tension $\bar{\sigma}_{ii}$, with wake, showing stress $\sigma_\mu(x)$ and strain $\varepsilon_\mu(x)$ (bottom) for element of material within this zone. Toughening increment R_μ given by area under $\sigma_\mu(\varepsilon_\mu)$ curve. (Cf. fig. 3.12.)

obtain $R_x/R_0 \approx 1.1$, a modest increase. Materials with greater expansion anisotropy (e.g. aluminium titanate, barium titanate), and therefore larger e^M, might be expected to exhibit a more substantial increment.

There has been considerable debate as to the role of microcrack clouds in the toughness of brittle materials. Direct confirmations, e.g. from *in situ* frontal-zone observations of extending primary cracks, are conspicuously lacking in monophase ceramics. And we have indicated that the conditions for microcrack activation are in any event stringent. For such microcracking to be viable at all the sources must be sufficiently large and closely spaced, requirements which in geometrically similar microstructures tend to be mutually exclusive.

Two-phase ceramics, on the other hand, are not bound by the same similitude restrictions. One may for instance increase the microcrack *density* independently of the *size* of the sources, by increasing the volume fraction of the second phase. Indeed, the most convincing reports of activated microcracking in ceramics are from observations on matrix/particle systems (Evans 1990). With due allowance for such independence between density and size of sources, and for the existence of elastic-mismatch in addition to expansion-anisotropy stresses (with consequent enhancement of the residual strain), analyses of two-phase systems may be developed along the same lines as above.

7.4 Frontal-zone shielding: phase transformations in zirconia

If there remains some doubt as to the efficacy of dislocations and microcracks as practical toughening agents in brittle ceramics, the same is certainly not true of phase transformations in zirconia. The realisation of modern zirconia ceramics with uncommonly high toughness ('ceramic steel') traces back to the pioneering work of Garvie and colleagues (Garvie, Hannink & Pascoe 1975). Contributions from later players, Clausen, Heuer, Lange, Swain, etc. have led to material refinements, with ultimate toughness levels $T_\infty \approx 20$ MPa m$^{1/2}$.

Central to the zirconia story is the displacive 'martensitic' transformation from tetragonal to monoclinic phase. Unconstrained, the transformation is spontaneous at room temperature, with relatively large dilational ($\approx 4\%$) and deviatoric ($\approx 7\%$) strains. The key is to inhibit the transformation by use of appropriate processing additives (MgO, CaO, Y_2O_3, CeO_2) and heat treatments, such that the resultant 'partially stabilised' zirconia (PSZ) contains a distribution of fine *metastable* tetragonal precipitates in an elastically constraining matrix, usually cubic phase. A classical precipitate structure is shown in fig. 7.17. The transformation can then be triggered hysteretically by imposed stresses within a crack frontal zone, in the manner envisaged above for microcracking.

So far, zirconia is the only truly functional transformation-toughened ceramic, although alumina and mullite have been successfully used as matrix materials for the tetragonal zirconia particles. The exploration of alternative martensitic systems and other transformation types continues. Again, we present only essential details of this important area of materials

Fig. 7.17. Transmission electron micrograph of Mg–PSZ showing untransformed tetragonal precipitates in cubic matrix. Width of field 1.75 μm. (Courtesy A. H. Heuer.)

development. The interested reader is referred to a monograph by Green, Hannink & Swain (1989) for an extensive coverage.

7.4.1 Experimental observations

In this subsection we present the results of selected observations on zirconia ceramics in relation to the transformation-zone micromechanics and attendant toughness properties.

We start with micrographic confirmation of the hysteretic tetragonal–monoclinic transformation in Mg-PSZ, fig. 7.18. The micrographs show the surface of a specimen with a notch-induced crack viewed in two-beam interference. The fringes represent contours of uplift associated with surface relaxation of residual dilation in the frontal-wake zone. From these fringe patterns we directly estimate the zone width and transient zone length at ≈ 1 mm. Also shown are tetragonal- and monoclinic-phase Raman microprobe signals from the same crack. The spatial variation of these signals relative to the crack plane correlates directly with that of the fringes.

Now consider fracture data on a similar Mg-PSZ with moderate toughness characteristics. Fig. 7.19 shows steady-state toughness and strength as a function of aging. Both data sets increase to a maximum, and thereafter decline, with aging time. The material before, at and after the maximum is referred to as under-, peak- and over-aged. Microstructural examinations by transmission electron microscopy (e.g. fig. 7.17) reveal the action of the aging to be a coarsening of the transforming precipitates. At first this coarsening enhances the toughness, apparently by reducing the critical stress for particulate transformation. But at a certain precipitate size the transformation occurs spontaneously, and the toughening is then lost (cf. critical grain size for microcracking, sect. 7.3.2).

Toughness vs crack-extension (T-curve) data for the same material are plotted in fig. 7.20 for different aging states. The rise in toughness is accentuated in the peak-aged state. This rise persists over extensions ≈ 1 mm, i.e. in accord with the transient zone length measured directly in fig. 7.18. Even the over-aged specimen shows some remnant T-curve behaviour, suggesting that the toughening may not be due exclusively to transformation; note the strongly deflected and bridged crack path (sect. 7.5) at coarsened precipitates in the heavily over-aged material in fig. 7.21.

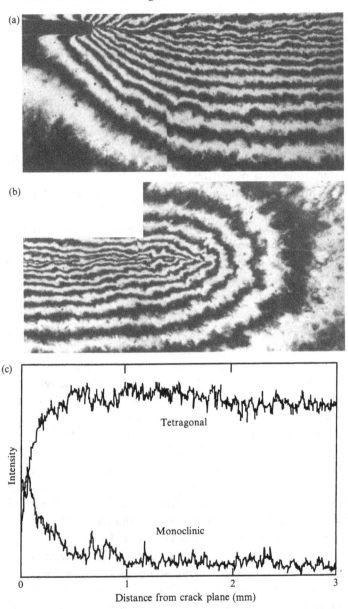

Fig. 7.18. Two-beam optical interference micrographs of surface of notch-induced crack in CT specimen of Mg-PSZ showing surface uplift associated with transformation-induced dilation at (a) notch and (b) crack tip. Each fringe represents height change of half-wavelength of light (≈ 0.25 μm). Broadening fringe contour pattern ahead of notch measures transient evolution of transformation zone, and thereby reflects shape of *T*-curve. Width of field 2500 μm. (c) Raman

Many microstructural and extraneous variables have been investigated in zirconia systems: volume fraction of stabilising additive, composition of matrix material, size of precipitates and matrix grains, aging temperature, are just a few examples (Green, Hannink & Swain 1989). There is much about our current knowledge of the transformation process that is incomplete.

7.4.2 Theoretical fracture mechanics

The discovery that zirconia can be substantially toughened by stress-activated phase transformation sparked a flurry of activity in fracture-mechanics modelling in the early 1980s by Evans and others (McMeeking & Evans 1982; Marshall, Drory & Evans 1983; Budiansky, Hutchinson & Lambropoulus 1983). Here is one area of materials research that has stimulated significant contributions from the solid mechanics community.

As with microcracking, we first investigate the condition for initiation of the constrained martensitic transformation, and then evaluate the ensuing toughness increment.

(i) *Initiation of transformation.* Again, one may anticipate a threshold particle size for *spontaneous* transformation. A loose treatment considers just the initial, tetragonal (t) and final, monoclinic (m) states. The energy difference between these 'end-point' states for a particle of radius a is

$$\Delta F = \tfrac{4}{3}\pi a^3 \Delta \mathcal{U}_E + 4\pi a^2 \Delta U_S \qquad (7.22)$$

with $\Delta \mathcal{U}_E = \mathcal{U}_E^m - \mathcal{U}_E^t$ the differential volume energy density (including chemical energy as well as elastic strain energy of constrained particle) and $\Delta U_S = U_S^m - U_S^t$ the differential surface energy (chemical and configurational components associated with the particle–matrix interface – cf.

spectroscopy signal as function of scan distance from crack plane at intermediate location between notch and crack tip, indicating strength of tetragonal-to-monoclinic transition. (After Marshall, D. B., Shaw, M. C., Dauskardt, R. H., Ritchie, R. O., Readey, M. & Heuer, A. H. *J. Amer. Ceram. Soc.* **73** 2659.)

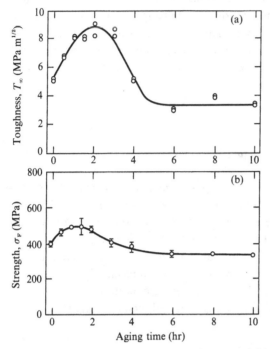

Fig. 7.19. (a) Steady-state toughness, and (b) strength, of Mg-PSZ, as function of aging time at 1400 °C. (Courtesy A. H. Heuer, R. Steinbrech & M. Readey.)

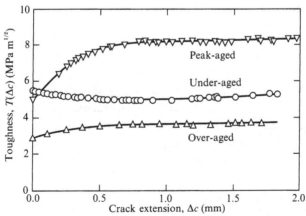

Fig. 7.20. Measured toughness curves for Mg-PSZ, under-aged, peak-aged, over-aged. CT specimens. (Courtesy A. H. Heuer, R. Steinbrech & M. Readey.)

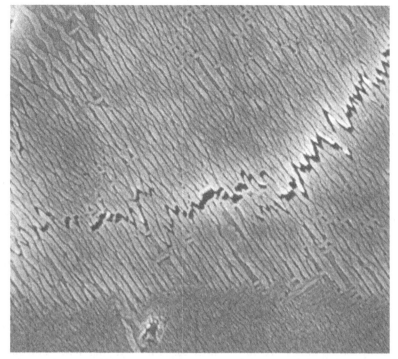

Fig. 7.21. Coarsened precipitates in over-aged Mg-PSZ. Note strong crack deflection and bridging. Width of field 12.5 μm. (After Hannink, R. H. J. & Swain, M. V. (1986), in *Tailoring of Multiphase and Composite Ceramics*, eds. G. L. Messing, C. G. Pantano & R. E. Newnham, Plenum Press, New York, p. 259.)

W_{AB} in (7.3) – plus energy of any twin interfaces). Then it is thermodynamically permissible for the transformation to proceed when $\Delta F < 0$, which determines the threshold radius

$$a_C = -3\Delta \mathcal{U}_E / \Delta U_S \qquad (7.23)$$

(noting that, in analogy to cracks, as volume energy is released, $\Delta \mathcal{U}_E < 0$, surface energy is gained, $\Delta U_S > 0$).

This argument is at best incomplete, for it makes no statement as to the 'reaction path' *between* the end points, specifically to the existence of intervening energy barriers. The condition $\Delta F < 0$ is therefore a necessary but not sufficient condition for the transformation, so (7.23) is a lower bound. Nevertheless, the notion of a threshold particle radius prevails. Indeed, size effects are characteristic of any process where the energetics balance volume and surface terms, and we shall come across them several more times in later chapters.

In analogy to microcracking, the field of a primary crack can *activate* transformations at $a < a_C$ within a frontal cloud. We therefore expect the threshold condition in (7.23) to manifest itself as a 'window' in precipitate size (cf. fig. 7.15), in accordance with the experimental data of fig. 7.19.

(ii) *Toughness increment.* Now consider the shielding contribution to the toughness from a well-developed transformation zone about a primary crack. Ignoring the influence of shear components in the transformation, the shape of the frontal-wake zone and the constitutive stress–strain history of a volume element adjacent to the advancing crack are as depicted earlier in fig. 7.16. Our analysis for the steady-state resistance increment $R_\mu^\infty = -G_\mu^\infty$ then follows identically from that of sect. 7.3.2:

$$R_\mu^\infty = 2\sigma_C^T \varepsilon^T w_C \qquad (7.24)$$

in analogy to (7.19), where σ_C^T is the critical dilation stress (again a function of particle size) and $\varepsilon^T \approx e^T V_f$ is the net wake transformation strain, with e^T = the unconstrained dilation strain of a single transformation particle and V_f the volume fraction of particles. Similarly, the 'weak-shielding' zone width is

$$w_C = \Omega_T (T_\infty/\sigma_C^T)^2 \qquad (7.25)$$

with $\Omega_T = (1/2\pi) \bar{f}_{ii}^2 (60°) \sin 60° = 0.062$ (cf. (7.20)). We obtain the steady-state resistance

$$R_\infty = R_0(1 + 2\Omega_T e^T V_f E'/\sigma_C^T), \quad \text{(weak shielding)} \qquad (7.26)$$

approximately, with E' the composite modulus (cf. (7.21)).

As an estimate for a peak-aged zirconia (fig. 7.19), insert $e^T \approx 0.04$ and $V_f \approx 0.20$ ($\varepsilon^T \approx 0.008$), $E' \approx 250$ GPa, $\sigma_C^T \approx 25$ MPa, $R_0 \approx 35$ J m^{-2} (cubic phase, $T_0 \approx 3.0$ MPa m$^{1/2}$), yielding $R_\infty \approx 400$ J m^{-2} ($T_\infty \approx 10$ MPa m$^{1/2}$). There would appear to be scope for even higher toughness levels, e.g. by increasing V_f.

Turn now to the transient state. The increase in toughness with crack extension for the zirconia in fig. 7.20 reflects the evolution of an active wake behind the advancing frontal zone. This evolution, inferred from direct observations like those in fig. 7.18(a), is depicted schematically in fig.

7.22. Recall from sect. 3.7.2 that an expanding dilation zone along the front of an initially stationary crack, state (a) in fig. 7.22, makes no contribution to the shielding (cf. immobile dislocation cloud, sect. 7.3.1). That arises because the dilated elements ahead of the crack ($\theta < 60°$) exert tensile (hoop) opening stresses on the crack plane, while adjacent and behind ($\theta > 60°$) exert compressive (radial) closure stresses (cf. 7.5)); the net effect at equilibrium is cancellation. At a critical zone size the crack begins to propagate, state (b), creating an imbalance in closure stresses from the irreversibly unloaded particles in the expanding wake, so the toughness begins to increase. Ultimately, after the crack has extended sufficiently that the wake is fully-developed, state (c), the intensity of closure stresses saturates, and steady state obtains. The transient problem has been treated by McMeeking & Evans (1982), in the approximation of an ever-constant wake width $w = w_C$, by integrating over the transformation zone boundary in a stress-intensity analysis of the toughness increment $T_\mu = -K_\mu$,

$$T_\mu(\Delta c) = \eta E \varepsilon^T w_C^{1/2} \phi(\Delta c / w_C) \tag{7.27}$$

where $\eta = 1/3^{1/4} 2\pi^{1/2}(1-v) \simeq 0.28$ and $\phi(\Delta c/w_C)$ is a (numerically evaluated) dimensionless function of crack extension. This last function is monotonic with limits $\phi = 0$ at $\Delta c = 0$ (no wake) and $\phi = 1$ at $\Delta c \gg w_C$ (steady state), as reflected directly in the data of fig. 7.20.

The above description of phase-transformation toughening smooths over many important issues. In the interest of obtaining simple closed-form toughness relations we have invoked the restrictions of weak shielding and invariant wake width. Self-consistency between (7.24) and (7.27) cannot in fact be obtained without modifying the approximate expression for w_C in (7.25) to allow for relaxation of the displacement K-field at the zone boundary by the transformation 'plastic' zone and an ever-widening transient zone. It is also implicit in our analysis that we are dealing exclusively with long cracks. For the zirconia in fig. 7.18 this means cracks of length $\gg w_C \approx 1$ mm. The validity of extrapolating such long-crack analyses to the domain of flaws and strength is then seriously at issue. A related matter is the uniqueness of the R-curve and associated nonlinearity in the macroscopic stress–strain characteristic that arises when the zone size becomes comparable to the specimen dimensions. Other issues not addressed here include: the role of nucleation forces in determining the martensitic transformation stress σ_C^T; shape of zone, as influenced by shear

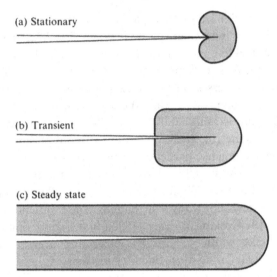

(a) Stationary

(b) Transient

(c) Steady state

Fig. 7.22. Development of frontal-wake zone with crack extension: (a) zero extension, equilibrium frontal zone, $R = R_0$ ($T = T_0$); (b) short extension, partial wake, $R = R_0 + R_\mu$ ($T = T_0 + T_\mu$); (c) long extension, steady-state wake, $R_\infty = R_0 + R_\mu^\infty$ ($T_\infty = T_0 + T_\mu^\infty$).

stresses and twinning, and non-uniformity of stress–strain distribution $\sigma_\mu(\varepsilon_\mu)$ within this zone (necessitating point-by-point integration of (3.37)); 'subcritical' vs 'supercritical' transformations in the constitutive relation. Phase-transformation toughening is a complex and still-evolving topic.

7.5 Shielding by crack-interface bridging: monophase ceramics

Many monophase polycrystalline ceramics exhibit pronounced resistance-curve behaviour in microstructural domains where frontal-zone mechanisms are non-operative. The principal source of toughening in these cases is intergrain contact bridging across the crack interface behind the advancing front. A basic understanding of this bridging process is of special interest in the microstructural design of flaw-tolerant ceramics, including two-phase and composite materials (Becher 1991).

Brief mention was made at the start of sect. 7.3 to the early precedent for bridging in the concrete literature. But the current understanding of bridging in monophase ceramics really begins with a clever experiment by Knehans & Steinbrech (1982) on notched alumina specimens. The essence of their experiment is illustrated in fig. 7.23. After propagating a crack

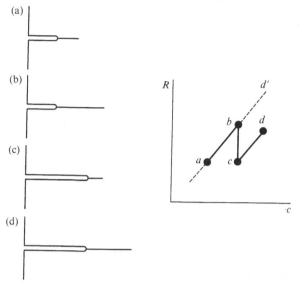

Fig. 7.23. Knehans-Steinbrech experiment. (a) Notched specimen, (b) propagation from notch, (c) sawcut, to remove wake zone from extended crack, (d) re-propagation. Sawcut removes shielding tractions, as indicated by decrement *b–c* in *R*-curve. (After Knehans, R. & Steinbrech, R. (1982) *J. Mater. Sci.* **1** 327.)

some distance up the *R*-curve from the notch (*a–b*) they removed interfacial material by sawcutting behind the tip (but with care not to remove the tip itself). They found the re-notched crack to revert to the base of the *R*-curve (*b–c*) and then to repropagate anew up this curve (*c–d*), rather than to continue along its original path in *R*-space (*b–d'*). The conclusion was unequivocal that the source of the toughening increment resides in the wake. Knehans & Steinbrech conceded that the results were not inconsistent with a frontal-wake process like microcracking. But they also cited frictional tractions across the interface as an alternative mechanism. The latter possibility was to strike a chord with those observant fracture experimentalists who noted that 'failed' specimens sometimes remain intact, as if the test halves are held together by ligaments.

As we shall demonstrate, bridging can be an effective mode of toughening in ceramics. On the other hand, increased toughness does not automatically translate into increased strength; again, a trade-off between long and short cracks is implied. It will be seen that certain source conditions required for effective frontal-zone processes, e.g. strong internal stresses and weak interfaces, also favour bridging. Indeed, in some

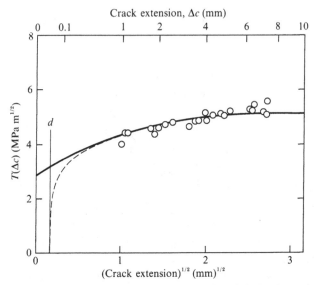

Fig. 7.24. Toughness curve, $T(\Delta c)$, for high-density polycrystalline alumina, grain size 35 μm. Compliance-calibrated CT data (starting notch length ≈ 14 mm). Solid curve is long-crack data fit to theory in sect. 7.5.2. Dashed curve is equivalent short-crack prediction in region $\Delta c > d$. (Data and analysis courtesy S. Lathabai & N. Padture.)

cases (e.g. fig. 7.21) bridging may act in concert with microcracking and transformation toughening.

7.5.1 Experimental observations

The first reports of resistance-curves or toughness-curves in monophase ceramics were a culmination of studies on alumina by the German school in the late 1970s and early 1980s (see Dörre & Hübner 1984). Some long-crack $T(\Delta c)$ data for an alumina of grain size 35 μm are plotted in fig. 7.24. There is a respectable rise in T, approaching a factor of two, over extensions Δc in excess of 1 mm. Generally, the steepness of such curves increases with grain size, confirming the importance of microstructure in the toughening. Similar results are reported by other schools on other simple ceramics using other test geometries. In cubic monophase ceramics, however, the increases are relatively minor, implying that internal residual stresses may be a vital factor in the underlying toughening process.

Definitive identification of bridging in monophase ceramics came only with *in situ* observations of crack propagation. Initial studies focussed on

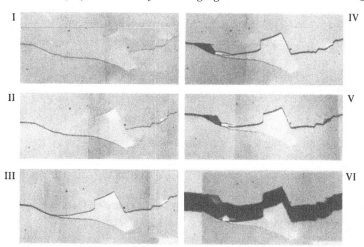

Fig. 7.25. Evolution of a grain-bridging site in pure polycrystalline alumina (same material as in fig. 7.4), at six stages of applied loading (wedge-loaded tapered DCB specimen). The crack tip lies (I) 0.1 mm, (II) 0.4 mm, (III) 0.8 mm, (IV) 1.3 mm, (V) 1.7 mm, (VI) ∞ (failure) to the right of field of view. Reflected light micrograph. Width of field 225 μm. (After Swanson, P. L., Fairbanks, C. J., Lawn, B. R., Mai, Y-W. & Hockey, B. J. (1987) *J. Amer. Ceram. Soc.* **70** 279.)

polycrystalline aluminas with pronounced R-curves (e.g. fig. 7.24) and characteristically intergranular fractures (fig. 7.4), using both short-crack (indentation) and long-crack (DCB) test geometries (Swanson *et al.* 1987). Subsequent extension to other ceramics confirmed the generality of the phenomenon (Swanson 1988). The most immediately conspicuous feature of the fracture in these observations was a spasmodic extension, two or three grain facets at a time, at steadily increasing load. Especially striking was the first 'pop-in' from indentation (or even natural) surface flaws, where the incremental extension could exceed the initial length itself. Notwithstanding these discrete jumps, the overall stability of crack growth was markedly enhanced relative to control runs on (monocrystalline) sapphire. This enhanced stability enabled exhaustive examinations of the entire propagating crack during its growth up the T-curve. Such examinations revealed a high incidence of persistent bridging grain–grain contacts at the primary crack interface behind the tip. No sign of microcrack cloud activity was observed in any of the materials examined, even with the crack-tip K-fields fully sustained.

Fig. 7.25 shows the evolution of one particularly active grain-bridging site in alumina at six stages in loading to failure. The sequence corresponds to progressive advance of the crack tip out of the field of view at right.

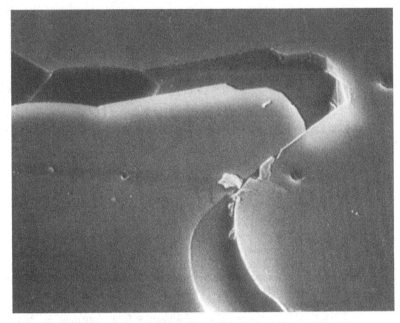

Fig. 7.26. Crack in same alumina as in previous figure, showing evidence for frictional contact at sliding grain–matrix facet. Note 'debris' at contacting facet. Scanning electron micrograph. Width of field 30 μm. (After Swanson, P. L., Fairbanks, C. J., Lawn, B. R., Mai, Y-W. & Hockey, B. J. (1987) *J. Amer. Ceram. Soc.* **70** 279.)

Observe how the primary crack appears to segment about the active grain in stage I, with increasing overlap in II. In stage III the primary crack links up over the top portion of the grain. Enhanced optical reflectively beneath this grain reveals that the segmented cracking extends below as well as along the specimen surface. Even this linking does not signify the end of activity: witness the evidence for continued, if minor, subsidiary damage at left in stage IV and at right in stage V, even though in the latter case the crack tip is almost 2 mm distant. Ultimately, in stage VI, the specimen fails. In the final stages the crack segment below the ruptured grain closes up, although incompletely, indicating part release of the locally stored strain energy.

Closer examination of active bridge sites by scanning electron microscopy reveals high frictional stresses at sliding grain boundary facets. The contact 'debris' in fig. 7.26 attests to the intensity of these stresses. At large grains the frictional build-up can generate transgranular fractures contiguous with the primary crack. In extreme cases interlocking grains may be rotated in their 'sockets' or even ejected from the surface (fig. 7.4).

These more disruptive features are most noticeable in those materials with strong *T*-curves, i.e. with coarse, elongate grain structures, weak grain or interphase boundaries, and high levels of internal stress.

Detailed observations of this kind provide quantitative information on key microstructural dimensions for ensuing fracture mechanics analysis. In monophase ceramics, the mean separation between bridging grains is typically two to three grain diameters. The distance behind the crack tip over which bridges remain intact is some orders of magnitude greater, corresponding to a typical crack-opening displacement $\approx 10\%$ grain diameter. Such quantities imply a long-range 'tail' to the constitutive stress–separation pullout function (cf. fig. 3.11) for individual grains.

We alluded at the close of sect. 7.1 to the pitfalls that attend conventional failure analysis by post-mortem surface fractography. Here is a classic case in point. Bridging often passes unnoticed because the very act of separating the specimen halves destroys the 'evidence', by rupturing the links between contacting grains across the crack interface.

7.5.2 Theoretical fracture mechanics

The above observations lay the groundwork for modelling bridging in monophase ceramics at the microstructural level (Mai & Lawn 1987; Bennison & Lawn 1989). *Deflection* and *residual stress* are important ancillary elements in this modelling: the first in directing the crack along weak intergranular boundaries, thereby creating favourable conditions for bridge formation; the second primarily in augmenting grain-contact friction tractions.

Consider an intergranular crack in a homogeneous, equi-axed polycrystal of grain size l, with bridging at the interface. Fig. 7.27(a) depicts the crack, length c, extending from a starter notch, length c_0, subjected to distributed tractions $p_\mu(X)$ within the bridging zone $0 \leqslant X \leqslant \Delta c$. Fig. 7.27(b) designates the bridging grains, spacing d, as those with *compressive* components of the residual field, $-\sigma_R$, at transverse facets. On debonding, the clamped grains resist pullout by Coulomb friction. Fig. 7.27(c) is the configuration for an incipient flaw prior to any bridge activation. The most active flaws are those that originate on facets subject to *tensile* components of the residual field, $+\sigma_R$ (e.g. those responsible for spontaneous microcracking, sect. 7.3.2). As we shall see, the crack system in fig. 7.27 contains all the necessary ingredients for a universal description of bridging, encompassing both long- and short-crack domains.

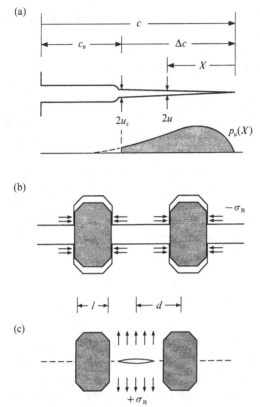

Fig. 7.27. Bridged crack. (a) Extension from notch activates bridging stresses (shaded). (b) Grain pullout at $c > c_0$. (c) Small penny-like cracks at radius $c < d$, subject to thermal expansion stress $+\sigma_R$. In homogeneous, geometrically similar microstructures bridge spacing d scales with grain size l, such that area density of bridges $l^2/2d^2$ is scale-invariant.

Once more, the constitutive relation for the restraining elements holds the key to the underlying role of microstructure. When (but only when) the crack intersects the first set of bridges in fig. 7.27 we replace the array of discrete closure forces over the crack area by a continuous interplanar stress–separation function, $p_\mu(u)$. Debonding by deflection of the primary crack along the transverse grain or interphase boundaries, although a necessary precursor to bridge formation, is omitted from this function because the ensuing frictional sliding occurs over a much greater crack-opening displacement (e.g. fig. 7.26), with correspondingly greater energy dissipation. The pullout constitutive relation takes the form

$$p_\mu(u) = p_C^B(1 - 2u/\xi^B), \quad (0 \leqslant 2u \leqslant \xi^B) \tag{7.28}$$

where $p_\mu = p_C^B$ is the critical bridging stress at first sliding ($u = 0$) and $2u = \xi^B$ is the wall–wall separation at grain disengagement ($p_\mu = 0$). The falloff in $p_\mu(u)$ reflects a diminishing grain–matrix contact area as the opposing walls separate. We acknowledge that (7.28), by the very (non-conservative) nature of frictional tractions, must be hysteretic.

Equation (7.28) may be related explicitly to microstructural variables by evaluating $p_C^B = P_C^B/d^2$ in terms of the restraining force P_C^B exerted by a *single* grain on the initially separating matrix walls. For systems dominated by Coulomb friction, this force is the product of the grain–matrix sliding contact area $\lambda l \xi^B$, with λ the number of contacting facets per bridging grain, the coefficient of friction μ at this interface, and the residual compressive stress $-\sigma_R$. Imposing the requirement that the net force from the internal stresses balance out over the extended primary crack plane, and defining a bridge-rupture strain ε^B, we obtain (Bennison & Lawn 1989)

$$\xi^B = \varepsilon^B l \tag{7.29a}$$

$$p_C^B = \varepsilon^B \lambda \mu (1 - l^2/2d^2) \sigma_R. \tag{7.29b}$$

For geometrically similar microstructures ε^B and d/l are scale-invariant quantities, so that a *grain size effect* is manifest in the disengagement displacement ξ^B but not in the critical stress p_C^B (contrast σ_C^M and σ_C^T in sects. 7.3.2 and 7.4.2). A graphical expression of this similitude is given by the plot of (7.28) at four grain sizes in fig. 7.28.

We may calculate the bridging contribution to toughness from the generic formalism of sect. 3.7.1. The usual difficulty in obtaining closed-form solutions for the full toughness curve is now compounded by the prospect of a subsidiary, *anti*-shielding contribution from those same thermal expansion anisotropy stresses responsible for the frictional tractions implicit in (7.28). For long straight cracks the influence of an alternating internal field over the crack plane rapidly integrates to zero beyond the first few bridge intersections. For short penny-like flaws, however, the initial evolution may be dominated by the local tensile component of this internal field (fig. 7.27). It is then convenient to make use of the superposibility of K-fields by evaluating the T-curve increment $-K_\mu = T_\mu = T_\mu' + T_\mu''$, with T_μ' from pullout once bridge intersection occurs and T_μ'' from local tensile fields prior to intersection:

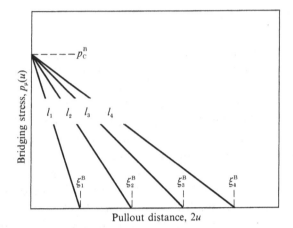

Fig. 7.28. Constitutive bridging-stress function $p_\mu(u)$ for grain–matrix pullout mechanism, grain sizes $l_1 < l_2 < l_3 < l_4$. Area under $p_\mu(u)$ curve gives energy absorbed in the bridge rupture process. Note critical stress at which pullout begins is constant, but pullout distance scales with l. Ordinarily, a precursor debonding stage (cf. fig. 7.30) occurs at relatively small interplanar separations, but is neglected here in the energetics.

(i) *Long straight cracks.* Consider a straight crack such that the internal stresses average to zero along an infinite front at first incremental extension. Then the only contribution $R'_\mu = -G'_\mu$ to the toughening is from the quasi-continuous bridging stresses in (7.28). Integration of (3.34) yields

$$\left.\begin{array}{l} R'_\mu(u_Z) = 0, \quad (2u_Z = 0) \\[4pt] R'_\mu(u_Z) = 2p_C^B u_Z(1 - u_Z/\zeta^B), \quad (0 \leqslant 2u_Z \leqslant \zeta^B) \\[4pt] R'_\mu(u_Z) = \tfrac{1}{2}p_C^B \zeta^B, \quad (2u_Z \geqslant \zeta^B) \end{array}\right\} \tag{7.30}$$

with u_Z the crack-opening displacement at the edge of the shielding zone. In the limit of long cracks ($\Delta c = c - c_0 \ll c_0$, fig. 7.27) we invoke (3.31)

$$T'_\mu(u_Z) = [E'R'_\mu(u_Z) + K_*^2]^{1/2} - K_* \tag{7.31}$$

to convert to T-notation.

Thus far the formalism makes no reference to crack geometry. A profile relation $u_Z(c)$ is now required to express the toughness component in (7.30) and (7.31) as a function of crack length. Here we simply use the Irwin near-field solution for slit cracks, in the same spirit of weak-shielding as in (7.20) and (7.25) for frontal zones, i.e. assuming the K_A-field to extend

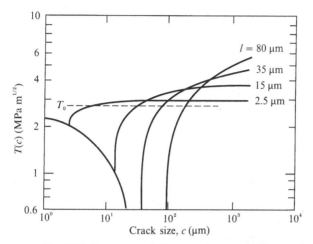

Fig. 7.29. Regenerated T-curves for penny-like cracks in alumina at four grain sizes using calibrated long-crack relations in (7.30)–(7.32) but including short-crack terms (7.31). At large c, the toughness is enhanced by the dissipative bridging stresses. At small c, prior to engagement of the first bridges, the counter effect of (tensile) matrix internal stresses is manifest, and $T(c)$ falls below the intrinsic boundary toughness T_0. (After Chantikul, P., Bennison, S. J. & Lawn, B. R. (1990) *J. Amer. Ceram. Soc.* **73** 2419.)

unperturbed to the boundary of the shielding zone (cf. fig. 3.8). For the notch of fig. 7.27(a) we insert $u = u_Z$ at $X = X_Z = \Delta c$ into (2.15),

$$u_Z(c) = [K_A(c)/E'](8\Delta c/\pi)^{1/2}, \tag{7.32}$$

which allows us to determine $T_\mu''(\Delta c)$.

The solution at $2u_Z \geqslant \xi^B$ in (7.30) is the steady-state limit, signifying disengagement of the bridges from the matrix at the notch edge. Thereafter, the shielding zone translates with the advancing crack tip. There is a fundamental correspondence between this bounding solution and its frontal-zone counterparts in (7.19) and (7.24): the shielding contribution to the crack-resistance energy is given by the area under the appropriate $p_\mu(u)$ curve. Note from (7.29a), however, that the scaling dimension ξ^B in (7.30) (unlike the frontal-zone dimension w_C in (7.20) and (7.25)) is a material quantity independent of any resistance parameter. Thus for bridging, R_μ and R_0 (or T_μ and T_0) are *not* multiplicative quantities.

Implementing the equilibrium requirement $K_A(c) = T_0 + T_\mu''(\Delta c) = T(\Delta c)$ in (7.32), (7.30)–(7.32) may be solved simultaneously to determine the

T-curve.[3] An analysis of this kind has been used to fit the straight-crack alumina data in fig. 7.24, with parameter adjustments $E' = 400$ GPa, $T_0 = 2.75$ MPa m$^{1/2}$ ($R_0 = 20$ J m^{-2}), $\varepsilon^{\mathrm{B}} = 0.040$, $d = l = 35$ μm, $p_{\mathrm{C}}^{\mathrm{B}} = 50$ MPa, ($\sigma_{\mathrm{R}} = 350$ MPa, $\lambda = 4$, $\mu = 1.7$). For the material in fig. 7.24, this parameter 'calibration' corresponds to a critical crack-opening displacement $\xi^{\mathrm{B}} \approx 2$ μm in (7.29a) and steady-state toughness $R_\infty = R_0 + R_\mu^\infty \approx 65$ J m^{-2} ($T_\infty \approx 5.2$ MPa m$^{1/2}$) in (7.30).

(ii) *Short penny cracks.* Now consider a small penny-like crack that begins its life at a tensile grain boundary sub-facet, fig. 7.27(c), as is the expectation for natural microstructural flaws (chapter 9). In addition to the above closure *K*-field from bridging, the crack experiences a countervailing opening *K*-field from the discrete internal tensile stress $+\sigma_{\mathrm{R}}$ over $0 \leqslant r \leqslant d$. To evaluate this second, negative contribution $K_\mu'' = -T_\mu''$ to the toughness we integrate the Green's function formula (2.22b) for *half-penny* cracks:

$$\left.\begin{array}{l} T_\mu''(c) = -\psi\sigma_{\mathrm{R}} c^{1/2}, \quad (c \leqslant d) \\ T_\mu''(c) = -\psi\sigma_{\mathrm{R}} c^{1/2}[1 - (1 - d^2/c^2)^{1/2}], \quad (c \geqslant d) \end{array}\right\} \qquad (7.33)$$

with $\psi = 2\alpha/\pi^{1/2}$ from (2.21). In the limit $c \gg d$ we have $T_\mu'' \to \psi\sigma_{\mathrm{R}} d^2/2c^{3/2} \to 0$, so the influence of microstructural discreteness is lost well before the system reaches steady state.

The short-crack *T*-curve is now given by $T(c) = T_0 + T_\mu'(c) + T_\mu''(c)$, with $T_\mu'(c)$ evaluated as for long cracks but with the 'notch' length in (7.32) determined by first bridge intersection at $c_0 = d$. A plot of this modified *T*-curve is included for the alumina in fig. 7.24. The effect of the local residual tensile field is to depress the toughness at $c \lesssim 10d$ in fig. 7.24, with consequential effects on the crack stability. Recalling the 'tangency' construction in fig. 3.10, a depression of this kind in $T(c)$ can accommodate initial pop-in and subsequent stabilised extension of microstructural-scale flaws, as observed in the previous subsection. Significantly, the region of these perturbations is that in which *strength* properties are decided.

As an illustration of the flexibility of the formulation we generate in fig. 7.29 complete short-crack *T*-curves for aluminas of several grain sizes, using the calibrated microstructural parameters. We assume that the microstructures satisfy similitude requirements, i.e. the quantities ε^{B}, λ and

---

[3] Strictly, these equations are implicit in $T_\mu(\Delta c)$, so must be solved iteratively. However, in the approximation $T_\mu(\Delta c) \ll T_0$, $K_{\mathrm{A}} \simeq K_* = T_0$, the weak-shielding limit $T_\mu(\Delta c) = E'R_\mu(\Delta c)/2T_0$ in (3.32b) affords a useful explicit solution.

d/l in (7.29) are scale-invariant, so that l is the only variable. The crossover in the ensuing curves simply reflects countervailing effects of the residual-stress driving K-field at small c and subsequent (ultimately dominant) bridge restraining K-field at large c. Increasing l enhances both these K-fields: the first by expanding the crack area over which the (scale-invariant) tensile stress $+\sigma_R$ operates; the second by increasing the frictional pullout distance ξ^B. At sufficiently large l the minimum in the T-curve at $c = d$ intersects the c-axis; the effective toughness is zero at this point, and the crack is on the verge of spontaneous extension to the stable branch at $c > d$. Thus *general microfracture* (sect. 7.3.2) is included as a limiting state in the bridging model. In scaling microstructures to improve long-crack toughness one has to be careful that intrinsic internal stresses in the short-crack domain do not cause the material to disintegrate.

We see that once the parameters in the bridging T-curve equations are suitably calibrated, the above model may be used as a predictive guide to the role of other important microstructural variables, such as grain boundary energy, internal residual stress, grain–grain sliding friction coefficient, etc. Extension of the analysis to grain shape (aspect ratio), second phases, and so on requires further, but minor, modifications to the fracture mechanics. In sect. 10.4 we shall give examples as to how this predictive capability may be put to use in material design.

It is well not to lose sight of the assumptions and approximations embodied in our analysis. The bridging zone is presumed small and weak, so application to materials with short cracks and strong T-curves should be made with special caution. The Irwin displacement relation (7.32), while decoupled from the toughness equations, makes no allowance for the modifications to the crack shape that inevitably attend the bridging tractions; even within the weak-shielding approximation the solution is not strictly self-consistent. Also lacking is direct experimental confirmation of the ideally linear tail in the constitutive relation (7.28). With all that, the model accounts for the highly distinctive features of crack response in simple bridging ceramics, and provides a useful starting point for the more general description of reinforcement in brittle composites.

7.6 Ceramic composites

We are now well placed to understand the elementary fracture mechanics of reinforced ceramic-matrix composites. In a perfectly brittle multi-phase system the sole contribution to the toughness comes from the intrinsic adhesion energies of the components, suitably 'weighted' according to some 'law of mixtures'. By optimising shielding processes of the type considered in the preceding sections one may ultimately hope to exceed this lower bound by perhaps as much as two orders of magnitude in R (one order in T). Ideally, the applied stress–strain curve for the composite would be highly nonlinear, with large peak stress (to maximise load-bearing capacity) and large strain to failure (to maximise energy absorption capacity). These prospects have led to a renewed outburst in research and development activity in ceramic-matrix composites over the past decade. The fruits of this activity are yet to be fully felt in the commercial arena. Complicating factors, e.g. failure modes other than single-crack propagation, chemical degradation of interfaces, especially at high temperatures, remain challenges to the composites designer.

Brittle composites can take many forms: (i) natural materials (e.g. bone, teeth, rocks); (ii) particulate second-phase ceramics (e.g. glass-ceramics, alumina/zirconia); (iii) continuous-fibre and whisker composites; (iv) ceramic-matrix/ductile-dispersion composites; (v) bi-material laminates. We focus on just two of the best studied and most topical of these systems here, (iii) and (iv). The more committed reader who undertakes to explore the various pathways through this diverse and still-evolving field will discover a literature rich in detailed solid mechanics analysis if somewhat deficient in experimental evaluation.

As indicated above, toughening in ceramic composites is attributable to shielding processes, notably *bridging* where the reinforcing phases are left intact as ligaments at the crack interface (Evans 1990; Becher 1991). In such cases the crack-resistance energy of the matrix is

$$R_0 = (1 - V_{\mathrm{f}})W_{\mathrm{BB}} \tag{7.34}$$

where $W_{\mathrm{BB}} = 2\gamma_{\mathrm{B}}$ is the intrinsic cohesion energy of the matrix material and V_{f} is the volume fraction (or, more specifically, area density at the crack plane) of reinforcing phase. Key to attaining effective toughening is the existence of suitably weak interfaces to allow debonding between the matrix and reinforcing phases, and hysteretic energy dissipation within the ensuing bridges at separation. In terms of processing strategy this means

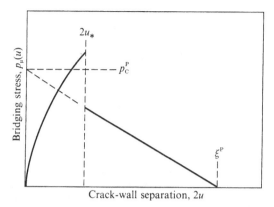

Fig. 7.30. Constitutive stress–separation function for fibre-reinforced brittle matrix. Between $0 \leqslant 2u \leqslant 2u_*$ the fibres debond and ultimately rupture. Thereafter at $2u_* \leqslant 2u \leqslant \xi^P$ the fibres pull out of matrix.

skilful tailoring of the reinforcing phases and associated interphase boundaries relative to the given matrix. The achievement of ultimate toughness in ceramic composites may require the incorporation of mixed frontal-zone and bridging processes of energy dissipation, preferably in mutually interactive (multiplicative) modes.

7.6.1 Fibre-reinforced composites

Reinforcement of ceramic matrices with high-strength fibres or whiskers can result in dramatic improvements in long-crack toughness. This area of materials development began with the early work of Kelly and co-workers (Kelly 1966; Aveston, Cooper & Kelly 1971), but then fell dormant for about a decade. The advent of a new breed of ultra-strong fibres in the 1980s, e.g. silicon carbide, alumina and zirconia, subsequently incorporated into matrices of glass, glass-ceramic and even cementitious materials, signalled a renaissance in fracture mechanics modelling (Marshall, Cox & Evans 1985; Mai 1988), which continues today. Most widely studied are uniaxial systems with arrays of continuous fibres aligned along the ultimate direction of tensile loading. An attractive feature of such systems is the capacity to pre-determine the all-important constitutive stress–separation function for the bridged crack by independent experiment, e.g. by tensile pullout or indentation push-in tests (sect. 8.6.3) on individual embedded fibres. It is on uniaxial configurations that we focus our attention here.

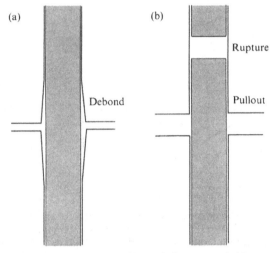

Fig. 7.31. Evolution of fibre reinforcement bridges at separating crack.
(a) Debonding (at $0 \leqslant 2u \leqslant 2u_*$), and (b) pullout (at $2u_* \leqslant 2u \leqslant \xi^P$).

Begin by considering the micromechanics of the stress–separation function in fig. 7.30 (Marshall, Cox & Evans 1985) in relation to the following microstructural parameters: composite Young's modulus $E = V_f E_f + (1 - V_f)E_m$, f and m denoting fibre and matrix; sliding-friction shear stress τ at the matrix/fibre boundary (with components from both residual-stress clamping and surface roughness); critical wall–wall displacement $2u_*$ (or strength σ_F) for fibre rupture; subsequent displacement ξ^P at fibre pullout; fibre radius r. We distinguish two stages in the $p_\mu(u)$ curve, depicted in fig. 7.31:

(i) *Debonding*, fig. 7.31(a). This is the essential precursor, bridge formation stage. The primary crack deflects along the matrix/fibre interface (cf. fig. 7.7), progressively transferring the applied load from the matrix to the fibre as the walls separate. The deflected crack develops a large component of shear ($K_I \approx K_{II}$, recall sects. 2.8, 7.1) as it spreads along the interface. Approximate analysis of this configuration gives the closure stresses on the matrix walls (Marshall, Cox & Evans 1985)

$$p_\mu(u) = p^D(2u/r)^{1/2}, \quad (0 \leqslant 2u \leqslant 2u_*) \tag{7.35}$$

with stress coefficient

$$p^D = (2\tau E_f)^{1/2} E V_f / E_m (1 - V_f). \tag{7.36}$$

(ii) *Frictional pullout*, fig. 7.31(b). At the critical displacement u_* (or critical stress σ_F) the fibre ruptures (or is debonded along its entire length), imposing an upper bound on the closure stress. The debonded fibre begins to slide out against the frictional restraint of the matrix walls, exerting a closure stress

$$p_\mu(u) = p_C^P(1 - 2u/\xi^P), \quad (2u_* \leqslant 2u \leqslant \xi^P) \tag{7.37}$$

in direct analogy to the bridging relation (7.28) for monophase ceramics, with p_C^P a critical sliding stress,

$$p_C^P = (2V_f \xi^P/r)\tau \tag{7.38}$$

in analogy to (7.29b). The quantity ξ^P scales with the distance of the fibre rupture relative to the crack plane. If the debond distance is small, as it will be for short or inclined fibres (e.g. whiskers), there is negligible pullout and the debond energy dissipates abruptly as acoustic waves.

Since the toughening increment is determined by the area under the stress–separation curve in fig. 7.30, it would seem that one should aim to maximise the quantities p^D, p_C^P, u_* and ξ^P in (7.35) and (7.37). The role of weak matrix/fibre interfaces in any such endeavour is paramount: the adhesion must not be too strong that the fibres rupture prematurely before debonding can ensue; nor must it be too weak that the frictional resistance to sliding is reduced effectively to zero. Achieving the proper balance in interfacial friction (e.g. by judicious incorporation of an appropriate interboundary phase) and in the other microstructural variables in (7.36) and (7.38) (plus any additional variables associated with residual stress and Poisson effects) poses a considerable challenge to composites designers. Matters are further complicated by statistical variations in fibre strengths, effectively prolonging the tail of the constitutive curve.

Even the simplest uniaxial systems can exhibit complex failure modes. In longitudinal loading, an initial matrix crack generates and runs through the specimen to the boundaries. Provided the strength of the reinforcing material exceeds that of the matrix, unbroken fibres bridge the separated walls, thereby sustaining the applied load. On further loading, the incidence of matrix cracking increases, and the general stress–strain curve begins to deviate markedly from linearity. At peak applied stress the fibres begin to fail. Thereafter the stress tails off as the remaining fibres pull out of the matrix. The system therefore exhibits the characteristics of 'ductility', with an ill-defined toughness. Transversely, however, the material may be relatively weak, so the applied stress state becomes an important factor. In bending, for instance, the composite can fail prematurely on the compression side by a variety of damage modes, e.g. microcrack coalescence, delamination, fibre buckling. Accordingly, special care has to be taken to ensure that strength in tension is not achieved at the cost of weakness in compression or shear.

According to this description, conventional fracture mechanics remains applicable until the first matrix crack breaches the specimen. The toughness increment $R_\mu = -G_\mu$ may be determined formally by substituting the stress–separation relations (7.35) and (7.37) into (3.34), exactly as in the bridging analysis in sect. 7.5 for monophase ceramics. Integration gives

$$R_\mu(u_Z) = (8u_Z/9r)^{1/2}p^D u_Z, \quad (0 \leqslant 2u \leqslant 2u_*) \tag{7.39a}$$

$$R_\mu(u_Z) = (8u_*/9r)^{1/2}p^D u_* + 2p_C^P u_Z[(1 - u_Z/\xi^P) \\ - (u_*/u_Z)(1 - u_*/\xi^P)], \quad (2u_* \leqslant 2u \leqslant \xi^P). \tag{7.39b}$$

Once more, a crack-profile relation $u_Z = u_Z(c)$ is needed to transform (7.39) into $R_\mu(c)$, and so determine the R-curve, $R(c) = R_0 + R_\mu(c)$. For a first approximation one may resort to the familiar unperturbed Irwin solution, as in (7.32), although the weak-shielding approximation on which this solution is predicated is especially suspect for the tougher composites. At steady state, $2u_Z = \xi^P$, we have, in the limit of dominant pullout ($2u_* \ll \xi^P$),

$$R_\mu = \tfrac{1}{2}p_C^P \xi^P. \tag{7.40}$$

Only modest pullout stresses and displacements, say $p_C^P \approx 100$ MPa and $\xi^P \approx 10$ μm, are required to produce substantial toughening increments, $R_\mu^\infty \approx 500$ J m^{-2}, $T_\mu^\infty \approx 12$ MPa m$^{1/2}$ ($E' \approx 300$ GPa).

An alternative approach uses a K-field analysis for $T_\mu = -K_\mu$ to determine the toughness-curve, $T(c) = T_0 + T_\mu(c)$ (Marshall, Cox & Evans 1985). That approach is especially powerful if superposed stress fields (cf. sect. 7.5) have to be taken into account.

As intimated earlier, theoretical activity in the fracture mechanics analysis of fibre-reinforced ceramic composites has not been matched by critical experimental evaluation. A notable exception is the work by Mai and co-workers (Mai 1988) on cementitious composites. Some of their T-curve data for a fibre-reinforced cement are plotted in fig. 7.32. The results depend strongly on specimen geometry, indicating large bridging zones

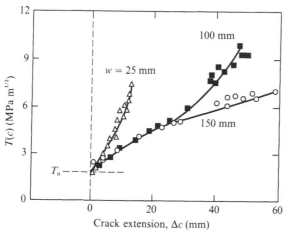

Fig. 7.32. *T*-curve data for a fibre-reinforced cement. Data for SENB specimens of different thickness w, with initial notch length $c_0/w = 0.3 =$ const. *T*-curve depends on w, reflecting large bridging zones relative to specimen widths. Solid curves are fits to bridging analysis (with due allowance for the specimen thickness effects). (After Foote, R. M. L., Mai, Y-W. & Cotterell, B. (1980), in *Advances in Cement-Matrix Composites*, ed. D. M. Roy, Materials Research Society, Pennsylvania, p. 135.)

relative to the specimen widths. Solid curves are analytical fits using a bridging constitutive law with large pullout tail $(2u_* \ll \xi^P)$, and with due allowance for specimen thickness effects in both the $K_A(c)$ and $K_\mu(c)$ terms. There is room for more data of this kind, on both model and practical systems, to test the limits of applicability of the existing toughness analyses.

7.6.2 Ductile-dispersion toughening

Another approach to tough ceramic-matrix composites is the incorporation of metal phases (e.g. alumina/aluminium, tungsten-carbide/cobalt 'cermets') (Evans 1990). The approach promises the best of two worlds: the brittle matrix preserves stiffness and lightness; the metal phase enhances toughness and energy absorption capacity. There are many variants of this composite type, relating to the continuity of the two phases. Here we examine the kind where the ceramic is continuous, so that no full-scale plastic zone develops about the crack tip. Then the shielding is associated predominantly with dissipation of plastic energy at metal bridges across the crack interface, as depicted in fig. 7.33.

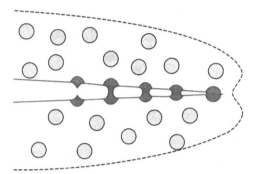

Fig. 7.33. Crack-interface bridging at metal particles in ceramic matrix.

The toughening increment may then be determined, using the J-integral, as the plastic work to cause the metal bridges to neck and ultimately rupture. Integration of (3.34) gives, at steady state

$$R_\mu^\infty = \alpha V_f \sigma_C^Y \varepsilon^Y a \tag{7.41}$$

with σ_C^Y the yield stress for plastic flow of the dispersed particles and ε^Y the corresponding rupture strain, a the particle radius; α is a numerical coefficient $\approx$ unity but depending sensitively on the detailed mode of particle detachment from matrix, work-hardening rate, etc. (as reflected in the shape of the stress–separation curve). Substituting reasonable values for an alumina/aluminium matrix/particle composite (Evans 1990), $V_f \approx 0.2$, $\sigma_C^Y \approx 200$ MPa, $a \approx 2$ μm, $\varepsilon^Y \approx 1.5$, $\alpha \approx 2.5$, we get $R_\mu^\infty \approx 300$ J m^{-2}, $T_\mu^\infty \approx 10$ MPa m$^{1/2}$ ($E' \approx 300$ GPa).

As indicated above, energy dissipation within any frontal plastic zone will augment the toughness increment. An additional contribution may come from the release of residual stresses in the particles (sect. 7.5).

The crack-stabilisation characteristics that attend the strong R-curve or T-curve behaviour outlined above offers the structural engineer the attractive prospect of flaw insensitivity in strength, and thence a well-defined design stress. To the ceramic scientist it offers an avenue to economic yet innovative processing, counteracting the need to eliminate all flaws by proper tailoring of reinforcing phases. In this latter context we recall from sect. 7.5 that wherever internal residual stresses are a critical element of the toughness equation, enhanced long-crack toughness may imply simultaneous degradation of short-crack properties. These issues will be addressed in chapter 10. A final evaluation of the ceramic composites revolution lies with tomorrow's scientists and engineers.

8
Indentation fracture

We turn now to a special kind of fracture, that produced by the contact of a hard indenter on a brittle surface. Indentation fracture, so-called, is of historical as well as practical interest. It dates back to 1880 with the celebrated studies of Hertz (see Hertz 1896) on conical fractures at elastic contacts between curved glass surfaces. The fully developed *Hertzian cone crack* is the prototypical stable fracture in brittle solids. Shortly after Hertz, Auerbach (1891) showed empirically that the critical load to initiate cone fractures in flat specimens is proportional to the radius of the indenting sphere, $P_C \propto r$. For 75 years, 'Auerbach's law' remained one of the great paradoxes in fracture theory: the notion that fracture should initiate when the maximum tensile stress in the Hertzian field just equals the bulk strength of the material implies an alternative relation, $P_C \propto r^2$. Resolution of the paradox awaited the advent of modern-day fracture mechanics (Frank & Lawn 1967). More recently, *radial–median cracks* produced in elastic–plastic fields by diamond pyramid indenters have assumed centre stage. The radial crack system is now arguably the most widely used of all fracture testing methodologies in the mechanical evaluation of brittle materials.

Indentation fracture has many facets: it yields valuable information on the fundamental processes of brittle fracture in covalent–ionic solids, and rare detail on subsidiary deformation processes at concentrated contacts; it provides 'controlled flaws' for systematically evaluating strength properties, with special insight into the stability of 'natural' flaws; it serves as a simple microprobe for determining material fracture parameters, toughness, crack-velocity exponent, etc.; for materials with R-curves (T-curves), it affords a much needed bridge between the short-crack domain of microstructural flaws and the long-crack domain of traditional toughness testing; it allows for the study of crack initiation as well as propagation, and thereby enables one to quantify 'brittleness', one of the most elusive of mechanical properties; it can be used to simulate service damage in ceramics, e.g. strength degradation from seemingly minute particle impingements and material removal in surface finishing, wear and erosion. The great appeal of the indentation methodology is its versatility, control and simplicity, requiring only access to routine hardness testing apparatus.

In this chapter we outline the basic principles of indentation fracture. We begin with a survey of classical contact stress fields, the starting point for fracture mechanics modelling. It is convenient to characterise indenters that produce elastic and elastic–plastic contacts as 'blunt' and 'sharp', respectively. Critical elements of cone and radial crack morphologies are then considered in relation to these two contact types. Distinction is made between propagation and initiation phases in the crack evolution: of these two phases, initiation is by far the more complex, because of extremely high stress gradients in the contact near-field and (at least for blunt contacts) a certain dependence on surface flaw states. Equilibrium and kinetic states are also distinguished, in particular relation to toughness and fatigue. Our ultimate goal is to establish a proper framework for basic fracture mechanics descriptions, first with just contact stresses present and then, as pertinent to strength, with uniform external applied stresses superimposed.

Our description will highlight *stress-field inhomogeneity* and *crack stability* as key elements of the contact problem. Nowhere else in fracture mechanics are these elements so compellingly demonstrated. Consequent breakdown of critical stress concepts, resulting in *size effects* in the initiation characteristics, will also be given special attention. In developing an analytical description we shall focus our attention on simple, ideal systems, referring the more serious reader elsewhere for greater analytical detail (Lawn & Wilshaw 1975; Evans & Wilshaw 1976; Lawn 1983; Cook & Pharr 1990). Specific materials applications, fundamental and applied, will be discussed in later sections.

8.1 Crack propagation in contact fields: blunt and sharp indenters

Indentation fracture starts with a knowledge of the contact stress field within which the cracks evolve (Lawn & Wilshaw 1975; Johnson 1985).

Such fields are determined principally by geometrical factors (indenter shape) and material properties (elastic modulus, hardness and toughness). As already indicated, it is expedient to classify indentations as 'sharp' or 'blunt', depending on whether or not there is irreversible deformation at contact. In the context of crack patterns it is the tensile component of the field that commands most attention. A characteristic feature of the tensile stress distribution is the existence of uncommonly high gradients in the near-contact region, especially at indenter corners and edges. Since our interest in this section is with the propagation of well-developed cracks in the far field, we defer detailed consideration of these extreme gradients to sect. 8.4 on subthreshold indentations.

Accordingly, we begin our discussion with the simplest configuration, the point-contact Boussinesq field. Explicit attention will be given only to normal loading, and to homogeneous, isotropic materials with single-valued toughness ($T = T_0 = K_C$), except where otherwise specified.

8.1.1 Contact stress fields

Consider a linear elastic half-space subjected to a normal point load P. This configuration corresponds to the axially symmetric *Boussinesq* field in fig. 8.1. In reality, the contact is accommodated elastically or plastically over a nonzero area of characteristic linear dimension a, to avert a stress singularity. Then the field may be characterised by two scaling quantities: in *spatial* extent, by the contact dimension a itself; in *intensity*, by the mean contact pressure, $p_0 = P/\alpha_0 a^2$, where α_0 is a dimensionless geometry constant (e.g. $\alpha_0 = \pi$ for circular contact of radius a). The far-field elastic stress distribution in spherical coordinates (ρ, θ, ϕ) about the load-point origin has the form

$$\sigma_{ij}/p_0 = (\alpha_0/\pi)(a/\rho)^2[f_{ij}(\phi)]_\nu, \quad (\rho \gg a) \tag{8.1}$$

with $f_{ij}(\phi)$ a well-defined function of the polar angle ϕ at Poisson's ratio ν. The (inverse-square) falloff in stress is responsible for the inherent stability of well-developed contact fractures. Again, the nature of the field at $\rho < a$ is of concern here only insofar as it may predetermine the spatial origin, and consequently the ultimate form, of the crack pattern.

The plots in fig. 8.1 represents *principal normal stresses* in the Boussinesq field: (a) trajectories (tangents to which denote direction); (b) contours (points on which denote magnitude). These stresses are so defined that

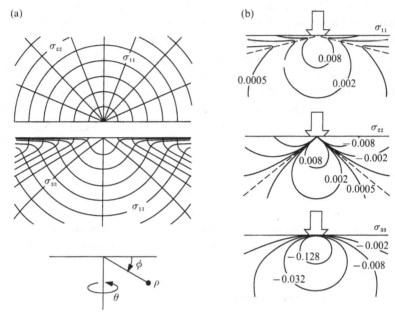

Fig. 8.1. Boussinesq field, for principal normal stresses σ_{11}, σ_{22} and σ_{33}. (a) Stress trajectories, half-surface view (top) and side view (bottom). (b) Contours, side view. Unit of contact stress is p_0, contact diameter $2a$ (arrow). Note sharp minimum in $\sigma_{11}(\phi)$ and zero in $\sigma_{22}(\phi)$, dashed lines. Plotted for $\nu = 0.25$. (See Johnson, K. L. (1985) *Contact Mechanics*. Cambridge University Press, Cambridge, Ch. 3.)

$\sigma_{11} \geqslant \sigma_{22} \geqslant \sigma_{33}$ nearly everywhere: σ_{11} is tensile at all points in the field, with maxima at the surface ($\phi = 0$) and along the contact axis ($\phi = \pi/2$); σ_{22} ('hoop' stress) is tensile subsurface; σ_{33} is everywhere compressive. We assert then that tensile stresses, however small, are generally unavoidable in contact fields. Recalling the tendency for brittle cracks to propagate along paths normal to the greatest tensile stresses (sect. 2.8), we may expect the fully developed cracks to lie on either quasi-conical σ_{22}–σ_{33} or median σ_{11}–σ_{33} trajectory surfaces. We shall encounter examples of both types below.

Other components of the contact field, shear, i.e. $\frac{1}{2}(\sigma_{11} - \sigma_{33})$, $\frac{1}{2}(\sigma_{11} - \sigma_{22})$, $\frac{1}{2}(\sigma_{22} - \sigma_{33})$, and hydrostatic compression, $-\frac{1}{3}(\sigma_{11} + \sigma_{22} + \sigma_{33})$, are not without interest. The intensities of these components substantially exceed those of the tensile stresses, typically by an order of magnitude. Hence wherever tension is suppressed (e.g. immediately under the contact circle, especially in sharp contacts), the material may be deformed irreversibly, leaving a residual, 'plastic' impression.

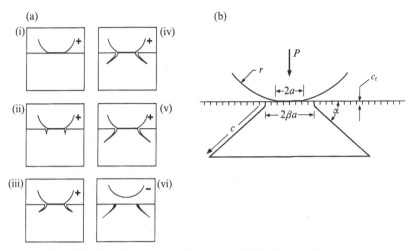

Fig. 8.2. Hertzian cone crack system. (a) Evolution of cone during complete loading (+) and unloading (−) cycle. (b) Geometrical parameters.

8.1.2 Blunt indenters

The most practical manifestation of the blunt indenter geometry is that of a hard spherical indenter loaded normally on a flat, thick elastic specimen. This configuration produces the classical *Hertzian cone crack*. Broad features of the crack evolution are depicted schematically in fig. 8.2(a): (i) pre-present surface flaws are subjected to tensile stresses outside the contact zone; (ii) at some point in the loading a favourably located flaw runs around the contact circle to form a surface 'ring' crack; (iii) on further loading, the embryonic ring crack grows incrementally downward in the rapidly weakening tensile field; (iv) at critical load the ring becomes unstable and propagates downward into the full frustum of the Hertzian cone (pop-in); (v) at still further loading the cone continues in stable growth (unless the contact circle expands beyond the surface ring crack, in which case the cone is engulfed in the compressive contact zone); (vi) on unloading, the cone crack closes.

Fig. 8.3 shows an example of a well-developed Hertzian crack in soda-lime glass, with load applied. The σ_{22}–σ_{33} conical surface forms an angle $\alpha \approx 22°$ to the free surface in this material. Such cracks tend to remain visible after unloading, indicating imperfect closure due to mechanical obstruction at the crack interface.

As indicated, the distinctive characteristic of blunt indenters is an *elastic*

Fig. 8.3. Cone crack in soda-lime glass. Photographed under load
($P = 40$ kN) from cylindrical punch, optical micrograph (block edge
length 50 mm). Crack makes angle $\simeq 22°$ to free surface. (After
Roesler, F. C. (1956) *Proc. Phys. Soc. Lond.* **B69** 981.)

contact, fig. 8.2(b). From the Hertz elasticity analysis the contact pressure
increases monotonically with expanding contact circle:

$$p_0 = P/\pi a^2 = (3E/4\pi k)(a/r) \tag{8.2}$$

with $k = \frac{9}{16}[(1 - v^2) + (1 - v_s^2) E/E_s]$ a dimensionless coefficient, E and v,
E_s and v_s Young's modulus and Poisson's ratio of the elastic half-plane
and spherical indenter respectively ($k = 1$ for $v = \frac{1}{3}$ and like materials).
The contact becomes inelastic only if p_0 exceeds some critical level for
irreversible deformation prior to development of the cone fracture. Note
the inverse dependence on r in (8.2) at fixed contact ($a = $ const), implying
higher contact pressures for 'sharper' indenters.

Now consider the fracture mechanics. A far-field solution was first
obtained for the cone crack by Roesler (1956), using dimensional analysis.
Here, in that the popped-in cone propagates on an ever-expanding circular
front, we simply regard the configuration as a centre-loaded (if distorted)
penny-like crack. Then (2.23b) gives immediately

$$K_P = \chi P/c^{3/2}, \quad (P > P_C, c \gg a) \tag{8.3}$$

with χ a dimensionless constant. The condition $K_P = K_C = T_0$ at $c = c_I$,
say, denotes a prevailing state of equilibrium at sustained load P. This state
is stable, because c_I increases with P, i.e. $P \propto c_I^{3/2}$. The quantity χ for cone
cracks depends on Poisson's ratio, $\chi = \chi(v)$ (recall (8.2)), and is therefore a
material constant. However, exact calculations of χ are not readily
available, and one usually calibrates this term empirically from fits to data
of the kind shown for soda-lime glass in fig. 8.4(a). Since the contact is
elastic, there are no residual stresses (other than those associated with the

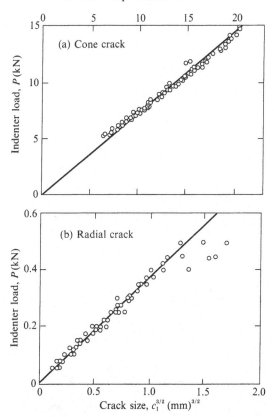

Fig. 8.4. Indentation load vs characteristic crack size for (a) cone cracks (spherical indenter with machined flat) and (b) radial cracks (Vickers indenter) in soda-lime glass, inert environment. (After Lawn, B. R. & Marshall, D. B. (1983), in *Fracture Mechanics of Ceramics*, eds. R. C. Bradt, A. G. Evans, D. P. H. Hasselman & F. F. Lange, Plenum, New York, Vol. 5, p. 1.)

incompletely closed crack) in the unloaded system. Observe that no information on the near-contact conditions is contained in (8.3).

Hertzian fracture has many variants. Two are illustrated in figs. 8.5 and 8.6. Fig. 8.5 shows the influence of crystallographic anisotropy on the cone crack morphology in silicon. The broad features of the surface ring remain, but with the cleavage tendency of the diamond structure (sect. 2.8) now imposed on the pattern symmetry. Fig. 8.6 shows the crack pattern when the indenting sphere is laterally translated across a glass surface. Friction at the contact significantly modifies the near field. Tensile stresses intensify at the trailing edge, resulting in the intermittent generation of 'partial' cone cracks. The friction has a relatively small effect on the *far* field,

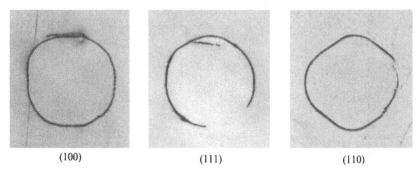

(100)   (111)   (110)

Fig. 8.5. Cone cracks in three monocrystal silicon surfaces showing effects of cleavage anisotropy. Surfaces etched after indentation and viewed in reflected light. Straighter segments of surface cracks lie along traces of {111} planes. Width of field 1500 μm. (After Lawn, B. R. (1968) *J. Appl. Phys.* **39** 4828.)

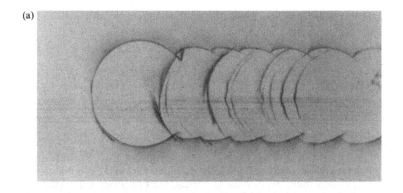

Fig. 8.6. Cone crack track in soda-lime glass, produced by sliding sphere (left to right), friction coefficient 0.1. (a) Surface view. Note partial formation of surface traces. (b) Profile view after section-and-etch through track. Increase in angle of crack with free surface (cf. fig. 8.3(c)) reflects tilting of resolved-load axis. Width of field 2000 μm. (After Lawn, B. R., Wiederhorn, S. M. & Roberts, D. E. (1984) *J. Mater. Sci.* **19** 2561.)

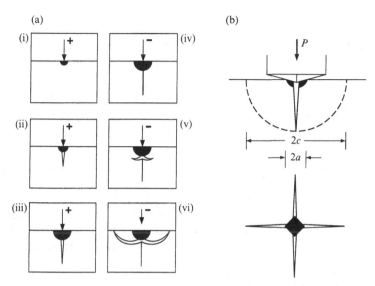

Fig. 8.7. Radial–median and lateral crack systems. (a) Evolution during complete loading ($+$) and unloading ($-$) cycle. Dark region denotes irreversible deformation zone. (b) Geometrical parameters of radial system.

reflecting primarily as an increase in resolved load P in (8.2). Thus the effect of translation is to increase the crack *density* rather than the crack *size*. Finally, in reactive environments, the cone cracks can extend further into the material at sustained loads according to some velocity function $v = v(K_P)$.

8.1.3 Sharp indenters

We alluded in the previous subsection to an increasing elastic contact pressure with diminishing indenter radius. In the sharp-point limit, (8.2) implies a stress singularity ($r \to 0$, $p_0 \to \infty$, at $a = $ const). Physically, the singularity is averted by irreversible (plastic) deformation beneath the indenter point until the contact is large enough to support the load.

Sharp indenters like the Vickers or Knoop diamond pyramids used in hardness testing produce two basic types of crack pattern: *radial–median* and *lateral* (Lawn & Wilshaw 1975). Fig. 8.7(a) depicts the evolution of these crack systems: (i) the sharp point induces inelastic, irreversible deformation; (ii) at a critical load one or more nascent flaws within the deformation zone become unstable, and pop-in to form subsurface radial

(a)

(b)

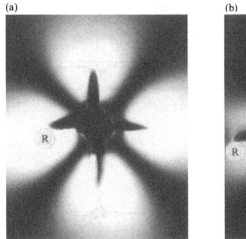
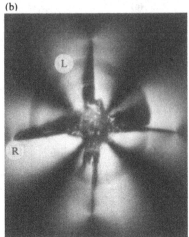

Fig. 8.8. Vickers indentations in soda-lime glass, showing evolution of radial (R) and lateral (L) crack systems as viewed in polarised, reflected light from below indenter during load cycle, at (a) full load ($P = 90$ N) and (b) full unload ($P = 0$). Note expansion of both crack systems as indenter is withdrawn, and strong remnant stress birefringence at completion of cycle, indicating substantial residual component of contact field. Width of field 700 μm. (After Marshall, D. B. & Lawn, B. R. (1979) *J. Mater. Sci.* **14** 2001.)

cracks on tensile median planes, i.e. planes containing the load axis (and, usually, some line of stress concentration, e.g. impression diagonal or cleavage-plane trace in the specimen surface); (iii) on increased loading, the crack propagates incrementally downward; (iv) on unloading, the median cracks close up *below* the surface but simultaneously open up in the residual tensile field *at* the surface as the contact recovers its elastic component; (v) just prior to removal of the indenter the residual field becomes dominant, further expanding the surface radials and initiating a second system of sideways spreading, saucer-like lateral cracks near the base of the deformation zone; (vi) the expansion continues until indenter removal is complete, both crack systems ultimately tending to half-pennies centred about the load point.

These sequential features in the crack evolution are confirmed by the micrographs of a Vickers indentation in soda-lime glass in figs. 8.8 and 8.9. Fig. 8.8 is a subsurface view of the contact in polarised light. The continued extension of the radial and lateral cracks during unloading and the persistence of the birefringence 'Maltese cross' in the glass specimen testify to the importance of residual contact stresses. Fig. 8.9, a view of the

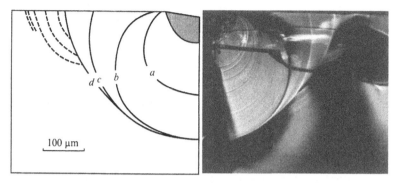

Fig. 8.9. Same indentation as fig. 8.8, showing view of median fracture plane in soda-lime glass after breaking specimen. Half-penny geometry of fully formed crack is apparent. Solid tracings at left delineate arrest markers from minor perturbations in the indentation loading cycle: (*a*) half load, (*b*) full load (cf. fig. 8.8(a)); (*c*) half unload; (*d*) full unload (cf. fig. 8.8b)). Dashed tracings delineate post-indentation markers in subsequent flexural loading, indicating a surface failure origin. Reflected light. (After Marshall, D. B. & Lawn, B. R. (1979) *J. Mater. Sci.* **14** 2001.)

median crack plane for the same indentation after specimen fracture, demonstrates the surface confinement of radial expansion during unloading and the ultimate penny-like geometry of the final configuration.

We acknowledge that the sequence of events in fig. 8.7 is ideal. Geometrical deviations occur in many materials (Cook & Pharr 1990). At light loads especially, the median stages (ii) and (iii) in fig. 8.7 may be suppressed, such that only surface radial ('Palmqvist') segments form. Much has been made of such deviants in the literature. We assert that even such partially formed cracks, insofar as they are centred about the contact point and tend subsequently to expand rapidly into the full median geometry in any post-indentation growth, retain the essential half-penny character.

Now examine the mechanics of contact. For rigid, fixed-profile pyramid (Vickers or Knoop) indenters where geometrical similarity prevails, we have an elastic–plastic contact of contact pressure

$$p_0 = P/\alpha_0 a^2 = H \qquad (8.4)$$

where H defines the 'indentation hardness'. For ideally homogeneous surfaces hardness is a material constant, so p_0 in (8.4) (in contrast to (8.2) for elastic contacts) is size-invariant.

An appropriate stress-intensity factor for the ideal radial–median

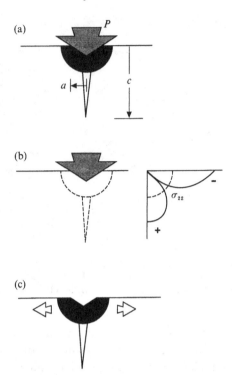

Fig. 8.10. Radial–median crack system. (a) Elastic–plastic configuration at full load, composed of superposable (b) reversible (elastic) in-plane (σ_{22}) stress and (c) residual (plastic) centre-opening force components. (After Lawn, B. R., Evans, A. G. & Marshall, D. B. (1980) *J. Amer. Ceram. Soc.* **63** 574.)

configuration may once more be written down from (2.23b) for centre-loaded penny-like cracks,

$$K_R = \chi P/c^{3/2}, \quad (P > P_C, \, c \gg a) \tag{8.5}$$

with χ a dimensionless constant. Again, writing $K_R = K_C = T_0$ at $c = c_I$, we obtain the (stable) equilibrium relation $P \propto c_I^{3/2}$, borne out by the data for glass in fig. 8.4(b).

However, if (8.5) has the same form as its cone-crack counterpart (8.3), the route to final equilibrium, and thus the physical significance of the coefficient χ, is totally different. The sharp-contact configuration may be analysed as an expanding cavity, fig. 8.10. A pressurised internal spherical volume (hardness impression) induces deformation in an annular surround region (deformation zone), the whole being constrained in an elastic matrix. The composite field at maximum load in (a) is the superposition of two components, elastic in (b) and residual in (c) (Lawn, Evans & Marshall 1980):

(i) *Residual component.* At full unload, configuration (c), a residual field arises from accommodation of the impression volume by expansion of the deformation zone against the constraining elastic matrix. This field is hoop-tensile and centre-symmetrical, in keeping with the final half-penny crack profile in fig. 8.9. An appropriate analysis of the accommodation gives

$$\chi = \xi_0 (\cot \Phi)^{2/3} (E/H)^{1/2} \qquad (8.6)$$

with Φ the indenter half-angle. The dimensionless constant ξ_0 in this relation depends on the nature of the deformation. It is smaller for volume-consuming ('anomalous') processes than for volume-conserving ('normal') processes, since in the former the indentation strain can be relaxed by compaction as well as by elastic compression within the deformation zone.

(ii) *Elastic component.* Because subsequent reloading traces the unload path, the field at full load may be reconstructed by superposition of an elastic field, configuration (b), onto the residual field. At the median plane this elastic field is approximated by the σ_{22} stress distribution in fig. 8.1, i.e. tensile below the surface and compressive near the surface. Thus a prevailing compressive restraint at maximum impression is effectively released by unloading, allowing the radials to complete their expansion in the surface region of fig. 8.9.

It is now apparent that the stress-intensity factor (8.5) relates specifically to the *residual* stress field at complete unload, not to the composite elastic–plastic field at full load. The quantity H/E in (8.6) emerges as an important parameter for elastic–plastic contacts; H quantifies the loading half-cycle, E the unloading half-cycle. Higher H/E (e.g. 'hard' ceramics) corresponds to a lower residual field intensity χ, implying a stronger elastic recovery and thus more pronounced radial extension during the unloading half-cycle.

Relations analogous to (8.5) and (8.6) may be derived for lateral cracks, although analysis is more demanding owing to interactions with the adjacent specimen free surface. The lateral system in turn interacts with the

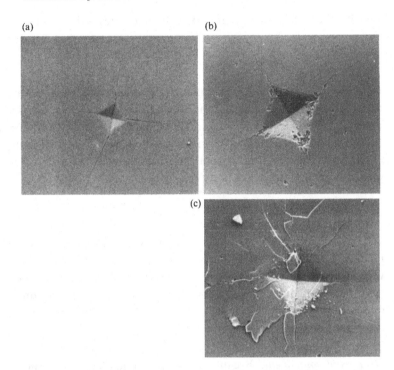

Fig. 8.11. Vickers indentations in alumina: (a) monocrystal (sapphire); (b) polycrystal, grain size 3 μm; (c) polycrystal, grain size 20 μm. Scanning electron micrographs. Width of field 175 μm. (After Anstis, G. R., Chantikul, P., Marshall, D. B. & Lawn, B. R. (1981) *J. Amer. Ceram. Soc.* **64** 533.)

radial system which precedes it, via residual-stress relaxation or by geometrical modification, as manifested, for example, by a reduction of χ in (8.5).

Additional facets of the Vickers crack system are illustrated in figs. 8.11 and 8.12. In fig. 8.11 we show three aluminas to demonstrate the influence of microstructure. Well-defined radial cracks are apparent in the mono-crystalline sapphire in (a) and the fine-grained polycrystalline alumina in (b). The coarse-grained polycrystalline alumina in (c) reveals irregular intergranular cracking. In fig. 8.12 we observe cracks in soda-lime glass at immediate completion of the contact cycle and after prolonged exposure to moist air to illustrate the pervasive influence of the residual contact field. The radials expand according to some velocity function $v = v(K_R)$, first rapidly then slowly as K_R diminishes with increasing c in (8.5), ultimately coming to rest as the system reaches threshold on the v–K curve.

(a) (b)

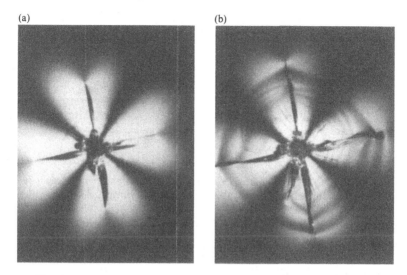

Fig. 8.12. Similar indentation to fig. 8.8, showing Vickers crack pattern in soda-lime glass (a) immediately on completion of indentation, (b) one hour later after exposure to moist atmosphere. Width of field 1000 μm. (Courtesy D. B. Marshall.)

8.2 Indentation cracks as controlled flaws: inert strength, toughness, and *T*-curves

The strength of a brittle solid subject to an applied tensile field is determined by the most severe flaw in the surface or bulk. By placing a *controlled* indentation flaw in the test surface one can predetermine the failure origin, and thus trace the micromechanics of flaw evolution. Some may argue that indentation flaws are 'artificial', and as such can not be representative of true strength properties. But, as we shall see in chapter 9, indentations do in fact contain many vital elements of 'natural' flaws, including local residual stress states. Moreover, since the flaw geometry is well-defined, indentation analysis enables us to decouple the material aspects from the geometrical aspects of the strength characteristic.

Consider an indentation half-penny crack, radius *c*, normal to an applied tensile stress, σ_A, as in fig. 8.13. The applied *K*-field from (2.20) is

$$K_A = \psi \sigma_A c^{1/2}.$$

In the most general case of an indentation with residual field and material

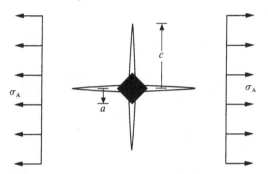

Fig. 8.13. Schematic of (Vickers) indentation half-penny crack, radius c and impression half-diagonal a, in applied tensile field σ_A.

with toughness-curve (T-curve), the net K-field $K_* = K_*(P, \sigma_A, c)$ at the crack tip is

$$K_* = K_A + K_R + K_\mu \qquad (8.7)$$

with $K_R = K_R(P, c)$ from the previous section and $K_\mu = -T_\mu(c,$ micro-structure) from chapter 7.

In this section we determine the failure stress for as-indented surfaces under essentially *inert* conditions; i.e. non-reactive environment and fast stressing rate, such that kinetic growth is negligible. One solves (8.7) for the 'inert strength' by imposing a state of equilibrium ($K_* = K_C = T_0$) and seeking the condition for unlimited unstable propagation ($\mathrm{d}K_*/\mathrm{d}c \geq 0$). Initially, we confine our attention to materials with single-valued toughness, e.g. glasses, monocrystals and fine-grained polycrystals. We then indicate how indentation-strength data may be inverted to evaluate the toughness of such materials. Finally, we extend the analysis to materials with T-curves.

It is understood here that the flaws are well-developed, i.e. the indentations are in the postthreshold domain. Failure from subthreshold flaws is addressed in sect. 8.5.

8.2.1 Inert strength

Now examine the mechanics of failure for blunt and sharp indenters in a material of constant toughness ($T = T_0$):

(i) *Blunt indenters*, cone cracks. In the absence of residual contact field

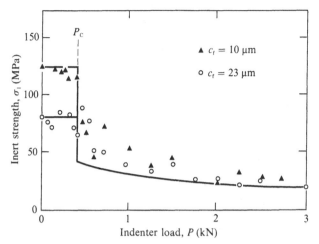

Fig. 8.14. Inert strength as function of load for blunt indenters (sphere radius $r = 1.6$ mm) on soda-lime glass. Surface pre-abraded with silicon carbide grit to produce flaws, size c_f. Curves are predictions from (8.8b). Note insensitivity of data to c_f above initiation threshold. (After Lawn, B. R., Wiederhorn, S. M. & Johnson, H. H. (1975) *J. Amer. Ceram. Soc.* **58** 428.)

($K_R = 0$) and microstructural shielding ($K_\mu = 0$), (8.7) reduces to $K_* = K_A$. Failure is *spontaneous* in the tradition of Griffith-like flaws at the post-indentation configuration $c = c_I$ (from (8.3) at $K_P = T_0$), $\sigma_A = \sigma_F = \sigma_I$:

$$c_I = (\chi P/T_0)^{2/3} \tag{8.8a}$$

$$\sigma_I = T_0/\psi c_I^{1/2}$$
$$= (T_0^4/\psi^3\chi P)^{1/3}. \tag{8.8b}$$

Inert strength data for soda-lime glass are plotted as a function of Hertzian contact load in fig. 8.14. Below threshold no cone cracks initiate (sects. 8.4, 8.5) and the strength is controlled by the pre-present flaw population, in this case abrasion flaws. Above threshold, flaw size has an insignificant influence, consistent with a far-field response. The curves represent the predictions of (8.8b), at $P < P_C$ using $c_I = c_f$, at $P > P_C$ using predetermined coefficients ψ (from theoretical analysis of the cone-crack geometry) and χ (experimental calibration from fig. 8.4(a)).

(ii) *Sharp indenters*, radial–median cracks. Residual stress is now present, so (8.7) gives $K_* = K_A + K_R$ for zero microstructural shielding ($K_\mu = 0$), with $K_R(c) = \chi P/c^{3/2}$ from (8.5). The function $K_*(c)$ is plotted in fig. 8.15:

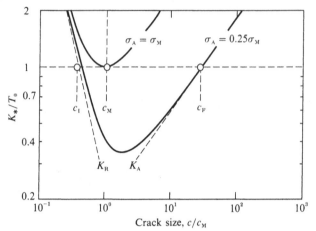

Fig. 8.15. Reduced plot of $K_*(c)$ for sharp indenters (zero microstructural shielding). Plots shown for two levels of uniform applied stress, σ_A. Inclined dashed lines are small-crack $K_R(c)$ and large-crack $K_A(c)$ asymptotic limits at $\sigma_A = 0.25\sigma_M$. Effect of increasing σ_A to σ_M is to propagate crack at stable equilibrium along $K_* = T_0$ from c_I to activated failure at c_M. (Cf. fig. 2.17.)

$K_R(c)$ dominates at small c (stable branch), $K_A(c)$ at large c (unstable branch). Failure is now *activated*, with precursor equilibrium growth ($K_* = T_0$) from immediate stable post-indentation state $c = c_I$ at $\sigma_A = 0$ ($\mathrm{d}K_*/\mathrm{d}c < 0$) to final instability $c = c_M$ at $\sigma_A = \sigma_F = \sigma_M$ ($\mathrm{d}K_*/\mathrm{d}c = 0$):

$$c_M = (4\chi P/T_0)^{2/3} \tag{8.9a}$$

$$\sigma_M = 3T_0/4\psi c_M^{1/2}$$
$$= \tfrac{3}{4}(T_0^4/4\psi^3\chi P)^{1/3}. \tag{8.9b}$$

This precursor stage attests to the potentially stabilising influence of extraneous forces in the flaw micromechanics. For the indentation flaw it is the consummate signature of the residual contact field.

Direct observations of precursor crack extension from indentations have been made in several brittle materials. Fig. 8.16 is an example, for Knoop indentations in silicon nitride specimens subjected to flexure. (The silicon nitride used in these studies was immune to environmental interaction, so the extension could not be attributed to spurious kinetic growth.) Control studies on similar specimens after a post-indentation anneal to remove the residual stresses show no such precursor growth, in which case the failure conditions revert to those of (8.8).

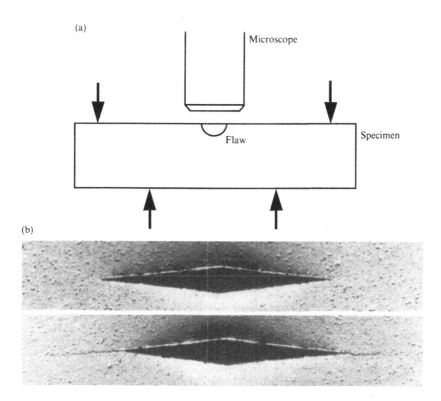

Fig. 8.16. Growth of radial cracks at indentations under applied stress. (a) Schematic of *in situ* arrangement for viewing crack evolution. (b) Micrographs showing Knoop indentation in hot-pressed silicon nitride before (above) and after (below) application of stress. Width of field 400 μm. (Courtesy D. B. Marshall.)

An alternative description of the 'activated' failure state is obtained by inserting $K_A = \psi\sigma_A c^{1/2}$ into the equilibrium relation $K_*(P, \sigma_A, c) = T_0$ in (8.7) and solving for the applied stress:

$$\sigma_A(c) = (1/\psi c^{1/2})[T_0 - K_R(c)]. \tag{8.10}$$

This function is sketched for as-indented surfaces ($K_R = \chi P/c^{3/2}$), as the lower curve in fig. 8.17. Stable extension of amount $c_M/c_I = 4^{2/3} \simeq 2.5$ (cf. (8.8a) and (8.9a)) occurs along path 1 over an 'energy barrier'. The $\sigma_A(c)$ data for silicon nitride in fig. 8.18, curve-fitted to (8.10) by adjusting ψ and χ, confirm the predicted response. Again, annealing or polishing away the

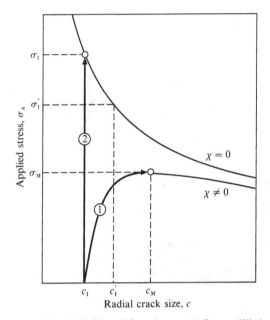

Fig. 8.17. Plot of function $\sigma_A(c)$ for equilibrium radial cracks *with* ($\chi \neq 0$) and *without* ($\chi = 0$) residual contact stress field. Path ① represents *activated* failure at σ_M, path ② *spontaneous* failure at σ_I. Post-indentation slow crack growth from c_I to c_I' diminishes spontaneous strength from σ_I to σ_I', but leaves activated strength σ_M unaffected.

deformation zone prior to the strength test ($K_R = 0$) causes $\sigma_A(c)$ to revert to the upper curve for Griffith-like flaws in fig. 8.17. Introduction of the residual contact term thereby corresponds to a strength reduction $\sigma_M/\sigma_I = \frac{3}{4}(c_I/c_M)^{1/2} \simeq 0.47$ (cf. (8.8b) and (8.9b)). If environmentally assisted post-indentation growth were to occur, from c_I to c_I' say, σ_M would be unaffected (other than through any relaxation in χ or ψ), but σ_I would be reduced to σ_I'.

It is instructive to re-interpret the strength formulation (8.10) in terms of the K-fields perceived by different observers (sect. 3.6). We do this in figs. 8.19(a), (b), (c) using data on soda-lime glass at two indentation loads:

(a) The $\sigma_A(c)$ data are plotted directly, and are again curve-fitted by adjusting ψ and χ.

(b) These same data are replotted from the perspective of a global K-field observer, i.e. as $K_A(c) = \psi\sigma_A c^{1/2} = T'(c)$, say. The dashed lines show $K_A(c)$ at the tangency points $\sigma_A = \sigma_M$. The solid curves represent the *pseudo*

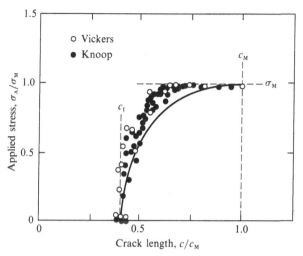

Fig. 8.18. Reduced plot showing stable crack extension from indentation flaws during stressing to failure in silicon nitride (same material as fig. 8.16). Data are experimental observations and solid curve is theoretical fit. (Courtesy D. B. Marshall.)

T-curve $T'(c) = T_0 + T_R(c)$, with $K_R = -T_R$ an *anti*-shielding term. Since $K_R(c)$ in (8.5) is a positive, ever-diminishing function, T' approaches T_0 asymptotically at large c.

(c) The data are now replotted from the perspective of a crack-tip enclave observer, i.e. directly as $K_*(c) = K_A(c) + K_R(c)$. The dashed curves are the computed functions $K_*(P, \sigma_M, c)$ at the prescribed P and for the tangency stresses $\sigma_A = \sigma_M$. The equilibrium path of stable radial crack growth at $K_* = T_0 = $ const is most apparent in this last case.

Generally, it is most convenient to plot indentation-strength data in the functional form $\sigma_M(P)$, in accordance with (8.9b), using the independent variable P to characterise the flaw severity. Fig. 8.20 shows such a plot for three relatively homogeneous brittle solids. The data cut off abruptly at low load, where failure occurs from natural rather than indentation flaws (cf. fig. 8.14). Apparent satisfaction of the $P^{-1/3}$ dependence in (8.9b) for the data sets may be taken as validating the assumption of single-valued toughness.

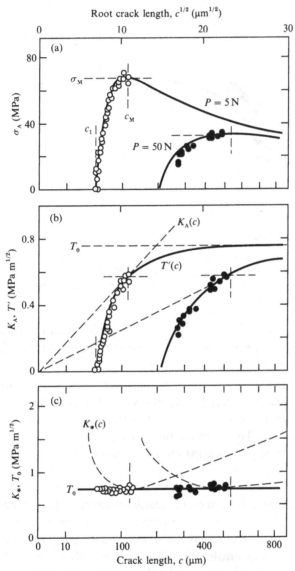

Fig. 8.19. Crack extension from Vickers flaws in soda-lime glass during application of applied stress to failure, inert test conditions: (a) stress vs crack size, at two indentation loads; (b) equivalent pseudo T-curves, $K_A(c) = T'(c)$; (c) equivalent $K_*(c) = T_0$ representation, after subtracting $K_R(c)$ from $T'(c)$. (After Mai, Y-W. & Lawn, B. R. (1986) *Ann. Rev. Mater. Sci.* **16** 415. Data courtesy D. B. Marshall.)

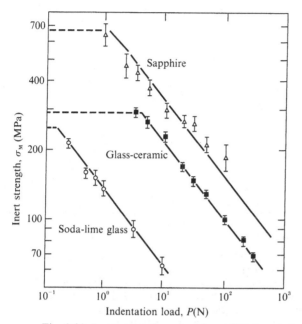

Fig. 8.20. Inert strength as function of Vickers indentation load for soda-lime glass, fine-grained cordierite glass-ceramic and monocrystal sapphire specimens with polished surfaces. All data points are failures from indentation sites. Error bars are standard deviation limits. Solid lines are fits with slope $-\frac{1}{3}$ in logarithmic coordinates. Dashed lines are natural flaw cutoffs. (Data courtesy R. F. Cook & D. B. Marshall.)

8.2.2 Toughness

Indentation provides a simple route to toughness evaluation. Sharp indenters are usually favoured because the tests can be carried out on routine hardness testing equipment. We discuss two procedures relative to independent long-crack toughness measurements in fig. 8.21 for selected glasses and fine-grained polycrystalline ceramics:

(i) *Direct measurement.* Immediate post-indentation radial crack size is measured as a function of load (e.g. fig. 8.4(b)) in the absence of applied stress ($\sigma_A = 0$) in an inert environment. Writing $K_R = T_0$ at $c = c_I$ in (8.5), together with $\xi = \xi_0(\cot \Phi)^{2/3}$ in (8.6), we obtain

$$T_0 = \xi(E/H)^{1/2}P/c_I^{3/2}. \tag{8.11a}$$

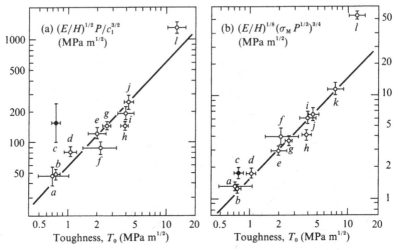

Fig. 8.21. Vickers indentation data illustrating toughness formulas (8.11) for (a) direct crack-size measurement and (b) indentation-strength methods. Vertical axis indicates indentation test variables, horizontal axis independently determined long-crack toughness. Selected glasses and fine-grained ceramics: *a*. silicon; *b*. soda-lime glass; *c*. fused silica ('anomalous'); *d*. barium titanate; *e*. silicon nitride (reaction-bonded); *f*. sapphire; *g*. cordierite glass-ceramic; *h*. alumina; *i*. silicon carbide; *j*. silicon nitride (hot-pressed); *k*. zirconia; *l*. tungsten carbide. (After Anstis, G. R., Chantikul, P., Lawn, B. R. & Marshall, D. B. *J. Amer. Ceram. Soc.* **64** 533, 539. With additional data courtesy R. F. Cook.)

Advantages: Extreme simplicity and economy, many indentations per surface.

Disadvantages: Often difficult to measure crack size, even on polished surfaces, at small and (in coarse-grained polycrystals) ill-defined indents; postthreshold crack growth leads to overestimates in c_I, underestimates in T_0; moderately dependent on H/E, which must be specified a priori; dependent on geometrical parameter ξ_0, which is sensitive to deformation mode (sect. 8.4.2 – witness strong deviation from fitted line in fig. 8.21(a) by fused silica, which deforms anomalously by compaction).

(ii) *Indentation-strength.* Inert strength is measured as a function of indentation load (e.g. fig. 8.20). Inverting (8.9b) and (8.6) and writing $\eta = (256\psi^3\xi^{2/3}/27)^{1/4}$, we have

$$T_0 = \eta(E/H)^{1/8}(\sigma_M P^{1/3})^{3/4}. \qquad (8.11b)$$

Advantages: No measurement of crack size required – radial system does not even have to be well-formed (e.g. fig. 8.11(c)); weak dependence on H/E, as well as on ψ and ξ_0 (note relatively small deviation from linear fit in fig. 8.21(b) by fused silica).

Disadvantages: Requires fabrication of flexure specimens; one strength data point per specimen.

The fitted solid lines in fig. 8.21 yield coefficients $\xi = 0.016$ and $\eta = 0.62$ in (8.11).

Inevitably, the method has its variants. One is a 'hybrid', entailing measurement of *both* crack size c_M and applied stress σ_M at instability: then, from (8.9), $T_0 = \frac{4}{3}\psi\sigma_M c_M^{1/2}$. This route is experimentally more demanding, but eliminates ξ_0 from the formulation, thereby avoiding any dependence on deformation mode. Another variant involves removal of the residual contact stresses, by annealing or polishing, and then determining crack size c_I and strength σ_I at (spontaneous) failure: from (8.10) at $K_R = 0$, $T_0 = \psi\sigma_I c_I^{1/2}$. That method relies on a first-principles analysis of the geometrical term ψ for surface half-pennies. Its drawback is the necessarily tedious specimen preparation (which can, moreover, alter the material microstructure). Also, computation of ψ is subject to considerable uncertainty; most attempts ignore interactions with co-existing orthogonal radial or lateral cracks.

As a means of measuring absolute toughness, indentation techniques are simple, but limited. From fig. 8.21 we see that systematic departures from the calibrated curves in excess of a factor of two may occur for the worst behaved materials. Relative measurements on a given material are typically reliable to much better than 20%.

8.2.3 Toughness-curves

So far we have investigated only materials with single-valued toughness. Many ceramics, especially those with coarse microstructures (e.g. fig. 8.11(c)), show significant *T*-curves (chapter 7). We may extend the indentation fracture analysis to such materials by incorporating a shielding term $-K_\mu = T_\mu$ into the mechanics. It is convenient to transpose (8.7) at $K_* = T_0$,

$$K_A'(c) = K_A(c) + K_R(c)$$
$$= T_0 + T_\mu(c) = T(c) \tag{8.12}$$

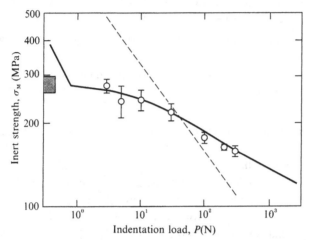

Fig. 8.22. Inert strength results for Vickers indentations in polycrystalline alumina, grain size 35 μm, as function of load. Data points represent confirmed failures from indentation sites. Error bars designate standard deviation limits. Shaded area at left indicates failures from natural flaws. Solid curve is theoretical fit to alumina data. Dashed line is comparative $\sigma_M \propto P^{-1/3}$ fit to monocrystal sapphire (fig. 8.20). Relative insensitivity of alumina data to load is measure of 'flaw tolerance'. (After Mai, Y-W. & Lawn, B. R. (1986) *Ann. Rev. Mater. Sci.* **16** 415. Data and analysis courtesy S. Lathabai & S. J. Bennison.)

so that K'_A is an *effective* applied K-field. (This is the relation pertinent to an observer between the indentation corner and crack tip.) Failure at $c = c_M$, $\sigma_A = \sigma_M$ is then determined from the instability requirement $dK'_A/dc = dT/dc$ in (3.33), with proper allowance for pseudo-failure states (see below). Generally, except for the simplest $K_\mu(c)$ functions, solving for $\sigma_M(P)$ in analogy to (8.9b) requires numerical analysis.

Indentation-strength results for a coarse-grained alumina with shielding by crack-interface *bridging* are plotted in fig. 8.22. Data points represent confirmed failures from indentation flaws, the shaded band at left failures from natural flaws. The solid curve is a fit obtained by adjusting microstructural parameters in $T_\mu(c)$ from sect. 7.5. The dashed $\sigma_M \propto P^{-1/3}$ line from a fit of (8.9b) to data for monocrystalline alumina (sapphire) establishes a comparative baseline. Marked departures from that baseline response are evident for the polycrystalline material. At decreasing P the

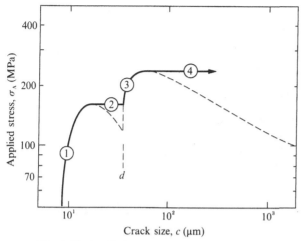

Fig. 8.23. Applied stress vs crack size for Vickers indentation in alumina, using parametric evaluations from fig. 8.22. Calculated for load $P = 2$ N. Note multiple stable and unstable branches of the equilibrium curve along path 1→2→3→4 to failure. (After Mai, Y-W. & Lawn, B. R. (1986) *Ann. Rev. Mater. Sci.* **16** 415. Data and analysis courtesy S. Lathabai & S. J. Bennison.)

data tend strongly to a plateau, asymptotic to the strength for failures from natural flaws. Herein lies the quality of *flaw tolerance*.

With $-K_\mu = T_\mu$ thus 'calibrated' from the alumina indentation data one may regenerate the function $\sigma_A(c)$ from (8.12) (e.g. by replacing T_0 with $T(c)$ in (8.10)). This function is plotted in fig. 8.23, at a load well within the plateau region (Mai & Lawn 1986). *Two* stabilising branches are now apparent (cf. fig. 8.17), one on either side of the first bridge activation at $c = d$ (fig. 7.27): that at $c < d$ relates to a dominant $K_R(c)$ function (short-range, positive diminishing); that at $c > d$ to a dominant $K_\mu(c)$ function (long-range, negative increasing). Path 1→2→3→4 traces the evolution to failure. The first instability at segment 2 corresponds to pop-in from the indentation flaw, in accordance with the observations reported in sect. 7.5. The second instability at segment 4 denotes final failure at σ_M. It is the diminished sensitivity to the load-dependent K_R term (which in turn controls the initial crack size at $\sigma_A = 0$ in fig. 8.23) in the vicinity of this latter barrier that ultimately accounts for the flaw tolerance in fig. 8.22.

The calibrated toughness function for a specific material may now be used to predict the effect of microstructural variation, e.g. grain size *l*, on the strength properties. We have already demonstrated the influence of *l* on the *T*-curve for alumina in fig. 7.29. The corresponding influence of *l*

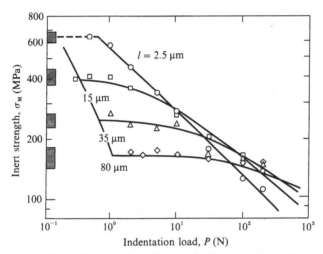

Fig. 8.24. Inert strength data for Vickers-indented aluminas of four grain sizes, *l*. All data points represent breaks from indentation flaws (error bars omitted for clarity). Shaded areas at left indicate failures from natural flaws. Curves are predictions of $\sigma_M(P)$ at each *l* using calibrated *T*-curve function. (After Chantikul, P., Bennison, S. J. & Lawn, B. R. (1990) *J. Amer. Ceram. Soc.* **73** 2419.)

on indentation-strength is plotted in fig. 8.24, together with experimental data. The results demonstrate a strongly enhanced flaw tolerance in the short-crack domain at coarser microstructures, albeit at the expense of decreased plateau strength, with consequent crossover in the curves in the large-crack domain. The crossover again demonstrates the need for extreme caution in using large-scale crack data to predict strength characteristics in polycrystalline ceramics.

8.3 Indentation cracks as controlled flaws: time-dependent strength and fatigue

8.3.1 Time-dependent strength

Strength is generally reduced in reactive environments by *kinetic* crack growth (chapter 5), the more so the slower the stressing rate. As with inert conditions, it is convenient to investigate the strength characteristics using controlled indentation flaws.

To set up the fracture kinetics for indentation flaws it is necessary only to replace the equilibrium condition for extension in the previous section with a suitable $v(K_*)$ or $v(G_*)$ velocity relation,

$$v = dc(t)/dt$$
$$= v\{K_*[P, \sigma_A(t), c(t)]\}. \tag{8.13}$$

For specified contact load P and stress–time function $\sigma_A(t)$, (8.13) reduces to a differential equation in $c(t)$. The most practical stressing configuration is $\dot{\sigma}_A = $ const, e.g. flexure tests at constant crosshead speed. One solves (8.13) by integrating c incrementally over t from initial stable equilibrium $(dK_*/dc < 0)$ at $t = 0, \sigma_A = 0, c = c_I$ (or c_I' in the event of post-indentation growth), to final unstable equilibrium $(dK_*/dc > 0)$ at $t = t_F, c = c_F$ (e.g. fig. 8.15), so determining the failure stress $\sigma_A = \sigma_F = \sigma_F(t_F)$.

Analytical solutions of (8.13) are available only in very special cases. One such is that of a power-law crack velocity function, $v = v_0(K_*/T_0)^n$ (cf. (5.22)), and material of single-valued toughness, $K_* = K_A + K_R$. Then at fixed indentation load and constant stressing rate one obtains a relation of the form (Marshall & Lawn 1980)

$$\sigma_F = A'(n'+1)\dot{\sigma}_A^{1/(n'+1)} \tag{8.14}$$

with n' for post-indentation-annealed $(\psi = 0)$ and as-indented $(\psi \neq 0)$ surfaces

$$n' = n, \quad \text{(annealed)} \tag{8.15a}$$

$$n' = \tfrac{3}{4}n + \tfrac{1}{2}, \quad \text{(as-indented)} \tag{8.15b}$$

and coefficient $A' = v_0 f(n', P, T_0)$. The empirical velocity curve may then be deconvoluted from appropriate data fits to (8.14).

Strength vs stressing-rate data in fig. 8.25 confirm the adequacy of (8.14) for Vickers-indented soda-lime glass in water. The slope of a linear fit over the data range for post-indentation-annealed specimens corresponds to $n = 17.9$ in (8.14) and (8.15a) (in agreement, within experimental scatter, with evaluations from long-crack soda-lime glass data in region I, e.g. fig. 5.11). The slope for as-indented specimens corresponds to $n' = 13.7$ in (8.14), which compares with $n' = 13.9$ from (8.15b). The parameter v_0 may similarly be evaluated from the A' intercepts. With these calibrations we regenerate (numerically) from (8.13) the solid curves in fig. 8.25 for high velocity cutoff at $K_* = T_0$, with corresponding asymptotic inert strength limits at high $\dot{\sigma}_A$. Enhanced fatigue susceptibility is clearly demonstrated

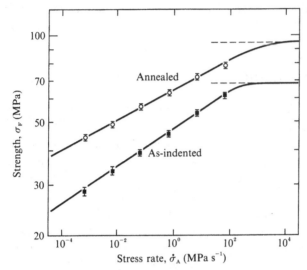

Fig. 8.25. Strength of Vickers-indented soda-lime glass in water, as function of stress rate. Data at load $P = 5$ N, for as-indented and post-indentation-annealed specimens. Error bars are standard deviation limits. Dashed lines indicate inert strength levels. Curves are regenerated solutions of (8.13), using n and v_0 by best-fitted (8.14) to data in linear region. (After Marshall, D. B. & Lawn, B. R. (1980) *J. Amer. Ceram. Soc.* **63** 532.)

for the flaws with local residual tensile stresses. This enhanced susceptibility is also apparent as an expanded pre-failure 'slow' growth region in the lower micrograph of fig. 8.26.

For most material–environment systems, however, the assumptions of power-law velocity function and single-valued toughness are unduly restrictive. Generally, we must adopt a more fundamental velocity function, to allow for a threshold (e.g. figs. 5.7–5.10), and incorporate a shielding term $-K_\mu = T_\mu$ into (8.13), $K_* = K_A + K_R + K_\mu$, to allow for T-curve behaviour (sect. 8.2.3). The cost of such elaboration is the necessity to resort totally to numerical solution of (8.13).

Accordingly, consider the data for constant stressing rate tests on a flaw-tolerant alumina in water, fig. 8.27. The solid curve is regenerated from (8.13) by adjusting parameters in the hyperbolic sine v–G_* function (5.27) for region I with region III cutoff, in accordance with the previously 'calibrated' bridging K-field term $K_\mu(c)$ from sect. 8.2.3. Note the tendency for the curve to a lower bound in strength at low stressing rates, in addition to the more familiar upper bound inert strength at high rates. From these fits one may thereby deconvolute the unique crack-tip *enclave* velocity

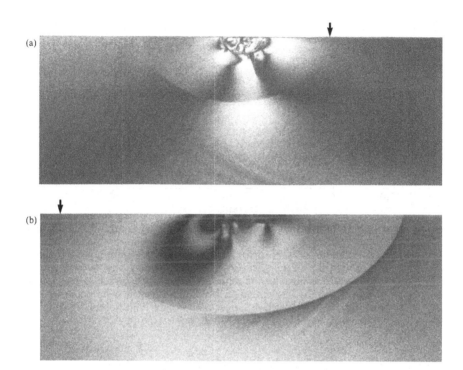

Fig. 8.26. Fracture surfaces of Vickers-indented ($P = 5$ N) soda-lime glass broken in water at a common stress rate ($\dot{\sigma}_A = 0.15$ MPa s^{-1}), (a) post-indentation annealed, (b) as-indented. Arrows denote failure origins at critical stress. Note more extensive pre-failure slow growth in the as-indented specimen. Width of field 700 μm. (After Marshall, D. B. & Lawn, B. R. (1980) *J. Amer. Ceram. Soc.* **63** 532.)

curve $v–G_*$ for the alumina (with threshold at $G_* = R_E = 2\gamma_{BE} - \gamma_{GB}$ for intergranular fracture, sect. 7.1). Recall from sect. 5.3 that the *global* $v–G_A$ curves are *non*-unique, i.e. are history dependent, as demonstrated in fig. 5.15. The latter curves displace further to the right on a crack velocity diagram at larger starting crack size and slower stressing rate, corresponding to larger pre-failure crack extensions and hence further progressions up the T-curve.

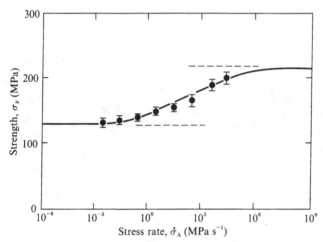

Fig. 8.27. Strength of Vickers-indented alumina of grain size 35 μm in water, as function of stress rate. Material is same as that in fig. 8.22. Data at load $P = 30$ N. Error bars are standard deviation limits. Curve is regenerated solution of (8.13), using microstructural K-field calibrated from fig. 8.22 and adjusted parameters in crack velocity function (5.18) by best-fitting (8.14) to data in linear region. Dashed lines indicate inert strength (upper) and fatigue limit (lower). (After Lathabai, S. & Lawn, B. R. (1989) *J. Mater. Sci.* **24**, 4298.)

8.3.2 Fatigue

Now let us explore how the results of the preceding subsection relate to fatigue. 'Fatigue' here implies a finite time to failure under any sustained stress, static or cyclic. Where fatigue is attributable solely to slow crack growth, one may integrate (8.13) over an appropriate v–K_* (or v–G_*) function (either directly determined, sect. 5.4, or deconvoluted from stressing-rate data, sect. 8.3.1) to predict 'lifetime' t_F at prescribed $\sigma_A(t)$.

Results of tests at $\sigma_A = $ const (static fatigue) on the same Vickers-indented soda-lime glass/water system as in fig. 8.25 are shown in fig. 8.28. Again, assuming a velocity function $v = v_0(K_*/T_0)^n$ and single-valued toughness, the solution of (8.13) at $\sigma_A = $ const is

$$t_F = A'/\sigma_A^{n'} \tag{8.16}$$

with n' and A' defined as in (8.14) and (8.15). The solid curves in fig. 8.28 are regenerated from (8.13) using the calibrated v–K_* function from fig. 8.25. Observe that annealing out the residual contact stresses increases the lifetime at any given applied stress by some three orders of magnitude.

Now consider the static fatigue data in fig. 8.29 for the same Vickers-

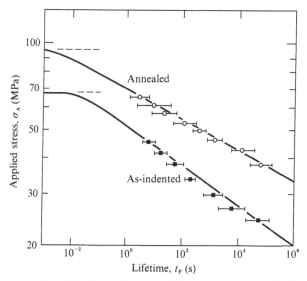

Fig. 8.28. Static fatigue of Vickers-indented soda-lime glass in water (same system as fig. 8.25). Data at load $P = 5$ N, for as-indented and post-indentation-annealed specimens. Error bars are standard deviation limits, dashed lines inert strength levels. Curves are regenerated solutions from (8.13). (After Chantikul, P., Lawn, B. R. & Marshall, D. B. (1981) *J. Amer. Ceram. Soc.* **64** 322.)

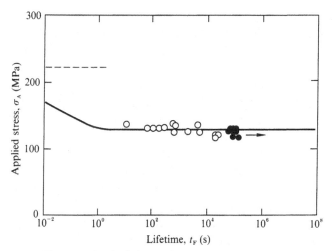

Fig. 8.29. Static fatigue of Vickers-indented ($P = 30$ N) alumina in water (same system as fig. 8.27). Data represent individual specimens: open symbols denote delayed failures, filled symbols survivors. Curve is regenerated solution from (8.13), dashed line is inert strength level. System is close to fatigue limit. (After Lathabai, S. & Lawn, B. R. (1989) *J. Mater. Sci.* **24** 4298.)

indentation alumina/water system as in fig. 8.27. Again, the solid curve is a prediction from (8.13) using the $v-K_*$ ($v-G_*$) function deconvoluted from fig. 8.27, along with the calibrated $K_\mu(c)$ function for alumina from sect. 8.2.3. The most striking feature of fig. 8.29 is the flatness of the curve at $t_F > 1$ s. This is indicative of a *fatigue limit*, marking a rapid approach to the $v-G_*$ threshold. Fatigue limits are of great importance in structural design. In the present case the limiting stress is significantly enhanced by the shielding effect of the K_μ bridging term on the crack-tip K-field in (8.7). We shall elaborate on this point when we examine the issue of flaw tolerance in the context of lifetime in sect. 10.4.

Accordingly, just two sets of routine indentation-strength data, $\sigma_M(P)$ under inert conditions to calibrate microstructural shielding parameters and $\sigma_F(\dot\sigma_A)$ under environmental conditions to calibrate velocity parameters, enable lifetime predictions relating to any material toughness characteristic (with or without T-curve), flaw state (specified geometry, with or without local residual stress), and loading mode (e.g. cyclic).

8.4 Subthreshold indentations: crack initiation

Our attention turns now to the critical conditions for *initiation* of indentation cracks. It is here that differences between blunt and sharp contacts are most striking. One of the most telling concerns the role of *flaw state*: with blunt contacts the critical load is considerably higher on pristine than on as-handled surfaces, indicating a dependence on pre-existing flaws; with sharp indenters the critical load is insensitive to surface state, implying that the indentation creates its own flaw population. Another difference relates to the underlying cause of *size effects* in the threshold stress-intensity factor: with blunt indenters the contact stress increases with load and the initial flaw size stays constant; with sharp indenters the initial flaw size increases with load and the contact stress stays constant.

We now examine the micromechanics of crack initiation for these two contact types. At the conclusion of the section we use the formalism for sharp contacts to define an 'index of brittleness'.

8.4.1 Hertzian cone cracks

Our broad aim in this subsection is to calculate the threshold conditions for initiation of a Hertzian cone crack beneath a spherical indenter on a flat, elastic–brittle surface. More specifically, we seek to resolve the Auerbach paradox referred to in the opening remarks of this chapter.

The Hertzian stress field through which the cone crack evolves is well-defined. It has a maximum tensile component, dependent on mean contact pressure p_0 and Poisson's ration v,

$$\sigma_T = \tfrac{1}{2}(1 - 2v)p_0 \tag{8.17}$$

at the contact circle. Generally, initiation occurs just outside the contact at radial distance βa ($\beta \geqslant 1$) from the centre at a favourably located pre-present flaw (fig. 8.2). In fig. 8.30 we plot the principal stress σ_{11} vs distance s downward along the conical σ_{22}–σ_{33} stress trajectory surface for $\beta = 1$ (sect. 8.1.1). We see that the tensile field falls off dramatically from its maximum value σ_T at the contact circle.

The Green's function expression (2.22a) for line cracks is used to obtain a stress-intensity factor for downward extension (Frank & Lawn 1967)

$$K_P(c/a) = p_0 a^{1/2} f(\beta, v, c/a) \tag{8.18}$$

with the crack-size-dependent dimensionless integral

$$f(c/a) = 2(c/\pi a)^{1/2} \int_0^{c/a} [\sigma_{11}(s/a, \beta, v)/p_0] \, \mathrm{d}(s/a)/(c^2/a^2 - s^2/a^2)^{1/2}.$$

The line-crack assumption is least suspect in the region of principal interest at $c/a \ll 1$, where the influence of cone curvature is small. An integration of (8.18) using $\sigma_{11}(s/a)$ from fig. 8.30 yields the curves in fig. 8.31, normalised to toughness T_0. At given load P, $K_P(c/a)$ has four branches, 1 and 3 unstable ($\mathrm{d}K_P/\mathrm{d}c > 0$), 2 and 4 stable ($\mathrm{d}K_P/\mathrm{d}c < 0$); branch 1 is asymptotic to the limit $K_P = \sigma_T(\pi c)^{1/2}$ for an undiminishing stress field $\sigma_{11} = \sigma_T$ at $c \ll a$; branch 4 is asymptotic to the limit $K_P = \chi P/c^{3/2}$ for the fully developed cone in the Boussinesq far field at $c \gg a$.

The construction in fig. 8.31 allows us to trace the cone-crack evolution from a surface flaw. Since both p_0 and a increase with P in (8.2), $K_P(c/a)$ in (8.18) intensifies as indentation proceeds. Accordingly, successive load increases $P' \to P'' \to P'''$ are represented as upward curve shifts in fig. 8.31.

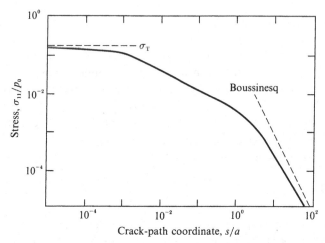

Fig. 8.30. Reduced plot of principal tensile stress σ_{11} vs distance downward along σ_{22}–σ_{33} stress trajectory surface in Hertzian field. Asymptotes (dashed lines) correspond to bounds of uniform field $\sigma_{11} = \sigma_T$ (Griffith flaw, $c \ll a$) and Boussinesq inverse-square field (Roesler cone, $c \gg a$). Note rapid stress falloff below surface, $\sigma_{11}/\sigma_T < 0.1$ at $s/a = 0.1$. Plots for $\beta = 1$, $\nu = \frac{1}{3}$. (After Frank, F. C. & Lawn, B. R. (1967) *Proc. Roy. Soc. Lond.* **A299** 291; Lawn, B. R. & Wilshaw, T. R. (1975) *J. Mater. Sci.* **10** 1049.)

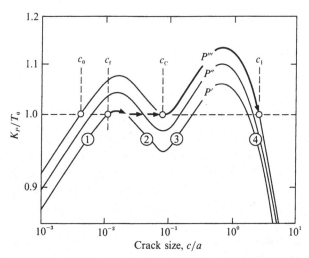

Fig. 8.31. Reduced plot of K-field as function of cone crack length, for increasing loads $P' < P'' < P'''$. Arrowed segments denote stages of stable ring crack extension from c_f to c_C (initiation), then unstable to c_I at $P = P_C = P'''$ (cone-crack pop-in). Within flaw size range $c_0 \leqslant c_f \leqslant c_C$ Auerbach's law is satisfied. (After Lawn, B. R. & Wilshaw, T. R. (1975) *J. Mater. Sci.* **10** 1049.)

Suppose the surface contains pre-present flaws of initial size within $c_0 \leqslant c_f \leqslant c_C$. Spontaneous growth of any such flaw into a surface ring crack from $c = c_f$ at branch 1 to branch 2 ensues at $P = P'$. Increasing the load to $P = P''$ extends the ring stably downward, along $K_P/T_0 = 1$. Ultimately, when the ring reaches depth $c = c_2 = c_3 = c_C$ ($\approx 0.1a$) at $P = P''' = P_C$ ($a = a_C$), the system becomes unstable and the crack pops-in to the full cone at $c = c_1$. Using (8.2) to eliminate p_0 and a, and invoking $K_P = T_0, f(\beta, \nu, c_C/a) = f_C$, (8.18) yields the critical condition

$$P_C = ArT_0^2(1 - \nu^2)/E$$
$$= ArR_0, \quad (c_0 \leqslant c_f \leqslant c_C) \tag{8.19}$$

with $R_0 = T_0^2(1 - \nu^2)/E$ (plane strain) and $A = 4\pi^2 k/3(1 - \nu^2)f_C^2 = A(\beta, \nu)$. *This result formally expresses Auerbach's law,* $P_C \propto r$, *with A an appropriate 'Auerbach constant'.* Absolute evaluation of A from the integral in (8.18) is generally not feasible, due to sensitivity of the stress field to ν and material anisotropy, curvature of the crack surface, uncertainty in location β of the critical flaw, and so forth.

The validity of Auerbach's law is contingent on the existence of flaws in the range $c_0 \leqslant c_f \leqslant c_C$. Within this range, P_C is *independent* of c_f, a direct manifestation of the condition of *activated* failure (cf. sect. 2.7). It is the critical depth c_C, not the initial depth c_f, that determines this failure condition. Experimentalists often abrade their test surfaces to *guarantee* a sufficient density of starting flaws in the Auerbach domain. For very small flaws, $c_f \ll c_0$, cone cracks initiate *spontaneously* at some higher load, $P_C = P''''$ (say) $> P'''$. Approximating the $K_P(c/a)$ curve in this region by the asymptotic solution $K_P = \sigma_T(\pi c_f)^{1/2}$ for branch 1 in fig. 8.31, and combining with (8.2) and (8.17) at $K_P = T_0$, we obtain an alternative relation $P_C \propto r^2/c_f^{3/2}$ for the critical load. Experimentally, $c_0 \approx 1\text{--}10 \, \mu\text{m}$ for typical brittle solids, so this last relation applies only to well-finished (e.g. highly polished) surfaces. The critical load can be exceptionally high on pristine surfaces (sect. 9.1.1).

Auerbach's law has been verified on several materials (Lawn & Wilshaw 1975). Data for a wide range of sphere radii on abraded soda-lime glass are shown in fig. 8.32. Such data enable a calibration of the Auerbach constant A in (8.19). We note especially the insensitivity to abrasion flaw size.

The derivation (8.19) does more than just confirm Auerbach's law. It provides a means for evaluating variations in intrinsic toughness, i.e. $P_C \propto R_0$. Fig. 8.33 plots P_C at fixed r within the Auerbach range as a function of temperature for two forms of SiO_2. The toughness for

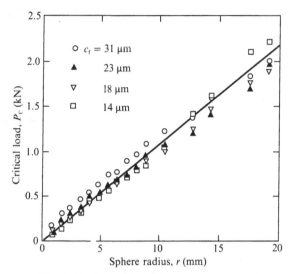

Fig. 8.32. Critical load for cone-crack initiation, as function of indenter radius on soda-lime glass. Glass surfaces pre-abraded to produce controlled flaws, sizes c_f indicated. (Each data point mean of at least ten tests.) Note Auerbach's law, $P_C \propto r$, insensitive to initial flaw size. (After Langitan, F. B. & Lawn, B. R. (1969) *J. Appl. Phys.* **40** 4009.)

crystalline quartz shows a strong decline relative to its amorphous silica counterpart at rising temperature. The accentuated minimum at the quartz $\alpha \to \beta$ transition temperature reflects an instability in the stacking of SiO_4^{4-} tetrahedra in the Si–O network, from an ordered to disordered state (Swain, Williams, Lawn & Beek 1973).

We alluded earlier to *size effects* in the crack initiation conditions. Such a size effect is manifest in Auerbach's law. Using (8.2) to evaluate the tensile stress $\sigma_T = \sigma_C$ in (8.17) at the critical initiation condition $P = P_C$ in (8.19), we obtain $\sigma_C \propto r^{-1/3}$, i.e. an increasingly higher fracture stress at diminishing sphere radius. Once more, the concept of an invariant critical stress is violated. Ultimately, at some sufficiently small indenter size, the cohesive stress in shear is exceeded before fracture and irreversible deformation ensues, signifying a transition to 'sharp' contact.

Environmental interaction and sliding friction strongly reduce the critical load for cone-crack initiation (Lawn & Wilshaw 1975). Kinetic crack growth can occur through the energy barrier in fig. 8.31 at sustained, subcritical loads, e.g. at $P'' < P_C$, leading to pop-in at $c > c_C$. Friction enhances the tensile stresses at the trailing edge of the contact, increasing $f(\beta, v, c/a)$ in (8.18) and, conversely, diminishing A in (8.19).

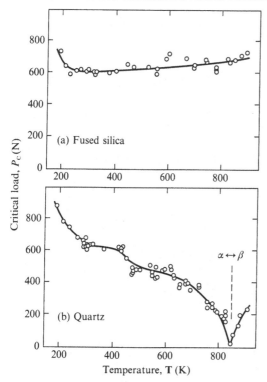

Fig. 8.33. Critical load for cone-crack initiation in (a) fused silica and
(b) quartz as function of temperature. Vacuum data on abraded
surfaces, tungsten carbide sphere, $r = 6.35$ mm. (Each data point
mean of at least ten tests.) (After Swain, M. V., Williams, J. S., Lawn,
B. R. & Beek, J. J. H. (1973) *J. Mater. Sci.* **8** 1153.)

8.4.2 Radial cracks

As with Hertzian fracture, there is a critical load for radial cracking.
However, the radial threshold is *independent* of whether the test surfaces
are pristine or abraded. This leads us to conclude that sharp indenters are
capable of generating their own starting flaws within the inelastic contact
zone. Whereas in metals such inelastic deformation is well understood in
terms of crystal dislocations, the same is not true of the 'harder' (high
H/E) covalent–ionic solids (sect. 8.1.3). For those materials, the contact
pressure is of the same order as the intrinsic cohesive strength (sect. 6.1), in
which case it makes more sense to regard the deformation as a cooperative
breakdown of the rigidly bonded structure.

Thus arises the notion of 'shear faulting', or 'block slip'. Although this
kind of deformation does tend to occur in directions of highest shear stress

(a)

(b)

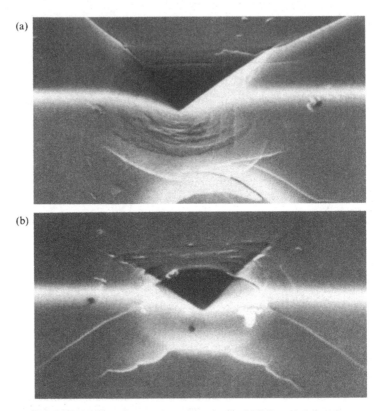

Fig. 8.34. Half-surface and section views of Vickers deformation zones (obtained by propagating a hairline pre-crack through the indent diameter) in (a) soda-lime, (b) fused silica glass. Scanning electron micrographs. Width of field 35 μm. (After Multhopp, H., Dabbs, T. P. & Lawn, B. R. (1984), in *Plastic Deformation of Ceramic Materials*, eds. R. E. Tressler & R. C. Bradt, Plenum, New York, p. 681.)

in the manner of dislocations, it is not confined to ordinarily preferred glide planes. Indeed, it is not even necessary for the material to be crystalline. Shear faults have been observed in silicate glasses, by Hagan (1980) and others. Surface–section views of the deformation zones in soda-lime glass and fused silica are shown in fig. 8.34. The faults are apparent as the discrete surface traces, akin to the slip lines of plasticity fields. Observe the somewhat deeper subsurface fault penetration in the volume-conserving ('normal') soda-lime glass; in the ('anomalous') fused silica part of the deformation is accommodated by volume compaction.

These observations suggest the model in fig. 8.35(a). The driving force for radial crack initiation comes from the residual field, evaluated in fig.

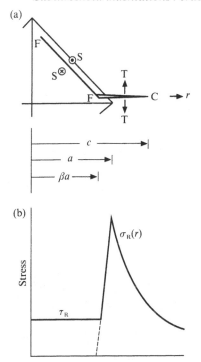

Fig. 8.35. Model for crack initiation at Vickers indentation. (a) Quadrant of indentation, with tensile (T) microcrack extension FC from edge of shear (S) fault FF. (b) Residual shear stress τ_R = const on fault, tensile stress $\sigma_R(r)$ on microcrack segment. (After Lathabai, S., Rödel, J., Lawn, B. R. & Dabbs, T. P. (1990) *J. Mater. Sci.* **26** 2157.)

8.35(b) from an elastic–plastic 'expanding cavity' representation of the deformation zone (sect. 8.1.3). We regard the fault FF as a constrained shear crack with uniform friction stress τ_R at its interface, extending radially from its edge at βa ($\beta \leqslant 1$) as a tensile segment FC into the outer field $\sigma_R(r)$. From similarity arguments (Marshall, Lawn & Chantikul 1979), the intensity of this stress field scales with the contact pressure H in (8.4), and the size of the fault scales with a. The stress-intensity factor for the fault–microcrack system in fig. 8.35 thereby assumes the form

$$K_R(c/a) = Ha^{1/2}f(\beta, v, c/a) \qquad (8.20)$$

with $f(c/a)$ a dimensionless function determinable by integration of τ_R/H and σ_R/H in fig. 8.35(b) over the crack area (cf. (8.18)). Insofar as H in (8.20) (unlike p_0 in (8.18)) is a material constant, any increase in K_R with load is now due *solely* to spatial scaling of the contact zone.

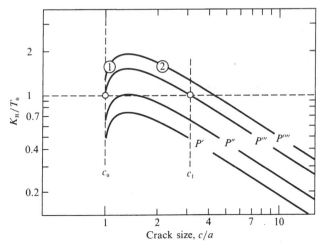

Fig. 8.36. Reduced plot of residual K-field vs crack length for radial-crack initiation, at increasing load $P' < P'' < P''' < P''''$, for $\beta = 1$. Initiation is spontaneous at $P = P_C = P'''$, from c_0 to c_1. In limit $c \gg a$, $K_R \propto c^{-3/2}$. (After Lathabai, S., Rödel, J., Lawn, B. R. & Dabbs, T. P. (1990) *J. Mater. Sci.* **26** 2157.)

Plots of $K_R(c/a)$ are given in fig. 8.36, for successive loads $P' \to P'' \to P''' \to P''''$. Because the fault is presumed well-developed at the instant of unload, we consider solutions only in the domain of microcrack extension, $c \geqslant c_0 = a$, $\beta = 1$. For equilibrium fracture, radial pop-in occurs spontaneously from $c = c_0$ on unstable branch 1 at $P = P''' = P_C$, $a = a_C$, $f(\beta, v, c_0/a) = f_C$. Inserting $K_R = T_0$ into (8.20) and (8.4), we obtain the critical conditions for *initiation* (Lawn & Evans 1977):

$$a_C = \Theta(T_0/H)^2 \tag{8.21a}$$

$$P_C = \Theta\alpha_0 H(T_0/H)^4 \tag{8.21b}$$

with dimensionless coefficient $\Theta = 1/f_C^2$. Again, absolute computation of Θ from the $f(\beta, v, c/a)$ integral is subject to much uncertainty. Recall also that the residual stress components τ_R and σ_R which determine this integral derive from an elastic-plastic field, so, strictly, Θ should include a term in H/E (sect. 8.1.3).

Once the radial cracks pop-in, they arrest on stable branch 2 at $c = c_1$ ($\approx 2a$–$3a$) in fig. 8.36. The stress-intensity factor in this region tends asymptotically to the solution (8.5) for fully developed radials, $K_R \propto c^{-3/2}$ ($c \gg a$), as required of any 'universal' model. Measurement of a_C and c_1 at pop-in allow an evaluation of adjustable parameters in $f(\beta, v, c/a)$.

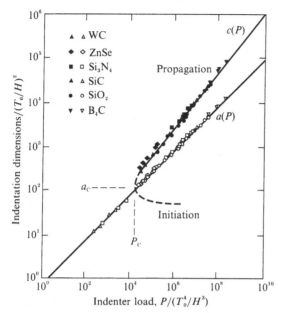

Fig. 8.37. 'Universal' indentation deformation–fracture diagram, constructed using Vickers data for several materials. Solid lines are representations of $a(P)$ from (8.4) and $c(P)$ from (8.5) at $K_R = T_0$. Extended, dashed $c(P)$ curve is function representing initiation condition (8.21). (After Lawn, B. R. & Marshall, D. B. (1979) *J. Amer. Ceram. Soc.* **62** 347. Data courtesy A. G. Evans & A. Arora.)

Environmental interactions can again lead to substantial reductions in the threshold load. An interesting manifestation of kinetic crack growth is 'delayed pop-in', i.e. abrupt initiation at some interval after completion of the indentation cycle at (say) $P'' < P_C$ in fig. 8.36 (Lawn, Dabbs & Fairbanks 1983). Such phenomena reinforce the vital role of residual contact fields in radial (and lateral) crack micromechanics.

8.4.3 Indentation threshold as index of brittleness

Intuitively, 'brittleness' measures the competition between deformation and fracture. It is manifest in the celebrated ductile–brittle transition in metals at low temperatures and high strain rates. But how may we *quantify* brittleness? Why is cracking so difficult to induce in metals and so easy in covalent–ionic solids?

The formalism for sharp contacts in the preceding sections provides a simple scheme for addressing these questions (Lawn & Marshall 1979).

Table 8.1. *Hardness H, toughness T_0 and brittleness H/T_0 (descending order), plus threshold contact size a_c and load P_c, for selected monocrystalline (mc) and polycrystalline (pc) solids.*

Material	H (GPa)	T_0 (MPa m$^{1/2}$)	H/T_0 (μm$^{-1/2}$)	a_c (μm)	P_c (N)
Diamond (mc)	80	4	20	0.3	0.004
Silicon (mc)	10	0.7	14	0.6	0.01
Magnesium oxide (mc)	9	0.9	10	1.2	0.02
Silica (glass)	6	0.75	8	2	0.1
Silicon carbide (mc)	19	2.5	8	2	0.2
Sapphire (mc)	20	3	7	3	1
Silicon nitride (pc)	16	4	4	8	0.8
Zirconia (pc)	12	3	4	8	60
Tungsten carbide (pc)	20	13	1.5	50	8
Zinc selenide (pc)	1.1	0.9	1.2	80	800000
Steel (pc)	5	50	0.1	12000	

T_0 for polycrystalline solids is short-crack (indentation) toughness.

Observe in the indentation threshold relations (8.21) that spatial dimensions (a, c) scale with the material quantity $(T_0/H)^2$, load (P) with $H(T_0/H)^4$. By appropriately normalising these variables in the hardness (H) relation (8.4) and toughness $(K_R = T_0)$ relation (8.5), we may represent deformation–fracture data for all materials on a universal diagram. This is done for Vickers data in fig. 8.37. The initiation configuration (a_C, P_C) is identified by the cutoff in radial-crack data. At $P > P_C$ (large-scale events) the indentation pattern is toughness-controlled, at $P < P_C$ (small-scale events) it is hardness-controlled.

Because of its governing role in the normalisation scheme of fig. 8.37, the ratio H/T_0 in (8.21) is proposed as an 'index of brittleness' (Lawn & Marshall 1979). Table 8.1 lists H/T_0, with a_C and P_C from (8.21) using a calibration value of Θ from fig. 8.37, for several materials. The extreme brittleness of ceramics relative to metals is clear. Such tabulations are useful for ranking the susceptibility of materials to strength-degrading surface damage.

The size effect embodied in the sharp-contact threshold can be resolved on dimensional grounds, noting that the work of indentation is partitioned into *deformation volume* and *fracture surface* (Puttick 1980). As remarked on several occasions in chapter 7, volume/surface size effects are commonplace in brittle fracture. It is therefore inevitable that the $a(P)$ and $c(P)$ functions in fig. 8.37 should intersect at some critical scaling dimension.

8.5 Subthreshold indentations: strength

Now examine the response of the sharp-contact flaw of fig. 8.35 in a material of single-valued toughness to an applied uniform stress σ_A (fig. 8.13). The crack-tip K-field is

$$
\begin{aligned}
K_* &= K_A + K_R \\
&= \psi\sigma_A c^{1/2} + Ha^{1/2}f(\beta, \nu, c/a)
\end{aligned}
\tag{8.22}
$$

with K_R from (8.20).

In this section we concern ourselves primarily with the transitional increase in *inert* strength on entering the subthreshold domain (Dabbs & Lawn 1985). Plots of $K_*(c/a)$ are shown in fig. 8.38 for indentation flaws in both the postthreshold $(P > P_C, a > a_C)$ and subthreshold $(P < P_C, a < a_C)$ domains. (The reader may view these plots as com-

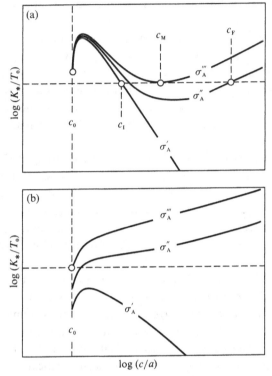

Fig. 8.38. Reduced plot of K-field as function of radial crack length in increasing applied field $\sigma'_A \rightarrow \sigma''_A \rightarrow \sigma'''_A$ ($\beta = 1$): (a) postthreshold indentation ($P > P_C$), activated failure; (b) subthreshold indentation ($P < P_C$), spontaneous failure. (After Lathabai, S., Rödel, J., Lawn, B. R. & Dabbs, T. P. (1991) *J. Mater. Sci.* **26** 2157.)

posites of figs. 8.36 and 8.15.) The effect of increasing σ_A through $\sigma'_A \rightarrow \sigma''_A \rightarrow \sigma'''_A$ is to shift $K_*(c/a)$ upward, the more so at larger c. At $P > P_c$ the radial crack extends incrementally at equilibrium from its initial well-developed state at $c = c_I$, $\sigma_A = \sigma'_A = 0$ to activated failure ($K_* = T_0$, $dK_*/dc = 0$) at $c = c_M$, $\sigma_A = \sigma'''_A = \sigma_F = \sigma_M$ (sect. 8.2.1). At $P > P_C$, on the other hand, the radial crack is in its embryonic state at $c = c_0$, $\sigma_A = \sigma'_A = 0$ and remains stationary until the applied stress meets the requirement for spontaneous failure ($K_* = T_0$, $dK_*/dc > 0$) at $\sigma_A = \sigma_F = \sigma'''_A = \sigma_0$. Whereas in the postthreshold region the strength is *propagation*-controlled, in the subthreshold region it is *initiation*-controlled.

To illustrate, inert strength data for Vickers-indented fused silica are plotted in fig. 8.39. We stress that all these data, even in the subthreshold region, represent breaks from indentation sites. The solid curve is a

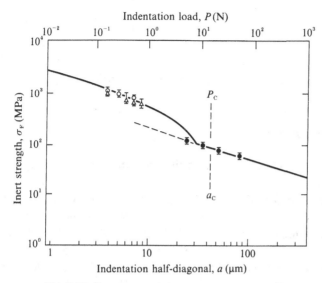

Fig. 8.39. Inert strength vs Vickers indentation size (lower axis) or load (upper axis) for fused silica. Bars are standard deviations for confirmed breaks from indentation flaws in subthreshold (open symbols) and postthreshold (filled symbols) domains. (Former surfaces subjected to pre-indentation acid etch to remove pre-present flaws and ensure breaks from indentation sites.) Solid curve is fit to theory. (Segment along the extrapolated postthreshold curve immediately below $P = P_C$ corresponds to activated pop-in *during* stressing to failure.) (After Lathabai, S., Rödel, J., Lawn, B. R. & Dabbs, T. P. (1991) *J. Mater. Sci.* **26** 2157.)

theoretical fit, by adjusting coefficients in $f(\beta, v, c/a)$ and solving (8.22) for the instability stress $\sigma_A = \sigma_F$ at each prescribed P. At $P > P_C$ the $\sigma_F(P)$ fit is close to linear in logarithmic coordinates with slope $-\frac{1}{3}$, consistent with (8.9b). At $P < P_C$ the strengths are substantially higher than one would predict by extrapolation from the postthreshold region. Discontinuities of this type are symptomatic of the bimodal strength distributions observed in optical quality glass fibres (sect. 10.3.1).

The indentation analysis for subthreshold flaws is readily extended to *time-dependent* strengths. In principle, one has only to incorporate an appropriate velocity equation, $v = v\{K_*[P, \sigma_A(t), c(t)]\}$, into the (calibrated) $K_*(c/a)$ function, and solve the ensuing differential equation in the usual way (cf. sect. 8.3). Tests on subthreshold flaws in silicate glasses in water reveal an enhanced fatigue susceptibility, along with a tendency to abrupt strength loss by premature radial pop-in from the upper to (extrapolated) lower curve at loads immediately below P_C in fig. 8.39.

8.6 Special applications of the indentation method

Indentation techniques have found many applications in the investigation of brittle fracture properties. We describe just three below.

8.6.1 Sharp vs blunt flaws

An extensive study by Mould in 1960 on the strength of abraded soda-lime glass surfaces set the scene for a long-standing debate on the intrinsic structure of brittle cracks. Mould measured strength increases $\approx 30\%$ on aging the abrasions in water for several days. Without the means for observing the responses of individual flaws, he concluded that the strength increase must be due to moisture-induced blunting (sects. 5.5.5, 6.7). Even higher strengths were measured after annealing, suggesting that elevated temperatures must accelerate the blunting process. However, the abrasions showed no sign of any further strength change with post-anneal aging.

A more recent, comparative study using Vickers indentation flaws (Lawn, Jakus & Gonzalez 1985) has demonstrated the need for revision of this earlier conclusion. As shown in fig. 8.40(a), the inert strength of as-indented surfaces, like Mould's abraded surfaces, increases with aging time in water, approaching a plateau after a day or more. Again, the strengths of surfaces annealed after indentation are higher than those of their as-indented predecessors, but remain invariant with subsequent aging time.

The singular advantage of indentations over abrasion flaws is their capacity for direct observation at every stage of the aging process. We have made much of the fact that fresh Vickers indentations (like abrasion flaws) are subject to stabilising residual stress fields, and hence to (decelerating) post-indentation extension in moist environments (sect. 8.1.3, fig. 8.12). As seen in fig. 8.40(b) such extension actually continues steadily through the aging period, in both radial *and* lateral systems, progressing down the $v–G$ curve (fig. 5.10) until, at threshold, the strength attains its long-term plateau.

These results provide a powerful case against the blunting hypothesis. Any dissolutive blunting would be expected to become most pronounced at the slowest velocities, reaching its maximum efficacy at threshold. But this is precisely the region where the strength saturates in fig. 8.40(a). We conclude that the aging effect has nothing to do with a fundamental change in crack-tip structure, but is due instead to progressive relaxation in the

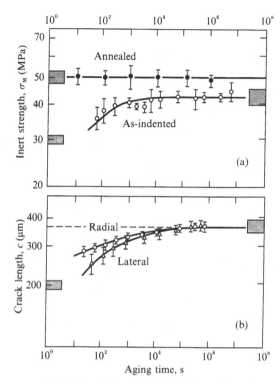

Fig. 8.40. Aging of Vickers indentation flaws ($P = 50$ N) in soda-lime glass in water: (a) ensuing inert strength, as-indented and post-annealed; (b) crack dimensions, radial and lateral. Shaded areas indicate limiting configurations. (After Lawn, B. R., Jakus, K. & Gonzalez, A. C. (1985) *J. Amer. Ceram. Soc.* **68** 25.)

residual stress term χ or geometry term ψ in (8.9b) by interaction between the strength-controlling radial crack and its relatively slowly expanding lateral neighbour. We draw attention once more to the conspicuous absence of any aging effect in the post-annealed specimens: the anneal *completely* removes the residual stress, thereby precluding further lateral extension and, in the manner of fig. 8.17, increasing the strength above that for the fully aged, as-indented specimens.

The indentation aging experiment is compelling evidence for the enduring atomic sharpness of brittle cracks.

(a)   (b)

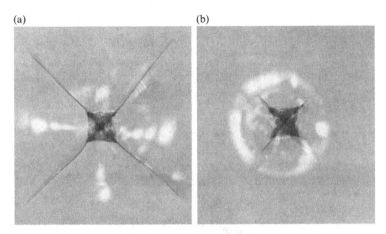

Fig. 8.41. Vickers indentations in (a) pre-annealed and (b) thermally
tempered soda-lime glass. Load same in both cases. Residual
compression in tempered surface inhibits radial crack extension. Width
of field 400 μm. (Courtesy D. B. Marshall.)

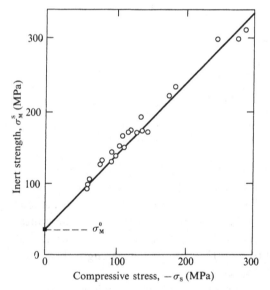

Fig. 8.42. Inert strength of surface-leached glass rods as function of
(independently measured) surface compressive stress. Point at $\sigma_M^S = \sigma_M^0$
($\sigma_S = 0$) is stress-free control. Data at fixed Vickers indentation load
($P = 100$ N). (After Chantikul, P., Marshall, D. B., Lawn, B. R. &
Drexhage, M. G. (1979) *J. Amer. Ceram. Soc.* **62** 551.)

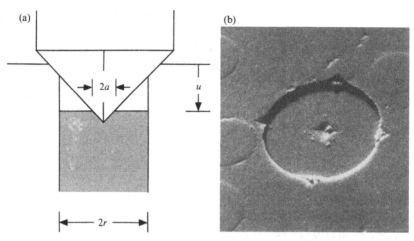

Fig. 8.43. Indentation push-in method for evaluating friction stress at sliding matrix–fibre interface. (a) Coordinate system for analysis. (b) Scanning electron micrograph of Vickers indentation ($P = 0.5$ N) on silicon carbide fibre in glass-ceramic matrix. Note residually depressed fibre, and remnant pyramid markings at rim of hole in matrix. Width of field 50 μm. (After Marshall, D. B. (1984) *J. Amer. Ceram. Soc.* **67** C–259.)

8.6.2 Surface stress evaluation

Surfaces of brittle materials may be subject to macroscopic residual stresses as a result of thermal, mechanical or chemical treatments. *Compressive* stress layers, by virtue of their inhibiting effect on flaw extension, afford effective surface 'protection' (provided the flaws do not penetrate into the countervailing subsurface tensile region). An indication of the benefit of surface compression is given by the Vickers indentations in pre-annealed and thermally tempered glass surfaces in fig. 8.41.

Indentation fracture may be used to determine the magnitude, σ_S, of the residual stress. Consider an indentation flaw in a uniformly compressed surface layer ($\sigma_S < 0$) of depth d ($> c$). Then we may define $K_S = \psi\sigma_S c^{1/2}$, which superposes onto $K_A = \psi\sigma_A c^{1/2}$ from any subsequently applied tensile stress σ_A to give a net 'effective applied' K-field $K'_A = \psi(\sigma_A + \sigma_S)c^{1/2}$. At unstable equilibrium ($K_* = K'_A + K_R = T_0$, $\mathrm{d}K_*/\mathrm{d}c = 0$), the inert strength σ_M in (8.9b) is simply replaced by

$$\sigma_M^S = \sigma_M^0 - \sigma_S \tag{8.23}$$

where superscript 0 denotes a control test without surface stress.

Fig. 8.42 shows results of indentation-strength tests on leached glass surfaces. The leaching produces a uniform compressive layer, with stress level dependent on (among other things) treatment time. The $\sigma_M^S(\sigma_S)$ data can be fitted by a straight line, with intercept at $\sigma_M^S = \sigma_M^0$, in accordance with (8.23).

Analyses for non-uniform stress layers are not so straightforward, but are available. Indentations (e.g. fig. 8.41) are ideal for probing point-by-point stress gradients *across* a given surface.

8.6.3 Matrix–fibre sliding interface friction

An ingenious application of the indentation methodology has been devised by Marshall (1984) to evaluate frictional tractions at sliding matrix–fibre interfaces. We recall from sect. 7.6 that friction is a key factor in the mechanism of toughening in fibre-reinforced ceramic composites.

Marshall's method is illustrated in fig. 8.43(a). A Vickers pyramid is loaded axially onto a fibre in a polished section. The fibre slides over some debond length, and remains depressed below the surface on unloading. An energy-balance argument leads to the fibre/matrix displacement

$$u = P^2/4\pi^2 r^3 E_f \tau - 2\Gamma/\tau \qquad (8.24)$$

with P the indenter load, r the radius and E_f the modulus of the fibre, τ the friction stress and Γ the (mode II) debond energy. In the micrograph of fig. 8.43(b) for an indented silicon carbide fibre in a glass matrix the displacement can be inferred from the marks left by the corners of the Vickers pyramid at the surface rim of the hole. From measurements at different loads, Marshall evaluates $\tau = 3.5$ MPa and $\Gamma = 40$ mJ m^{-2} for the system in fig. 8.43(b).

Several elaborations of this technique have been developed in the ceramic composites literature.

8.7 Contact damage: strength degradation, erosion and wear

The earlier sections in this chapter establish a foundation for modelling surface damage due to impinging contacts, either inadvertent from particle impact during service (erosion and wear) or deliberate from abrasive finishing (grinding and polishing). Controlled indentations and particulate

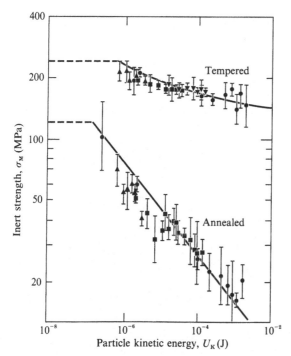

Fig. 8.44. Inert strength of soda-lime glass, (a) pre-annealed and (b) tempered, measured after impact with silicon carbide particles. (Each symbol a different silicon carbide grit size.) Dashed lines are natural flaw cutoffs. (After Wiederhorn, S. M., Lawn, B. R. & Marshall, D. B. (1979) *J. Amer. Ceram. Soc.* **62** 66, 71.)

contacts produce basically the same damage pattern, distinguished only by cosmetic differences in geometrical detail.

We investigate here two kinds of contact damage process of practical interest, strength degradation and erosive wear.

8.7.1 Strength degradation

The incidence of small particles onto brittle surfaces may induce radial or cone cracks of sufficient depth as to constitute dominant surface flaws. The strength properties are thereby degraded.

As a worst case, imagine a surface in *impact* with one or more *sharp* particles. Assuming that the incident kinetic energy of a given particle is consumed entirely as work of penetration, contact mechanics allows us

(a)

(b)

(c)

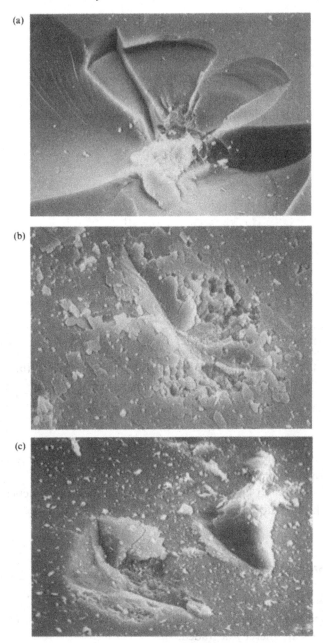

Fig. 8.45. Damage on ceramic surfaces after impact with 150 μm silicon carbide particles at 90 m s^{-1}: (a) soda-lime glass, width of field 100 μm; (b) fine-grained alumina, width of field 50 μm; (c) hot-pressed silicon nitride, width of field 50 μm. Scanning electron micrographs. Note grain-scale debris in (b) and (c). (Courtesy B. J. Hockey.)

to calculate the impulsive load (Lawn, Marshall, Chantikul & Anstis 1980)

$$P = [9\alpha_0 H(U_K \tan \Phi)^2]^{1/3} \tag{8.25}$$

with Φ an average 'indenter' half-angle. Substitution into $\sigma_M(P)$ in (8.9a) yields the functional dependence $\sigma_M(U_K)$ for materials with single-valued toughness. One may therefore make a priori predictions of strength degradation for moving components in hostile particulate environments.

Inert strength data for pre-annealed and tempered soda-lime glass after impact in a grit-injected gas stream are shown as a function of particle kinetic energy in fig. 8.44. The solid curves are predictions, using 'calibrated' constants χ and ψ from static indentation tests (e.g. fig. 8.19) and $\sigma_S = -130$ MPa in (8.23) for the tempered surfaces. The benefits of surface compressive stresses are again evident.

Similar strength studies on polycrystalline ceramic materials, especially those with strong toughness-curves, are only now being made. In such cases one might expect the surfaces to suffer less strength loss because of the flaw tolerance (sect. 7.5).

8.7.2 Erosion and wear

Repetitive particulate contacts can cause surface removal, resulting in erosion and wear. In brittle ceramics, lateral cracks from sharp contacts are the most potent agents of such removal. Examples of incipient surface erosion from sharp particle impacts are shown in fig. 8.45.

Erosion and wear formulas may be derived directly from contact fracture mechanics. Consider again just sharp particles and constant toughness materials. The potential removal rate at individual contact event i is determined by the volume of material V_i above the lateral crack. In terms of crack radius c_i, from (8.5) at $K_R = T_0$, and indentation depth a_i, from (8.4), we obtain

$$
\begin{aligned}
V_i &\approx c_i^2 a_i \\
&= \omega P_i^{11/6} / T_0^{4/3} H^{1/2}
\end{aligned}
\tag{8.26}
$$

with ω a 'wear coefficient'. In the approximation of non-interacting contacts the total removal rate is simply the sum $V = \Sigma V_i = NV_i$ over all N events. Then (8.26) suggests that the greatest resistance to wear is to be found in materials with the greatest toughness and, to a lesser degree, hardness.

Relations like (8.26) are useful *guides* to wear rates. They are inevitably limited, however, to fracture-generated removal processes, and to simple brittle solids. Below threshold (sect. 8.4.3), removal is due to more complex deformation processes. This latter is the subtle realm of 'plastic erosion' and 'polishing'. Moreover, in polycrystalline ceramics there is the complication of toughness-curves, particularly in those ceramics with strong bridging (sect. 7.5). Associated internal residual stresses may enhance fracture-controlled removal at the microstructural level (negating the benefits of flaw tolerance). Note, for instance, the grain-scale debris around the impact sites in fig. 8.45(b,c). Thus it is in some non-cubic ceramics that an increase in grain size promotes an abrupt transition from deformation- to fracture-controlled wear removal, as cumulative deformation-induced stresses augment σ_R in (7.17) and activate premature microcracking at the larger sub-facet flaws. We shall present a specific example of such transitional behaviour in our consideration of material reliability in sect. 10.5.

8.8 Surface forces and contact adhesion

The surface-forces apparatus described in sect. 6.5 can be used to measure adhesion energies between two contacting elastic bodies. Adhesive forces modify the Hertzian contact relations. Characteristic features are a nonzero radius at zero applied load and a tensile pulloff force to rupture the contact. An example of an adhesive contact, between mica and silica glass, is given in fig. 8.46. Such phenomena are of great interest in colloid chemistry and elsewhere.

The problem is properly analysed by computing the system energy for a sphere of radius r on a like half-solid of Young's modulus E and Dupré work of separation W. There are two (nonlinearly superposed) components, the elastic strain energy in the volume of the bodies and the adhesion energy at the surfaces. Imposing the condition $dU/dA = (1/2\pi a)\,dU/da = 0$, A the area and a the radius of contact, one obtains the equilibrium relation (Johnson, Kendall & Roberts 1971)

$$a^3 = (4kr/3E)\{P + (3\pi Wr)[1 + (1 + 2P/3\pi Wr)^{1/2}]\} \qquad (8.27)$$

where k is the dimensionless constant of sect. 8.1.2. At zero adhesion, $W = 0$, (8.27) reverts to $a = (4kPr/3E)^{1/3}$, as contained in the relation (8.2)

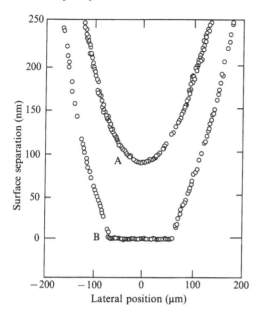

Fig. 8.46. Surface profiles for contact between crossed-cylinder surfaces of mica and silica in air. Plots are from digitised images of interference fringes of equal chromatic order. (Note difference in scale between the two axes.) On approach, A, the surfaces attract and snap into contact, B. The region immediately outside the contact has the essential character of a sharp crack. A pulloff force is needed to re-separate the two surfaces. (Courtesy D. T. Smith & R. G. Horn.)

for ideal Hertzian contacts. At zero contact force, $P = 0$, (8.27) predicts a nonzero contact $a = a_0$:

$$a_0 = (8\pi k W r/E)^{1/3}. \qquad (8.28)$$

On reversing the contact load, the system remains stable until a critical pulloff instability is attained, at $dP/da = 0$, $a = a_C$, $P = -P_C$ in (8.27):

$$P_C = \tfrac{3}{2}\pi r W. \qquad (8.29)$$

Pulloff tests are widely used in adhesion energy studies.

An essential complementarity exists between adhesive contacts and brittle cracks. This fact was exploited by Maugis & Barquins (1978), who treated contact adhesion as a classical fracture problem, with dU/dA a configurational force analogous to the mechanical-energy-release rate G. It is interesting to observe that, with $W = R_0$, (8.29) for adhesive rupture is identical in form to (8.19) for cone fracture. Since the contact interface is not generally coherent, the pertinent W term is that for healed cracks in angular misorientation (sect. 6.5). Thus, for contacts formed in interactive (moist) environments, we have $W = {}^{h'}W_{BEB}$ (like solids – cf. bottom data set for mica–mica in fig. 5.22) or $W = {}^{h'}W_{AEB}$ (unlike solids – fig. 8.46). The healed crack profile in fig. 6.20(d) is then a suitable representation of the adhesion interface immediately adjacent to the contact circle.

9
Crack initiation: flaws

All discussions on the strength of brittle solids in the preceding chapters are predicated on the existence of flaws. But what are the underlying *origin* and *nature* of such flaws? What flaw geometry and material parameters govern the micromechanics of *initiation* into the ultimate well-developed crack? How do persistent nucleation forces influence the *stability* of flaws during initiation and subsequent development?

The decades following 1920 witnessed a preoccupation of silicate-glass researchers with the search for 'Griffith' flaws. But observing such flaws by any direct means proved elusive. Typically, characteristic flaw dimensions in these and other homogeneous covalent–ionic solids like monocrystalline silicon, sapphire and quartz range from 1 nm (pristine fibres and whiskers) to 1 μm (aged, as-handled solids), and usually occur at the surface. As we indicated in sect. 1.6, planar defects on this scale are likely to lie below the limit of detectability by optical means. More recently, with the advent of modern heterogeneous polycrystalline ceramics, we have seen a shift in focus to microstructural fabrication flaws, the existence and character of which can be all too obvious. These latter range from 1 μm (high-density, fine-grained, polished materials) to 1 mm and above (refractories, concrete), and occur in both the surface and the bulk.

The small scale of typical flaws highlights the sensitivity of brittle solids to seemingly innocuous extraneous events and treatments (recall the one-hundred-fold reduction in strength on aging glass fibres, sect. 1.6). Almost invariably, the defects introduced during material processing, finishing and service are widely distributed in size, location and orientation. Hence the term 'flaw population', a pre-eminent constituent of reliability analyses (chapter 10). The coexistence of more than one flaw type gives rise to a bimodal or even multimodal population. Distributions for a given material are difficult to characterise statistically because one is usually more

Table 9.1. *Typical 'Griffith' flaw sizes and strengths of brittle materials.*

Flaw size	Material	Strength
nm	Pristine optical fibres, whiskers	10 GPa
	Ultra-fine-grain ceramics	
		1 GPa
μm	As-handled glasses, single crystals Fine-grain polycrystalline ceramics	
	Coarse-grain polycrystalline ceramics Composites	100 MPa
mm	Rocks	
	Refractories	10 MPa
	Concretes	
m	Earth's crust, glaciers	1 MPa

concerned with extreme rather than average values: the concept of the 'critical flaw' emerges as the most singular element of all traditional strength descriptions.

In this chapter we discuss several kinds of flaws, for both homogeneous and heterogeneous materials. The driving forces for formation derive primarily from concentrated mechanical fields, but chemical, radiation and thermal fields are other effective factors. Insofar as the flaws subsequently develop as sharp *microcracks*, classical fracture mechanics may be retained to describe their evolution to full propagation.

The almost universal prevailing notion of flaw response in brittle solids is that of unlimited instability at a critical applied stress, in the manner envisaged by Griffith. However, as we saw in discussing microstructural shielding in chapter 7 and residual stresses around contacts in chapter 8, crack growth from flaws can be highly stabilised by internal fields,

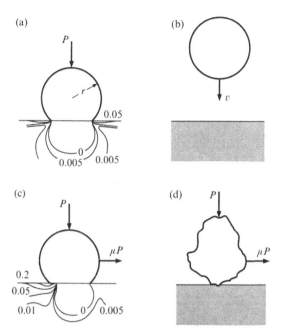

Fig. 9.1. Contact damage model: (a) static blunt particle, sphere radius *r* and load *P* (Hertz); (b) impacting sphere, at normal velocity *v*; (c) sliding sphere, with tangential force μP (μ = friction coefficient) superposed on normal force; (d) irregular particle. Equi-stress contours (units of mean normal contact pressure) indicated in (a) and (c).

implying significant departures from the Griffith ideal. Thus, microscopic flaws may be activated into a macroscopic 'pop-in' state, and thereafter grow stably to final instability at increasing applied stress. As we saw with our comparisons of subthreshold and postthreshold indentations in the preceding chapter, strength properties can be profoundly affected by any such precursor stages in the flaw evolution.

9.1 Crack nucleation at microcontacts

Pristine surfaces of homogeneous brittle solids may contain defects from mechanical preparation, e.g. fracture steps in cleavage. More pernicious, however, are spurious Hertzian and elastic–plastic contact flaws from subsequent exposure to quartz and other hard dust particles in the atmosphere. As a rule of thumb, we may expect damage of the same scale

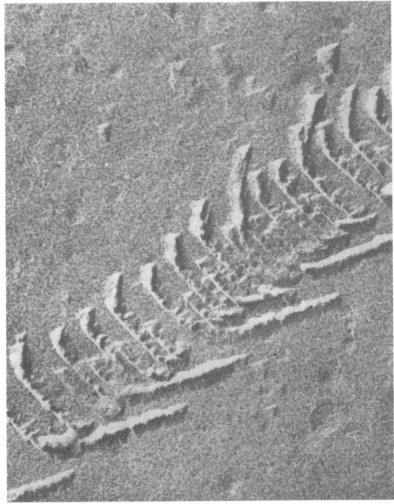

(a)

Fig. 9.2(a). For legend see facing page.

as the contacting particle, typically ≈ 1 μm. As foreshadowed in sect. 8.7, indentation fracture provides a sound basis for the analysis of such flaws.

9.1.1 Microcontact flaws

Consider the microcontact of a small particle on a brittle surface. Fig. 9.1 indicates how one may build up a practical model, starting with an idealised elastic Hertzian contact, progressing to more complex impact and

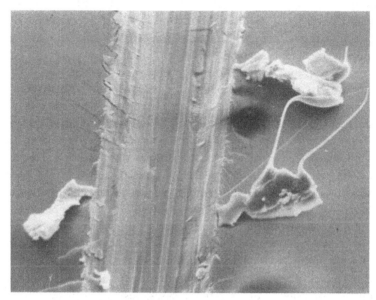

Fig. 9.2. Sliding damage on brittle surfaces. (a) Contact chatter track
on natural diamond (111) surface. Direction of sliding from bottom
left to top right. Electron micrograph of carbon replica. Width of field
1.0 μm. (After Lawn, B. R. (1967) *Proc. Roy. Soc. Lond.* **A299** 307.)
(b) Polishing groove on glass produced by translating diamond tool at
10 m s^{-1}. Direction of sliding from bottom to top. Note extrusion layer
inside groove and molten 'stringers' outside. Width of field 75 μm.
(After Schinker, M. G. & Doll, W. (1985), in *Strength of Glass*, ed.
C. R. Kurkjian, Plenum, New York, p. 67.)

sliding configurations, through to the extreme example of a translating
sharp particle. Let the axially loaded sphere in fig. 9.1(a) be of radius r,
Young's modulus E and Poisson's ratio v; and let the flat surface be of like
material and free of pre-existing defects, such that the system remains
perfectly elastic until the maximum tension at the contact circle, σ_T,
exceeds the theoretical cohesive tensile strength, p_Th. Such an ideally elastic
response is most likely to be realised in the harder, more covalent solids;
i.e. high H/E, sect. 8.1.3. Since we are now operating in the domain of
'small flaws', i.e. well below the limits of validity of Auerbach's law
(sect. 8.4.1), a cone crack initiates spontaneously. Solving the elasticity
relations (8.2) and (8.17) directly for the critical load gives

$$P_\mathrm{Th} = [2\pi/(1-2v)]^3 (4kr/3E)^2 p_\mathrm{Th}^3. \tag{9.1}$$

For a quartz dust particle $r = 1$ μm, say, on glass, with $E = 70$ GPa, $v = \frac{1}{3}$, $k = 1$ (sect. 8.1.2), $p_{\mathrm{Th}} = E/10$ (sects. 1.5, 6.6.1), we calculate $P_{\mathrm{Th}} \approx 1$ N. From extrapolation of the data in fig. 8.4(a), we estimate a corresponding pop-in cone-crack size ≈ 8 μm.

A contact load of 1 N is orders of magnitude greater than the weight of any quartz dust particle, but might be attained if the particle were to impact at high velocity (fig. 9.1(b)) or slide in a constrained surface–surface contact (fig. 9.1(c)). The 'chatter' track on a natural diamond seen in fig. 9.2(a) (cf. fig. 8.6) demonstrates that microcontact damage of the latter kind is possible on even the hardest surface. The most severe microcontact events are those involving sharp particles (fig. 9.1(d)). We recall from sect. 8.4.2 that even subthreshold contacts can produce deformation-induced degrading flaws, e.g. shear faults augmented by residual stresses. Subtle effects like smearing by surface melting during sliding, as evident at the scratch on the glass surface in fig. 9.2(b), can obscure subsurface damage of this type; damage which may nevertheless be severe enough to cause crack pop-in, thereby interfering with certain wavelengths in an optical component, or, in extreme cases, destroying the component (e.g. laser damage, sect. 9.3.3).

9.1.2 Flaw distributions

As with most natural flaw types, surface microcracks due to spurious contacts and similar surface stress concentrations are subject to considerable variability. This variability is strongly manifest in the strength of as-handled glass. It is the grounds for statistical theories of structural design, which we shall address in the final chapter.

An interesting device for determining distributions of surface flaws on glass and other homogeneous materials is a macroscopic Hertzian probe. One takes a spherical indenter of radius $r \gg$ flaw size c_{f}, and records the critical loads P_{C} and radial distances R_{C} at which full cone cracks initiate. Then, in the 'small-flaw limit' $c_{\mathrm{f}} \ll c_0$ of fig. 8.31 (sect. 8.4.1), one estimates c_{f} by assuming that cone fracture occurs when the tensile stress σ_{T} on the critical flaw at $R_{\mathrm{C}} \geq a$ exceeds the bulk tensile strength σ_{F}. Combining (8.17) with (8.2), together with the Hertzian elasticity result $\sigma_{\mathrm{R}} = \sigma_{\mathrm{T}}(a/R)^2$ (cf. (8.1)), gives

$$c_{\mathrm{f}} = (\beta R_{\mathrm{C}}^2/P_{\mathrm{C}})^2 \qquad (9.2)$$

with β a material constant. Several hundred probes can usually be made on a single test surface, and the flaw count $n(c_{\mathrm{f}})$ for any specified size range thereby determined.

An estimate of the surface density of flaws within any such specific size range may be obtained by dividing the appropriate flaw count by the sum of areas $A_i(c_{\mathrm{f}})$ 'searched',

$$\lambda(c_{\mathrm{f}}) = n(c_{\mathrm{f}})/\sum_{i=1}^{N} A_i(c_{\mathrm{f}}), \quad (N \text{ indentations}). \tag{9.3}$$

We define $A_i(c_{\mathrm{f}})$ for load P_i as that surface area within which a flaw of size c_{f} (here taken as the mid-range value) would have caused cone fracture. Note that this definition does not necessarily require that fracture actually occurs, or even that there be any flaws within the searched area. The searched area is an annulus of major radius R_i and minor radius R_0; R_i has a meaning analogous to R_{C} in (9.2), i.e. a flaw of size c_{f} at $R = R_i = c_{\mathrm{f}}^{1/2} P_i/\beta$ just becomes critical at $P = P_i$, *provided* the flaw is not first encompassed by the compressive zone within the contact circle at the minor radius $R = R_0 = a(P_0) = \alpha P_0^{1/3}$ (from (8.2)) at $P = P_0$. Thus

$$\sum_{i=1}^{N} A_i(c_{\mathrm{f}}) = 0, \quad (R_i \leqslant R_0, P_i \leqslant P_0)$$

$$\left. \begin{aligned} \sum_{i=1}^{N} A_i(c_{\mathrm{f}}) &= \sum \pi(R_i^2 - R_0^2), \quad (R_i > R_0) \\ &= \sum \pi c_{\mathrm{f}}^{1/2} P_i/\beta - \sum \pi \alpha^6 \beta^2/c_{\mathrm{f}} \\ &= (\pi c_{\mathrm{f}}^{1/2}/\beta)\sum P_i - (\pi \alpha^6 \beta^2/c_{\mathrm{f}})N', \quad (P_i > P_0). \end{aligned} \right\}$$

The searched area is then determined by summing the N' ($\leqslant N$) indenter loads P_i in excess of P_0.

A set of experimental data for a glass plate is reproduced as the histogram in fig. 9.3. The mean flaw size ≈ 1 μm and spacing ≈ 20 μm evaluated from this particular data set are typical of as-handled surfaces.

We would re-emphasise that the identification of surface tensile stress σ_{R} with bulk strength σ_{F} in this calculation is contingent on the uniformity of stress over the full depth of the microcrack. We took pains in sect. 8.4.1 to point out the dramatic downward gradients that prevail in Hertzian fields, especially close to the contact circle, so that even micrometre-scale flaws

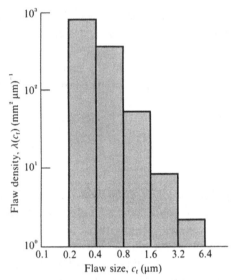

Fig. 9.3. Histogram showing variation of flaw density with flaw size for 'as-received' plate glass. Results from Hertzian tests, tungsten carbide ball $r = 0.35$ mm, $N = 99$, $N' = 97$. (After Poloniecki, J. D. & Wilshaw, T. R. (1971) *Nature* **229** 226.)

may suffer highly non-uniform stresses in normal contact loading. The above analysis can therefore be expected to underestimate c_i, especially toward the right of fig. 9.3. The issue of stress gradients has been largely overlooked by flaw statistics analysts.

It is apparent that exposed surfaces of brittle solids are susceptible to damage by the most minute of extraneous mechanical agents. We have seen the deleterious side of such damage with regard to strength degradation and, in more severe cases, erosion and wear (sect. 8.7). We have also encountered some benefits, in the controlled abrasion of surfaces for reproducible strength testing (e.g. sects. 8.2.1, 8.4.1). Preventative measures may be taken to minimise the incidence of microcontact damage by applying a protective coating to pristine or chemically polished surfaces.

9.2 Crack nucleation at dislocation pile-ups

Certain 'soft' ceramics, notably ionic solids with the rocksalt structure, may undergo limited room temperature plasticity in homogeneously applied stress fields prior to the onset of fracture, in a manner akin to those more brittle metals with restricted numbers of primary slip systems. The failure of such 'semi-brittle' solids then relates less to surface flaw population than to yield stress, as reflected for instance in the tendency for the strengths in compression to be of the same order as those in tension. The flow processes assume a controlling role in the crack initiation by augmenting and sometimes even creating the crack nuclei. Consequently, the component of resolved shear stress on the slip planes is at least as critical in the flaw micromechanics as the tensile stress on the ensuing crack planes.

Examples of this kind of crack nucleation are evident in the micrographs of deformed magnesium oxide bicrystals in fig. 9.4. Upon loading beyond the yield point, dislocation sources operate on {110} primary glide planes and cross slip to form discrete slip bands. The bands *pile up* at the grain boundary and concentrate the stress, fig. 9.4(a). If the misorientation between the adjoining grains is too severe the shear bands are unable to penetrate across the boundary (sect. 7.1.1), so microcracks generate, fig. 9.4(b).

The original pile-up model by Zener (1948) and subsequent derivatives by Stroh, Cottrell and others were developed principally in the context of metals, but the basic concept extends naturally to the soft ionic ceramics (as well as to the harder covalent ceramics above the brittle–ductile transition temperature). All such models are essentially based on the common hypothesis of a stress concentration at the edge of a shear-activated dislocation array, where some internal microstructural barrier, e.g. second-phase particle, grain boundary (fig. 9.4), adjacent slip band, microtwin, etc., constrains further penetration of the array. Our intent here is to present a generic, if simplistic, description of the pile-up model in terms of essential dislocation and barrier parameters.

Accordingly, consider the system in fig. 9.5. Under the action of the resolved shear stress σ_{xy} the source S generates n edge dislocation loops on the slip plane. These dislocations pile up against the barriers B, B′, distance d apart. The slip relaxes the mean shear stress to the lattice friction level $\sigma_{\rm C}^{\rm D}$ (sect. 7.3.1), corresponding to a relaxation of the elastic strain to $(\sigma_{xy} - \sigma_{\rm C}^{\rm D})/\mu$, with μ the shear modulus. This relaxation is accommodated by the plastic strain nb/d, with b the magnitude of the dislocation Burgers vector. Thus the number of dislocations needed to maintain equilibrium is

$$n = (\sigma_{xy} - \sigma_{\rm C}^{\rm D})\, d/\mu b. \tag{9.4}$$

Now examine the condition for microcrack nucleation in relation to

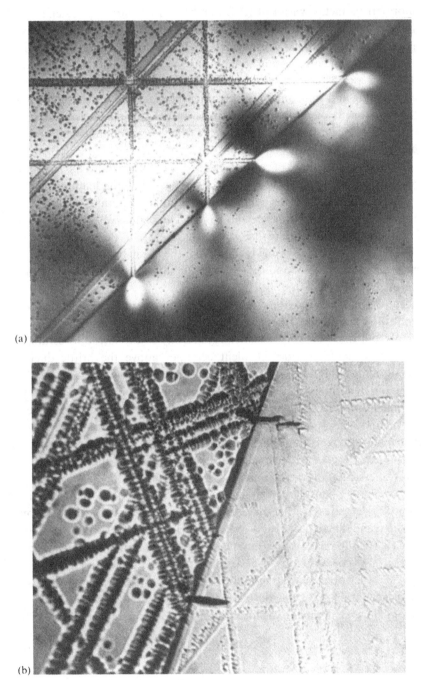

(a)

(b)

Fig. 9.4. For legend see facing page.

cohesive-bond rupture in the region of high stress concentration (Petch 1968). From dislocation theory it can be shown that the applied load exerts an effective force $(\sigma_{xy} - \sigma_C^D) b$ on each dislocation in the pile-up. The total force exerted by all the dislocations on the obstacles is therefore $(\sigma_{xy} - \sigma_C^D) nb$; that is, the pile-up acts as a single 'superdislocation' of Burgers vector nb, concentrating the shear stress at the slip-plane extremities by a factor n. A formal treatment of the pile-up problem indicates that the concentrated field contains a component of tension comparable with that of shear, acting on a plane inclined to that of the slip. Then the condition for microcrack nucleation may be written approximately

$$(\sigma_{xy} - \sigma_C^D) n = p_{Th} \tag{9.5}$$

with p_{Th} the theoretical cohesive strength in tension. From (6.6) we may eliminate p_{Th} in favour of surface energy γ_B to obtain

$$\sigma_{xy} = \sigma_C^D + \pi \gamma_B / nb. \tag{9.6}$$

We see that the applied stress to maintain the pile-up in equilibrium reduces toward the lattice stress σ_C^D as the number of dislocations increases.

Eliminating n from (9.4) and (9.6) and identifying $\sigma_F = \sigma_{xy}$ with the strength of the material with active dislocation sources results in

$$\sigma_F = \sigma_C^D + (\pi \mu \gamma_B / d)^{1/2} \tag{9.7}$$

which is the so-called 'Petch relation'. The $d^{-1/2}$ dependence signifies a defect-controlled property. In this sense (9.7) may be viewed as a shear counterpart of the Griffith relation (1.11) for tensile cracks, with the pile-up length d equivalent to microcrack length c in that earlier relation. It is this kind of 'equivalence' that has led theoretical elasticians like Eshelby, Frank & Nabarro and Bilby, Cottrell & Swindon to use hypothetical dislocation arrays to represent the stress fields in crack problems.

The clear implication from the Petch relation is that the strength of semi-

---

Fig. 9.4. Pile-up of slip band on {110} planes at grain boundary in magnesium oxide. (a) Showing stress concentrations after deformation. Transmitted polarised light. Width of field 3000 μm. (After Ku, R. & Johnston, T. L. (1964) *Phil. Mag.* **9** 231.) (b) Showing ensuing microcrack formation in etched crystal. Width of field 150 μm. (After Johnston, T. L., Stokes, R. J. & Li, C. H. (1962) *Phil. Mag.* **7** 23.)

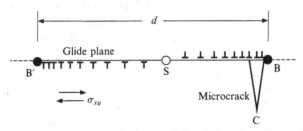

Fig. 9.5. Nucleation of microcrack by dislocation pile-up. Under shear stress σ_{xy} source S generates dislocation loops which pile-up at barriers B, B' in glide plane. Stress concentration at barrier B nucleates microcrack BC.

Fig. 9.6. Microcracks on (100) planes of sodium chloride, formed at edge of crystalline deposit on polished (001) surface. Note cube-like voids in deposit. Width of field 400 μm. (After Stokes, R. J., Johnston, T. L. & Li, C. H. (1960) *Trans. Met. Soc. A.I.M.E.* **218** 655.)

brittle solids may be improved by refining the microstructure. However, while widely applicable to metals, (9.7) does not carry over universally to brittle ceramics. It presumes that the condition for nucleating a microcrack corresponds identically with that for failure, and thereby ignores any flaw stabilisation that may occur from shielding and residual nucleation stresses (chapters 7, 8), as we shall discuss in sect. 9.6.

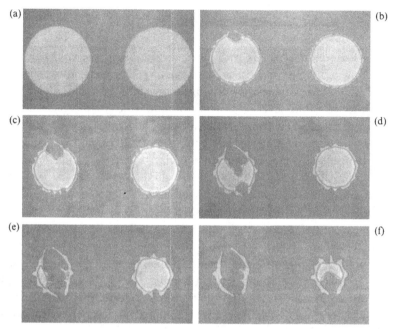

Fig. 9.7. Healing of cracks in sapphire. Basal-plane cracks were introduced by lithographically etching a disc-shaped depression into one of two matching crystal halves, and then sintering the two halves together for 1 hr at 1370 °C. Sequence shows evolution of 'flaws' after annealing at 1800 °C for (a) 0, (b) 4, (c) 8, (d) 14, (e) 22, and (f) 35 hr. Width of field 600 μm. (After Rödel, J. & Glaeser, A. M. (1990) *J. Amer. Ceram. Soc.* **73** 592.)

9.3 Flaws from chemical, thermal, and radiant fields

Solids in an initially pristine, defect-free state, such as freshly drawn glass fibres, whiskers and dislocation-free covalent monocrystals, may develop Griffith flaw populations on exposure to chemical, thermal or radiant fields. Aged solids with pre-existing defects may evolve populations of greatly enhanced severity (especially if pop-in ensues) from interactions with such fields. Occasionally the flaw severity may be reduced, by healing. We describe a few examples of such extraneously induced flaw types here. Many others undoubtedly exist.

9.3.1 Chemically-induced flaws

The interaction of solids with chemical species in the environment can generate a diversity of surface flaws. At very high temperatures corrosion layers form, in extreme cases driving fissures into the underlying substrate. At a more subtle level, crystal dissolution and growth generate stress-raising etch pits and grooves at dislocation and grain boundary outcrops. Under superimposed external loads these and other incipient surface inhomogeneities, e.g. embedded particles or microcontact deformation sites, may evolve into ellipsoidal (or elliptical) cavities by preferential dissolution in regions of stress concentration, creating 'Inglis flaws' (sect. 1.1) from which sharp cracks may ultimately initiate.

One means of 'decorating' surface flaws in alkali-silicate glasses is by ion exchange (Ernsberger 1960). Exposure to specific chemical environments promotes exchange of Na^+ ions in the open network for smaller species, e.g. H^+ (in acid solution) or Li^+ (in molten salt). This leaves the glass surface layers in a state of residual tension, causing pre-present flaws to extend laterally into a shallow craze ('mudflat') pattern. The question as to whether some members of the final flaw population are actually created by the chemistry is not easily resolved. We will provide an illustrative example of such surface cracking in the next subsection.

In certain instances chemistry can change the very nature of the failure mode. A celebrated case is the 'Joffe effect', discovered in 1924. Sodium chloride monocrystals, brittle in air, behave in a ductile mode when tested under water. The latter mode is explained by dissolution of defective surface layers in solution, thereby removing critical flaws. What is not so easily explained is the reversion to brittle behaviour when the crystals are removed from the water, dried and re-tested in air. A thin layer of saturated solution adheres to the crystal surfaces, leaving a highly defective deposit of dried matter which generates high stress concentrations. The end result is the parallel array of severe surface microcracks seen in fig. 9.6.

9.3.2 Thermally-induced flaws

Heating cycles can significantly modify flaw states. Dislocations and grain boundaries are subject to thermal etching at surface sites, which produces surface pits and grooves similar to chemical attack. Coalescence of point defects or impurities at internal heterogeneities generates bubbles and

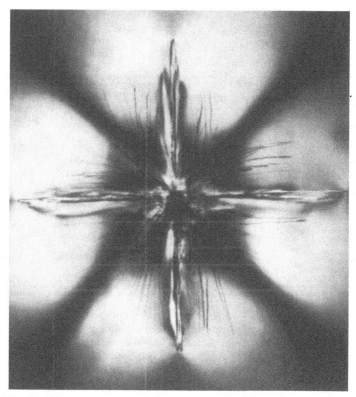

Fig. 9.8. Damage pattern at defect centre in lithium fluoride after radiation with YAG laser (power density 30 MW mm^{-2}, pulse interval 8 ns). Penny-like radial cracks generate along {100} planes. Transmitted polarised light. Width of field 1 000 μm. Cf. indentation crack, fig. 8.8. (After Wang, Z-Y., Harmer, M. P. and Chou, Y. T. (1989) *J. Mater. Sci.* **24** 2756.)

second-phase inclusions. Thermal aging enhances devitrification in glass and phase changes in polycrystalline ceramics, setting up favoured sites for microcrack generation and development (sect. 9.4). In extreme cases the combined action of elevated temperatures and applied stresses can generate entirely new flaw populations, especially if creep occurs.

Under the right conditions thermal cycling can actually be beneficial, by healing any existing microcracks. It is well known from studies of sintering in ceramics and pressure heating in rocks that fissures can be virtually eliminated by matter transport (surface–volume diffusion, evaporation–condensation) to the interface. An example of this kind of healing at artificially produced internal cracks in sapphire is shown in fig. 9.7. Such

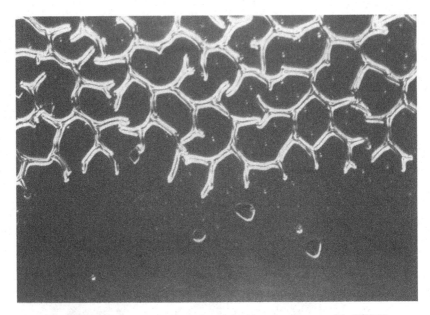

Fig. 9.9. Soda-lime glass surface after bombardment with 480 kV protons, dose 200 ions nm^{-2}. (Lower region masked from incident flux.) Mudflat craze pattern of shallow microcracks is due to lateral propagation of mutually intersecting surface flaws. Optical micrograph. Width of field 150 μm.

microcracks will clearly become less deleterious with annealing time, but may never be fully eliminated by heating alone.

9.3.3 *Radiation-induced flaws*

Radiation can inject substantial doses of energy into solids, with consequent adverse effects on flaw characteristics. We illustrate here with two examples, one using photon and the other particle radiation sources.

The first example considers the damage induced in a nominally transparent lithium fluoride monocrystal by a laser pulse. Fig. 9.8 shows a radial crack pattern resulting from adiabatic energy absorption at a submicroscopic *internal* inhomogeneity. The stress birefringence reveals a strong residual stress field around the 'zapped' inhomogeneity site. The similarity in the birefringence pattern with that from inelastic indentations (e.g. fig. 8.8) is striking, suggesting that the localised energy absorption generates a centre of expansion (sect. 8.1.3). Analogous popped-in crack

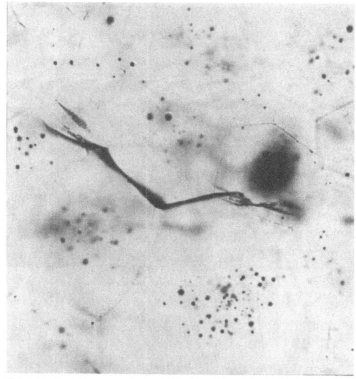

Fig. 9.10. Spontaneous microcracking associated with thermal expansion anisotropy stresses in alumina. Initiation occurs at tensile grain facets, arrest at compressive facets. Note small pores within individual grains. Transmitted light. Width of field 125 μm. (Courtesy P. Chantikul & S. J. Bennison.)

systems have been observed at impurity centres in glasses. It is clear that even minuscule defects have the potential to evolve into severe microcracks during pulsed in-service radiation.

Our second example demonstrates the effects of ion bombardment on pre-existing *surface* flaws. Fig. 9.9 shows a soda-lime glass specimen after exposure to a flux of protons. The irradiation has produced a damage state of residual tensile stress in the glass surface, generating a mudflat surface crack pattern of the type referred to in the preceding subsection. The implanted proton species may remain as mobile hydrogen, with persistent damage evolution in the form of chemically assisted slow crack growth.

Radiation damage of an even more extensive kind can result from prolonged exposures to intense fluxes, such as those in nuclear reactors. Specific processes include the cumulation of point defects (with consequent

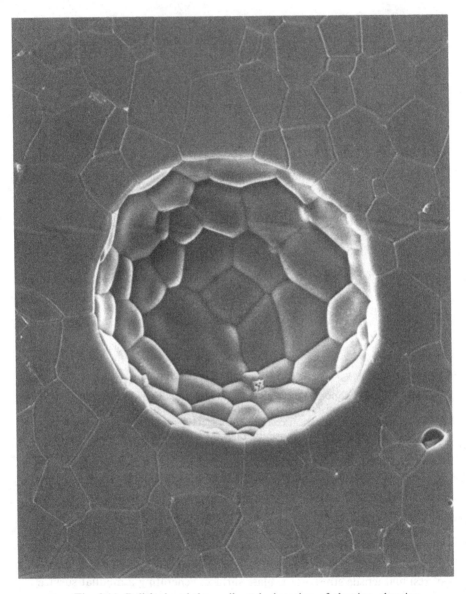

Fig. 9.11. Polished and thermally etched section of alumina, showing large pore. Scanning electron micrograph. Width of field 100 μm. (Courtesy J. S. Wallace.)

formation of pressurised gas bubbles), displacement spikes and cascades, and nuclear transmutations.

9.4 Processing flaws in ceramics

Modern tough ceramics tend to have complex microstructures. But the same microstructure responsible for high toughness can also be the source of an assortment of stress-concentrating flaws. As a rule of thumb, the size of microstructural flaws scales with some characteristic dimension of the microstructure itself. In coarser materials, therefore, the defect that leads to failure is more likely to originate from the fabrication procedure than from any subsequent surface finishing. The detailed processing route, powder preparation (including impurity content), consolidation, sintering, and aging, is then key to the critical flaw population (Davidge 1979; Dörre & Hübner 1984; Lange 1978). Some of the more common processing flaw types are:

(i) *Microcracks.* Polycrystalline ceramics are vulnerable to sub-facet defects at grain and interphase boundaries, especially triple-point junctions, at the surface and in the bulk. Internal thermal expansion and elastic mismatch stresses augment microcrack nucleation during the cooling stages of processing, as per sect. 7.3.2. Recall from (7.17) for monophase ceramics that such microcracking is expected above a critical grain size

$$l_\mathrm{c} = \Phi(T_0/\sigma_\mathrm{R})^2 \tag{9.8}$$

where T_0 is the grain boundary toughness, σ_R is the internal stress, and $\Phi = \pi/4\beta$ is a dimensionless constant with β the initial defect-to-grain size ratio. For alumina, with $T_0 \approx 2.5$ MPa $\mathrm{m}^{1/2}$, $\sigma_\mathrm{R} \approx 250$ MPa, $\beta \approx 0.5$ (say), we compute $l_\mathrm{c} \approx 150\ \mu\mathrm{m}$. Actually, this estimate is just a mean value, because in real polycrystals the grain boundary misorientations, and hence the σ_R (and even β) values in (9.8), inevitably vary from facet to facet. Accordingly, we may expect isolated microcracks to initiate at $l < l_\mathrm{c}$. An example is shown for an alumina with $l = 80\ \mu\mathrm{m}$ in fig. 9.10. The microcracks initiate at tensile grain facets, and arrest at adjacent compressive (bridging) facets after extensions through ≈ 2–3 grain diameters. Subsequently applied external fields can exacerbate the flaw severity by further extending such popped-in microcracks through the bridging field. For grain sizes $l > l_c$ the density of microcracking becomes sufficiently high that coalescence between neighbours ensues, with virtually total loss in strength.

Analogous cracks can result from sintering stresses, by shrinkage around abnormally large or agglomerated grains.

(ii) *Pores*. Pores can assume several forms. Small sintering pores at grain boundary triple points are preferred sites for microcrack initiation. Those contained wholly within grains (e.g. fig. 9.10), are relatively innocuous. It has been proposed that the major influence of porosity P on strength properties of flaw-sensitive ceramics is via a diminished elastic modulus, according to the empirical relation

$$\sigma_F = \sigma_0 \exp(-bP) \tag{9.9}$$

with σ_0 and b adjustables. This means that for a 'typical' value $b \approx 7$ the strength reduces to about one-half at 10 % porosity.

Large voids of the type seen in fig. 9.11 can arise from burnout of sintering aids or imperfect powder packing in the green compacts. These voids tend to be more prevalent in materials with non-uniform grain sizes and phases. They concentrate applied stresses over distances large compared to a single grain diameter, and are therefore common failure origins. Re-entrant facets provide favoured nucleation sites for extension. For a spherical void of radius a with an annular crack of effective radial width $c-a$ in tensile loading σ_A, the stress-intensity factor is

$$K_A(c) = \psi \sigma_A (c-a)^{1/2} f(a/c), \quad (c \geqslant a). \tag{9.10}$$

The function $f(a/c)$ decreases monotonically with increasing c, from $f = 2$–3 (depending on Poisson's ratio) at $c = a$ to $f = 1$ at $c \gg a$ (as required for (9.10) to restore to $K_A(c) = \psi \sigma_A c^{1/2}$). (At ellipsoidal voids, the function $f(a/c)$ includes an eccentricity concentration factor, cf. sect. 1.1.)

(iii) *Inclusions*. Inclusions from second-phase particles and impurity agglomerates can be even more effective than pores as sources of failure. The severity of these flaw sources is aggravated by the presence of local residual stress fields, the nature and intensity of which depend on

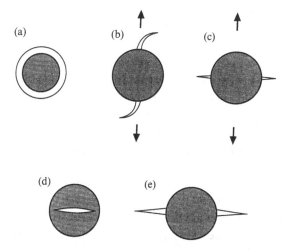

Fig. 9.12. Modes of inclusion-induced fracture, according to value of thermal expansion coefficient (α), elastic modulus (E) and toughness (T) of particle (P) relative to matrix (M). In order of increasing severity: (a) highly contracting, weakly bonded rigid particle ($\alpha_P \gg \alpha_M$, $E_P \gg E_M$), void-like defect; (b) contracting, strongly bonded, stiff, tough particle ($\alpha_P > \alpha_M$, $E_P > E_M$, $T_P > T_M$), polar tensile field in applied tension from elastic mismatch; (c) similar, but compliant particle ($\alpha_P > \alpha_M$, $E_P < E_M$), equatorial tensile field in applied tension; (d) similar again, but weak particle ($\alpha_P > \alpha_M$, $E_P \approx E_M$, $T_P < T_M$), particle failure; (e) contracting particle ($\alpha_P < \alpha_M$), with spontaneous radial crack pop-in above critical size ('most severe case'). (After scheme by A. G. Evans.)

differences in thermal expansion coefficients and elastic moduli between inclusion and matrix. Some of the possible modes of inclusion-induced cracking (cf. fig. 7.8) are depicted in fig. 9.12. A detailed fracture mechanics consideration of the most severe mode will be given in the following section.

The co-existence of two or more of the above flaw types gives rise to a multimodal flaw distribution. In cases of high flaw densities, neighbouring members may interact and coalesce. This can be highly deleterious if the material has no inbuilt stabilising influence, e.g. no toughness-curve (see sect. 9.6), in which case refinement of processing routes to eliminate large defects (Lange 1989) is advocated.

9.5 Stability of flaws: size effects in crack initiation

In practically all flaw systems considered thus far the elemental forces responsible for crack *nucleation* persist to drive the ensuing *propagation*. Cottrell (1958) was one of the first to analyse the influence of residual internal stresses on flaw instability, in the context of failure from dislocation pile-ups (fig. 9.5). The strong falloff in the residual tensile field about the microcrack origin is manifested as a precursor stage of extension prior to failure in subsequent applied loading. We have encountered analogous stabilising behaviour in our analysis of the role of residual stress fields on the evolution of indentation cracks (chapter 8). Accordingly, any general theory of strength should include provision for incorporating flaw-localised driving forces: the spontaneous failure of the ideal Griffith microcrack may be the exception rather than the rule.

We illustrate the general principle here with an analysis of a specific flaw system, that in fig. 9.13 of a spherical inclusion, radius a, exerting an outward pressure, σ_R, on a matrix of (single-valued) toughness T_0 ('worst' case in fig. 9.12). A uniformly applied tensile stress σ_A is superimposed on the system. Annular cracks of outer radius c $(c > a)$ extend radially from the particle circumference on a diametral plane normal to the applied stress. Although detailed solutions of this problem are available (Lange 1978; Green 1983), the physical principles are most adequately demonstrated in a simplified derivation, courtesy D. B. Marshall.

Analysis begins with expressions for the stress field on the crack plane. From (7.5), the internal tangential tensile component is

$$\left. \begin{array}{l} \sigma_I = \sigma_t = \tfrac{1}{2}\sigma_R(a/r)^3, \quad (a \leqslant r \leqslant c) \\ \sigma_I = 0, \quad (0 \leqslant r \leqslant a) \end{array} \right\}. \tag{9.11}$$

We emphasise at the outset that σ_R is a *material constant*, e.g. a thermal expansion anisotropy stress (7.4) or critical dilation phase transformation stress (sect. 7.4). This internal stress contributes to a residual K-field K_R. Similarly, the applied stress

$$\left. \begin{array}{l} \sigma_I = \sigma_A, \quad (a \leqslant r \leqslant c) \\ \sigma_I = 0, \quad (0 \leqslant r \leqslant a) \end{array} \right\} \tag{9.12}$$

contributes to a K-field K_A. The net K-field, $K_* = K_A + K_R$, is found by

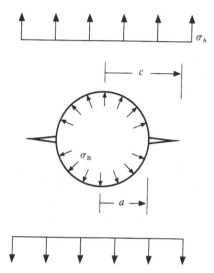

Fig. 9.13. Initiation of annular crack from particle with superposed outwardly directed internal residual pressure σ_R and applied tensile stress σ_A, showing essential coordinates.

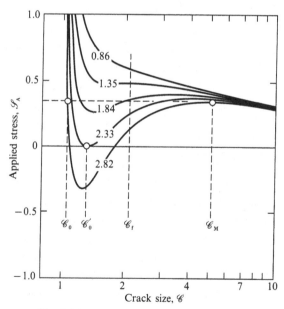

Fig. 9.14. Reduced plot of applied stress $\mathscr{S}_A$ as function of annular crack radius $\mathscr{C}$ from spherical particle, for indicated values of residual K-field parameter $\mathscr{S}_R$ in (9.14). Ensuing stability of system depends on size $\mathscr{C}_f$ of incipient flaw relative to $\mathscr{C}_0$ and $\mathscr{C}_M$.

inserting (9.11) and (9.12) directly into (2.22b) for penny-like cracks and integrating:

$$K_* = 2\sigma_A (c/\pi)^{1/2} (1 - a^2/c^2)^{1/2} + \sigma_R (a^2/\pi^{1/2}c^{3/2}) (1 - a^2/c^2)^{1/2}.$$
(9.13)

We note in passing the familiar limiting forms: small extensions, $\Delta c = c - a \ll a$, $K_* \propto \Delta c^{1/2}$ (negligible gradient); large extensions, $c \gg a$, $K_A \propto c^{1/2}$ (uniform stress over entire crack area), $K_R \propto c^{-3/2}$ (centre-loaded penny).

One may now solve for the applied stress as a function of crack size at equilibrium, $K_* = T_0$. Introducing normalised variables, crack length $\mathscr{C} = c/a$, applied stress $\mathscr{S}_A = 2\sigma_A a^{1/2}/\pi^{1/2}T_0$, and residual stress $\mathscr{S}_R = \sigma_R a^{1/2}/\pi^{1/2}T_0$, the equilibrium solution assumes a dimensionless form,

$$\mathscr{S}_A = 1/(\mathscr{C} - 1/\mathscr{C})^{1/2} - \mathscr{S}_R/\mathscr{C}^2.$$
(9.14)

Fig. 9.14 plots $\mathscr{S}_A(\mathscr{C})$ for various $\mathscr{S}_R$, in relation to an incipient flaw size $\mathscr{C}_f$. In general, the curves have three branches, two unstable and one stable, delineated by a maximum and minimum. Then, depending on $\mathscr{S}_R$ and $\mathscr{C}_f$, the stability configurations lie in one of three domains:

(i) *No pop-in, spontaneous failure* $(0 < \mathscr{S}_R < 1.35, \mathscr{C}_f > 1)$. The residual field is small and the curve has no extrema. The crack nucleus $\mathscr{C}_f$ propagates unstably and spontaneously to unlimited failure at critical $\mathscr{S}_A = \mathscr{S}_F$ in the manner of Griffith flaws.

(ii) *Activated pop-in, activated failure* $(1.35 < \mathscr{S}_R < 2.33, \mathscr{C}_0 < \mathscr{C}_f < \mathscr{C}_M)$. The residual field is of intermediate intensity and the curve has a minimum. However, the minimum occurs at $\mathscr{S}_A > 0$, so there can be no pop-in without the superposition of an applied field. On increasing $\mathscr{S}_A$ the flaw pops-in to the stable branch and extends with further loading to $\mathscr{C} = \mathscr{C}_M$, corresponding to ultimate failure at $\mathscr{S}_A = \mathscr{S}_M = \mathscr{S}_F$. Observe that the maxima in the curves in fig. 9.14 generally occur at large crack sizes, $\mathscr{C} \gg 1$, implying that once the crack reaches the stable branch the residual far field approaches that of a central point load.

At very small $(1 < \mathscr{C}_f < \mathscr{C}_0)$ and very large $(\mathscr{C}_M < \mathscr{C}_f < \infty)$ flaw sizes the system reverts to the spontaneous failure of case (i).

(iii) *Spontaneous pop-in, activated failure* $(2.33 < \mathscr{S}_R < \infty, \mathscr{C}'_0 < \mathscr{C}_f < \mathscr{C}_M)$. The residual field is of high intensity, so that the minimum in the curve is depressed below the $\mathscr{S}_A$ axis. The flaw now pops-in spontaneously without any applied load, and failure is again determined by precursor extension to the maximum at $\mathscr{S}_M$.

For somewhat smaller flaws, $\mathscr{C}_0 < \mathscr{C}_f < \mathscr{C}'_0$, initiation must be activated, case (ii). Again, at very small $(1 < \mathscr{C}_f < \mathscr{C}_0)$ and very large $(\mathscr{C}_M < \mathscr{C}_f < \infty)$ flaws, failure is spontaneous, case (i).

The above description opens the way to some very general conclusions concerning the nature of *thresholds* in crack initiation. One way conveniently classifies flaws as under-, well- and over-developed, depending on whether $\mathscr{C}_f$ intersects the first, second or third branch in fig. 9.14. Then a minimum requirement for spontaneous initiation, given the most favourable availability of under-developed nuclei, is that the intensity of the residual stress field for the inclusion should be sufficiently large that the minimum in $\mathscr{S}_A(\mathscr{C})$ occurs at $\mathscr{S}_A = 0$, i.e. corresponding to $\mathscr{S}_R = 2.33$ in fig. 9.14. Insofar as σ_R is a material constant in $\mathscr{S}_R = \sigma_R a^{1/2}/\pi^{1/2}T_0$, this requirement implies a critical particle radius

$$a_C = 5.52\pi(T_0/\sigma_R)^2. \tag{9.15}$$

Critical size relations of this kind have a fundamental universality (Puttick 1980), as the reader may affirm by recalling the corresponding relations (9.8) for grain-boundary microcracks and (8.21a) for indentation radial cracks $(\sigma_R \propto H)$.

The stabilising effect of the local residual field on the flaw evolution is reflected in the development of the extrema in $\mathscr{S}_A(\mathscr{C})$ at larger $\mathscr{S}_R$ in fig. 9.14. Flaws within the 'window' $\mathscr{C}_0 < \mathscr{C}_f < \mathscr{C}_M$ thereby undergo a characteristic stage of precursor extension immediately prior to final instability at $\mathscr{C} = \mathscr{C}_M$, $\mathscr{S}_A = \mathscr{S}_M$. Simple analytical solutions for these critical configurations may be determined from (9.14) in the approximation $\mathscr{C} \gg 1$ (consistent with our earlier observation that the maxima in fig. 9.14 tend to occur in this far-field domain). In this approximation, invoking $d\mathscr{S}_A/d\mathscr{C} = 0$ $(d^2\mathscr{S}_A/d\mathscr{C}^2 < 0)$ and recalling the definitions $\mathscr{C}_M = c_M/a$, $\mathscr{S}_M = 2\sigma_M a^{1/2}/\pi^{1/2}T_0$, we obtain

$$c_M = (4\sigma_R a^2/\pi^{1/2}T_0)^{2/3} \tag{9.16a}$$

$$\sigma_M = \tfrac{3}{8}(\pi^2 T_0^4/4\sigma_R a^2)^{1/3}. \tag{9.16b}$$

These solutions are of exactly the same form as the indentation-strength relations (8.9) for radial cracks in residual elastic–plastic fields, with $P \approx \sigma_R a^2$ an 'effective contact force'.

The possibility that flaws may evolve subcritically during applied stressing reinforces the need for proper caution when using non-destructive evaluations of flaw characteristics to pre-determine strength properties: σ_M in (9.16) is governed by the intrinsic crack size c_M, *not* the extrinsic starting flaw size c_f. Finally, the coexistence of under-developed and well-developed inclusions constitutes yet another source of bimodal flaw populations.

9.6 Stability of flaws: effect of grain size on strength

It is well documented that the strength of polycrystalline ceramics tends to diminish with increasing coarseness of the microstructure. Plotted against the inverse square root of grain size, strength data for homogeneous monophase ceramics can be fitted as two linear branches: at large grain sizes, an 'Orowan' branch through the origin; at small grain sizes, a 'Petch' branch intercepting the strength axis. The simplest interpretation of the Orowan branch is that of spontaneous failure from bulk 'Griffith' flaws whose initial size is directly proportional to the grain size (e.g. as assumed for sub-facet flaws in the derivation of the microcracking relation (9.8)). The Petch branch has been loosely attributed to the modifying influence on flaw growth by some internal residual field (e.g. as expressed by (9.7) in the pile-up model). But many variant micromechanical models of failure have been proposed for both branches; so that, despite its continued use over the past thirty years or so, the Orowan–Petch plot has remained a topic of considerable conjecture in the ceramics literature.

This state of conjecture is attributable to a conspicuous lack of direct observations of failure evolution in any specific flaw system. *In situ* experiments using controlled flaws are one recent exception. As demonstrated earlier (sect. 7.5.1), observations of radial crack extension from indentations have helped to identify grain-interface bridging as a primary cause of flaw stabilisation in aluminas and other polycrystalline ceramics. Such stabilisation profoundly reduces the sensitivity of strength to starting crack dimensions (sect. 8.2.3): it confers flaw tolerance. The implication is that the role of grain size as a scaling factor may be felt more strongly in the bridging micromechanics, i.e. in the toughness-curve, than in the initial flaw characteristics.

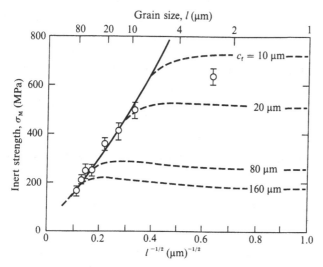

Fig. 9.15. Strength as function of inverse square root grain size for alumina (same material as figs. 7.29, 8.24), showing Orowan (left) and Petch (right) branches. Data are breaks for as-polished surfaces. Curves are calculations for natural flaws using calibrated T-curves: solid curve is prediction for microstructural, sub-facet grain boundary flaws; dashed curves are predictions for extraneous flaws of specified initial flaw sizes c_f. (After Chantikul, P., Bennison, S. J. & Lawn, B. R. (1990) *J. Amer. Ceram. Soc.* **73** 2419.)

As an illustration, consider the strength/grain-size plot in fig. 9.15 for a series of aluminas of different grain size l. Data are from the alumina study of fig. 8.24, but represent failures from natural flaws. Curves are corresponding strength predictions using the calibrated T-curve analysis from sect. 8.2.3: solid curve for sub-facet flaws of sizes $c_f = 0.5l$ (as used in evaluating (9.8)); dashed curves for extraneous flaws of sizes c_f specified. An Orowan–Petch transition is apparent at each value of c_f. From the convergence of the curves at large grain sizes we conclude that the Orowan branch is an *intrinsic material function*, totally independent of c_f, reflecting the flaw tolerance. This branch does not conform exactly to the classical $l^{-1/2}$ dependence, indicating a fundamentally complex relation for T-curve materials. Conversely, the Petch branch is predicted to be strongly dependent on initial flaw size. Only one data point in fig. 9.15, that for the finest grain size, lies in this latter region, suggesting the existence of extraneous flaws $\approx 15\,\mu m$ in that material (e.g. small pores, surface finishing defects). As may be noted from fig. 7.29, the T-curve is not pronounced for alumina of this grain size, accounting for the flaw sensitivity of the Petch branch.

There are materials processing implications concerning the scaling of ceramic microstructures that may be drawn from the Orowan–Petch diagram. Provided one could guarantee immunity from spurious damage, it would appear justifiable to refine the grain size so as to move as far up the Orowan branch as incidental processing and surface flaws allow. However, if the material were ever to suffer severe damage, say 100 μm extraneous flaws from particle impact (sect. 8.7), then the strength would actually decline with diminishing grain size along a dominant Petch branch. Proper attention to service conditions then becomes an important aspect of microstructural design.

10
Strength and reliability

In turning to engineering aspects of brittle fracture the focus shifts from toughness to *strength*. However, structural design is concerned not just with strength but also with *reliability*. How may we *guarantee* the strength of a brittle component? Or, more realistically, *how well* may we guarantee the strength and for *how long*? Much of our current methodology for quantifying the reliability of intrinsically brittle materials can be traced back to the endeavours of Evans, Wiederhorn, Davidge, Ritter and others in the early 1970s to address such questions in relation to fracture mechanics, specifically in the context of failure from Griffith flaws.

Reliability inevitably embodies a probabilistic element.[1] Designers reconcile themselves to the notion of a 'risk' of failure over a 'lifetime', acceptable values of these quantities depending on specific applications. The classical form in which risk is expressed for any mechanical structure (including, interestingly, the human body) is the 'bathtub' curve of fig. 10.1, representing the 'hazard' (mortality) rate as a function of time. Such a curve is certainly representative of the strength characteristics of ceramic components: component failure is most frequent during manufacture and initial screening or after prolonged wear and tear in stringent service environments.

It is in the recognition that variability in strength and lifetime is unavoidable that 'flaw statistics' enters as a central element of reliability analysis in brittle materials. One regards individual flaws as members of some determinable distribution. The most common flaw distribution is that due to Weibull, based on the notion of the weakest link. Despite its amenability to fracture mechanics, the flaw statistics approach continues

---

[1] More than one scientist has expressed unease with resort to the notion of 'rolling dice'. But until the advent of 'super' materials with ultra-high and, above all, non-degradable strength properties, statistics will remain an indispensable factor in engineering design.

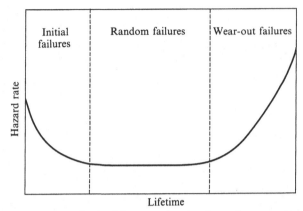

Fig. 10.1. 'Bathtub' curve, showing hazard rate against lifetime. 'Mortality' is more probable during and immediately after 'birth' (pre-existing defects), or after prolonged wear-and-tear in service (accumulated defects).

to reflect an empirical tradition: structural design remains more the practitioner's art than the theorist's science. Ceramics engineers still tend to regard quantities like 'strength' and 'toughness' as enduring material constants. As we have seen, it is only under the most exceptional circumstances that such quantities are single-valued and invariant; and, indeed, that the single-valuedness of one implies the variability of the other.

Whereas conservative attitudes to brittle design continue to prevail in the ceramics engineering industry, new philosophies are emerging from the materials community. With this backdrop we identify three main philosophies, in decreasing order of orthodoxy.

(i) *Flaw detection.* The conventional approach is to detect 'Griffith' flaws above a 'critical size', by proof testing and non-destructive evaluation (NDE). Then one may remove the component from service, or repair it. For ceramics, this approach has several drawbacks: even the largest flaw may be submicroscopic; flaw populations may deteriorate after laboratory screening; internal stresses (e.g. from local defect centres or the indigenous microstructure) may alter the very nature of flaw evolution. A large proportion of structural design is nonetheless still based single-mindedly on this strategy.

(ii) *Flaw elimination.* A more recent philosophy is to identify potential defect sources in materials preparation, and to eliminate these sources systematically by stepwise refinements in processing and finishing. This approach offers the prospect of relatively high and reproducible strengths and lifetimes. On the other hand, the component is now extremely susceptible to degradation from the chance inception of even a single large post-manufacture defect. There is therefore a need for strong 'protection' against spurious stress-concentrating encounters in service.

(iii) *Flaw tolerance.* The most forgiving approach is to learn to live with cracks by designing flaw tolerance into the microstructure. One again adopts a material processing route, but now to optimise microstructural resistance to crack propagation rather than to initiation: i.e. one incorporates toughness-curve (*T*-curve) behaviour. As we have seen, *T*-curve behaviour is common in non-cubic, coarse-grained ceramics, and is particularly pronounced in ceramic-matrix composites. This philosophy appeals because of its insensitivity to the subsequent development of any large, otherwise degrading service flaws.

The above three philosophies provide an appropriate foundation for the layout of this chapter. We begin by indicating how fracture mechanics may be used to quantify some of the more empirical notions of reliability in strength and lifetime prediction, with special reference to the Weibull flaw statistics. Then we indicate how microstructural design, via innovative processing routes, may be used to negate or combat the ordinarily strong flaw sensitivity of ceramic materials, either by *avoiding* heterogeneities to minimise crack initiation or by *including* heterogeneities to restrain propagation. In the latter case the traditional presumption of a fundamental connection between reliability and a critical flaw size is shown to be unduly restrictive. We finish by touching on other mechanical properties, thermal shock, wear, cyclic fatigue, damage accumulation and creep, that can be governing factors in ceramics reliability.

10.1 Strength and flaw statistics

To establish a proper footing for introducing flaw statistics, let us remind ourselves of the basic fracture mechanics for brittle failure under equilibrium conditions. In its most general form, the crack-tip *K*-field is expressible as $K_* = K_A + \sum K_i$, with K_i shielding contributions. The simplification implicit in most basic treatments is to ignore the K_i terms, in

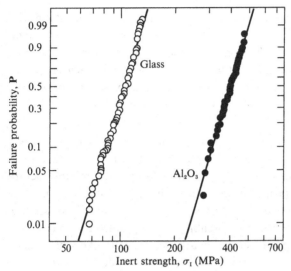

Fig. 10.2. Weibull diagram for soda-lime glass and a vitreous-bonded polycrystalline alumina (10 μm grain size, 4 vol.% additive), tested under inert conditions. (Glass data courtesy S. M. Wiederhorn. Alumina data from Gonzalez, A. C., Multhopp, H., Cook, R. F., Lawn, B. R. & Freiman, S. W. (1984), in *Methods for Assessing the Structural Reliability of Brittle Materials*, eds. S. W. Freiman and C. M. Hudson, A.S.T.M. Special Technical Publication 844, Philadelphia, p. 43.)

line with the Griffith flaw hypothesis (zero residual driving force) and supposed material homogeneity (single-valued toughness T_0). Then at equilibrium $K_* = K_A = T_0$, in conjunction with (2.20), the condition $\sigma_A = \sigma_F = \sigma_I$, $c = c_f$ for spontaneous instability defines the *inert strength*

$$\sigma_I = T_0/\psi c_f^{1/2}. \qquad (10.1)$$

Inert strength data for the simplest brittle materials indicate a wide variability, even for seemingly identical specimens. Moreover, strength tends to decrease as the stressed area (sometimes volume) increases. Given specimen-to-specimen constancy in T_0, this variability reflects a distribution in flaw sizes. Typical data sets (cf. sect. 9.1.2) suggest a large population of small flaws. Commensurate with the first design philosophy outlined earlier, we are concerned with detecting the 'worst flaw' in any component and ensuring that this flaw never exceeds a 'critical size'.

Thus it is that statistical theories of strength have emerged as a cornerstone of structural design with brittle ceramics. Their attraction (and at the same time their limitation) is that they allow the designer to

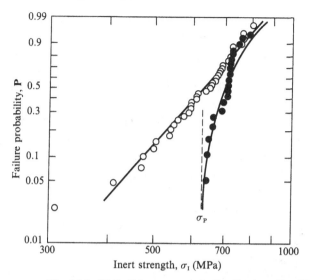

Fig. 10.3. Weibull inert strength distributions for silicon nitride (hot-pressed, magnesium-doped) before (open symbols) and after (filled symbols) proof testing. Post-proof curve determined from initial distribution using (10.6). (After Wiederhorn, S. M. & Tighe, N. J. (1978) *J. Mater. Sci.* **13** 1981.)

predict survivability without intimate knowledge of flaw or even material properties.

10.1.1 Weibull distribution

The most widely adopted function for describing the distribution of flaws in brittle materials is that proposed by Weibull in 1939 (see Weibull 1951). It is based on the concept of the weakest link, the domain of 'extreme-value statistics'. In terms of inert strengths σ_I, the Weibull failure probability is defined in its simplest form by the two-parameter relation

$$P = 1 - \exp[-(\sigma_I/\sigma_0)^m] \qquad (10.2)$$

where m, the Weibull modulus, and σ_0, a scaling stress, are regarded as adjustables. For a data sample of N strengths, cumulative probabilities are calculated by ranking values in ascending order and evaluating $P_n = n/(N+1)$ for all $1 \leqslant n \leqslant N$. A plot of $\ln\{\ln[1/(1-P)]\}$ against $\ln\sigma_I$ should then give a straight line of slope m and intercept $-m\ln\sigma_0$. Such a plot is known as a Weibull diagram.

Weibull plots for inert strength data on soda-lime glass and a vitreous-bonded polycrystalline alumina are shown in fig. 10.2. The linear fits are seen to be reasonable representations of the data, including the all-important low-strength 'tail'. Suppose that an acceptable failure rate for components fabricated from either material in fig. 10.2 is 1 in 100 ($P = 0.01$); then it would be necessary to allow for a safety factor relative to a central stress level ($P = 0.50$) of about two. The corresponding Weibull modulus for both materials is $m \approx 10$, which is typical of conventional as-finished ceramics. From an engineering perspective, it is clear that high Weibull modulus can be as important as high strength.

Considerable care needs to be exercised when using probability diagrams as a vehicle for design. For confidence in assessing safety margins in the extreme tail of the distribution it is essential to ensure that the data sample is sufficiently large (e.g. $N > 100$ for allowable failure rate $P = 0.01$). It is also important to devise test geometries that relate to actual component configurations. Further, the data sets represented in fig. 10.2 are somewhat ideal, indicative of a well-behaved flaw population. Bimodal populations, leading to strong deviations from a linear Weibull plot, are not uncommon. Other versions of the Weibull function, e.g. those with three or even more adjustables, or some other extreme-value function altogether, may be more appropriate for certain materials.

10.1.2 Proof testing

Proof testing is a potentially powerful methodology for eliminating weak members from a set of prospective components. The components are subjected to a short-term stress cycle of magnitude σ_P, in excess of that anticipated in service. Those with the largest flaws fail and are thereby eliminated from the distribution. This truncates the inert strength on the Weibull diagram at $\sigma_I = \sigma_P$, so establishing a well-defined stress level for design.

Fig. 10.3 shows inert strength data for a silicon nitride ceramic before and after proof testing. The original strengths plot as a simple Weibull distribution. Proof testing removes all specimens below σ_P and therefore biases the distribution toward the high-strength region of the plot. The curve through the remaining strengths is determined from the original function in (10.2) as the attenuated distribution

$$\mathbf{P'} = (\mathbf{P} - \mathbf{P_P})/(1 - \mathbf{P_P}) \tag{10.3}$$

with P_P the original failure probability at $\sigma_I = \sigma_P$. We see from fig. 10.3 that the risk of spurious failure is substantially reduced by the proof cycle.

Effective proof testing demands inert testing environments and rapid loading–unloading, to prevent strength degradation by environmentally enhanced flaw extension. As we shall indicate (sect. 10.2), subcritical extension is not easily avoided, even in vacuum conditions (region III, sects. 5.4, 5.5), so truncation may not always be well-defined. It is also desirable to run proof tests on actual components rather than dummy specimens, simulating the service stress state as faithfully as practicable.

10.1.3 Non-destructive evaluation (NDE)

A more direct route to screening potentially defective brittle components is NDE. The principle is simple: map out defect structures in the material surface and bulk using some imaging method (optical, x-ray, electron, acoustic), and remove or repair those components with unacceptably large flaws. To the engineer, this approach is alluring because of the potential for automation. Consequently, much effort has been expended in developing an impressive array of NDE techniques. For ceramics, however, the simplicity of the principle is belied by the difficulty of the practice. We recall the small scale of the typical flaw, usually $< 100\,\mu m$, which is on the bounds of investigative microscopy. There is an intrinsic limitation of resolution that tends to render defect imaging impracticable in the brittle materials of interest to us here.

Of the currently available NDE techniques, acoustic wave scattering is perhaps the most promising. Results of a demonstrative experiment using back-scattered surface (Rayleigh) waves from controlled flaws in silicon nitride are summarised in fig. 10.4. The cracks are penny-like Knoop indentation cracks, initial radius $150\,\mu m$, oriented normal to the incident acoustic beam. Tests are run on specimens in the as-indented and post-indentation-annealed states (sect. 8.2.1). Despite the clear detection capability of the technique, quantitative interpretation of the results in fig. 10.4 is not straightforward. Signal intensity at any given applied stress is considerably stronger for the *as-indented* specimen, reflecting a much enhanced residual crack opening from the irreversible contact field. In this case the monotonic rise in signal with increasing stress correlates directly with an expansion in crack radius from the same, stabilising contact field (cf. fig. 8.18); unloading–reloading does not retrace the loading curve, attesting to the irreversibility of this expansion. For the *annealed* surface,

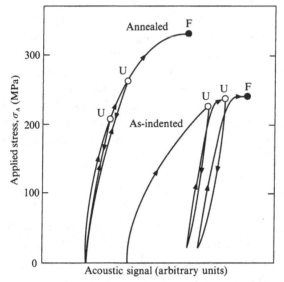

Fig. 10.4. Plot showing increase in acoustic signal with applied stress to failure (F) for backscattered Rayleigh waves (8.6 MHz) from Knoop indentations (crack plane normal to incident waves) in polished silicon nitride. Note stronger signal for as-indented relative to annealed indentations, despite initially identical crack size. Note also strong hysteresis at unload (U) points on as-indented but not on annealed specimen. (Reproduced from data in Tien, J. J. W., Khuri-Yakub, B. T., Kino, G. S., Marshall, D. B. & Evans, A. G. (1983) *Proc. Roy. Soc. Lond.* **A385** 461.)

the crack is initially held open only by spurious wall–wall asperities, and does not expand at all prior to failure. The increasing signal is then attributed to progressive disengagement of the interface asperities as the stress is applied; unloading–reloading now closely retraces the loading curve. Hence even for relatively simple and artificially large defect systems the acoustic signal by no means provides a direct measure of flaw size.

At present, the resolution of NDE techniques appears to be inadequate, and the complexity of the attendant algorithms for deconvoluting the imaging information too complex, to warrant the cost and effort of screening ceramic components for potentially critical flaws on a routine basis.

10.2 Flaw statistics and lifetime

The flaw detection philosophy of the previous section carries over to time-dependent failure at sustained stresses below the inert strength level, i.e. 'fatigue'. It is appropriate to recall the analytical groundwork for fatigue laid down in sect. 8.3. Central to most engineering analyses is an empirical power-law crack velocity function, taken in conjunction with a crack-tip stress-intensity factor for Griffith flaws in a material with no toughness-curve (zero internal driving force) and a fixed applied stress ('static fatigue') (Evans & Wiederhorn 1974):

$$\left.\begin{array}{l} v = v_0 (K_*/T_0)^n \\ K_* = \psi \sigma_A c^{1/2} \\ \sigma_A = \text{const} \end{array}\right\} \tag{10.4}$$

which constitutes a differential equation in $c(t)$. In this special case the 'lifetime' may be determined by direct integration between the initial flaw size, c_i, and the crack size at final instability, c_F:

$$\begin{aligned} t_F &= \int_{c_i}^{c_F} \mathrm{d}c/v[K_*(c)] \\ &= A/\sigma_A^n, \quad (n \gg 2) \\ &= B\sigma_I^{n-2}/\sigma_A^n, \quad (\sigma_A < \sigma_I) \end{aligned} \tag{10.5}$$

with $A = 2T_0^n/(n-2)\,\psi^n v_0\, c_i^{n/2-1}$ and $B = 2T_0^2/(n-2)\,\psi^2 v_0$ material–environment–flaw constants. We note the strong sensitivity of t_F: (i) to (inverse) applied stress σ_A; (ii) to intrinsic toughness T_0 or inert strength σ_I; (iii) to (inverse) initial (*not* final) crack size, this last reflecting the fact that the flaw spends most of its extension time in the slowest (low K_*) region of growth.

Granted that the operative flaw origins in fatigue are the same as those in inert strength tests, one may express lifetime distributions in terms of the Weibull function of (10.2). Direct substitution of (10.5) gives

$$P = 1 - \exp\left[-(t_F/t_0)^M\right] \tag{10.6}$$

with $t_0 = B\sigma_0^{n-2}/\sigma_A^n$ and $M = m/(n-2)$ (Davidge 1973, 1979; Evans & Wiederhorn 1974). Since $n-2 > m$ for most ceramics, $M < 1$ usually, corresponding to a large spread in lifetime values. In principle, one may evaluate (10.6) a priori from independent determinations of the inert strength Weibull parameters m and σ_0 and the material–environment parameters T_0, n and v_0.

A typical Weibull lifetime diagram is shown for a polycrystalline alumina in fig. 10.5. The alumina is the same as that represented in fig. 10.2. (The stress level at which the fatigue tests were conducted, $\sigma_A = 253$ MPa, lies well below the inert strength data range in fig. 10.2.) The spread in lifetimes covers several orders of magnitude, attesting to the sensitivity of the fracture kinetics to flaw size in (10.5). From the slope of the best-fit plot we obtain $M = 0.25$ in (10.6); this value compares with an independent evaluation $M = m/(n-2) = 0.18$, using $m = 9.8$ (fig. 10.2) and $n = 55$ (from a fit to independent stressing-rate data).

Proof testing may be used to eliminate the low-lifetime tail of the lifetime distribution function (Evans & Wiederhorn 1974). It is again imperative to carry out such tests in such a way as to avoid any degradation of the flaw population. With those precautions, testing at a proof stress σ_P guarantees a minimum lifetime

$$t_P = B\sigma_I^{n-2}/\sigma_P^n, \quad (\sigma_P < \sigma_I) \tag{10.7}$$

in (10.5), and the modified Weibull distribution is determined by combination of (10.6) with (10.3). A Weibull diagram for an alumina ceramic illustrating lifetime truncation after proof testing is shown in fig. 10.6.

Computational and graphical schemes for fatigue lifetime predictions, some highly elaborate, have been fashioned from the kind of fracture mechanics outlined above (Davidge 1973, 1979; Evans & Wiederhorn 1974; Ritter 1974). The value of any such predictive scheme is contingent on the following:

(i) *Validity of starting equations.* We reiterate that the constitutive equations in (10.4) are restrictive. Consider first the empirical power-law velocity function. Among other limitations, this function has no inbuilt provision for a velocity threshold (chapter 5), and thereby precludes a fatigue limit. The reader may recall the strong tendency to such a limit in the static fatigue data for indented alumina in fig. 8.29. In fact there is evidence, from the presence of 'survivors' in the alumina data of fig. 10.5, of a similar limit for natural flaws. Despite the most obvious implications of threshold stresses in the characteristic lifetime response, expressed so

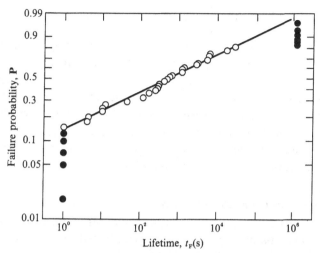

Fig. 10.5. Weibull diagram for vitreous-bonded alumina, showing distributions in lifetimes for sustained static stress $\sigma_A = 253$ MPa. Premature 'ramp' failures at lower end of distribution and 'survivors' at top end of distribution shown as filled symbols. Latter are suggestive of a fatigue limit. (Data from same source as fig. 10.2; analysis by Wiederhorn, S. M. & Fuller, E. R. (1985) *Mater. Sci. Eng.* **71** 169.)

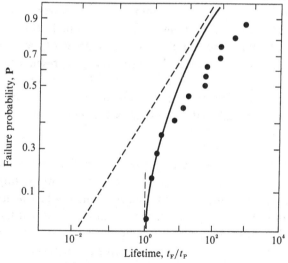

Fig. 10.6. Weibull lifetime distributions after proof testing (reduced relative to minimum lifetime t_P after proof test) for an alumina (grain size 15 μm). Dashed curve is initial distribution, solid curve is modified post-proof distribution from (10.6). (After Xavier, C. & Hubner, H. W. (1981) *Science of Ceramics* **11** 495.)

strongly in the metals literature, design schemes for ceramics continue to ignore this potentially important factor.

Now consider the K-field equation in (10.4). While arguably adequate for simple flaw types in homogeneous materials like annealed glass and monocrystals, the same can not be said of most ceramic materials, where shielding terms from internal residual stresses (sect. 9.5) and R-curve processes (sect. 9.6) are commonplace. As we shall see (sect. 10.4), the stabilising influence of such additional factors can significantly lessen the importance of initial flaw size in the failure mechanics, thereby changing the complexion of variability in strength and fatigue.

(ii) *Accuracy of calibrated parameters.* Compounding any restrictions in the starting equations are uncertainties in determinations of the associated material–environment parameters (T_0, n, v_0), as well as in the inert strength Weibull parameters (m, σ_0). We have already noted the sensitivity of t_F in (10.5) to such parameters. Extreme caution must therefore be exercised in any predictive scheme based on independent parameter calibrations, especially when extrapolating beyond the laboratory data range to long service times. This is especially so when employing traditional long-crack measurements to evaluate the crack velocity parameters. There can be no guarantee that long-crack data truly represent the domain of natural flaws. The indentation methods of chapter 8 certainly take us closer to the requisite short-crack domain; but, as intimated above, residual fields can strongly influence the fatigue susceptibility (cf. figs. 8.25, 8.29). Which indentation type, as-indented or annealed, more closely represents the natural flaw state for any given material system? Because of such reservations, many advocate the safer (if more tedious) route of parameter calibration from fatigue strength data on actual specimens in their as-finished state (Ritter 1974).

(iii) *Efficacy of proof test.* If not executed properly, proof testing is far from benign. In practice, it is not easy to avoid kinetic growth of existing flaws during the proof cycle, so lifetimes may fall substantially below the predicted truncation of (10.7). Indeed, in certain instances (notably if the proof cycle involves slow unloading in moist environments) the remnant flaw distribution can be more deleterious than the original. In extreme cases the proof cycle may initiate pop-in from subthreshold flaws.

It is implicit that the component is 'well-aged' prior to the proof test, so that the screened flaw population is in a steady state when it leaves the laboratory. If subsequent service conditions are severe (e.g. extraneous fields of the type considered in sect. 9.3), the flaw population may undergo additional deterioration. Repeat proof testing (or NDE) at strategic (shutdown) phases of operation may then be called for.

10.3 Flaw elimination

The second philosophy for improved reliability is to refine ceramic processing to produce homogeneous, defect-free components. Rather than detecting flaws in the finished product, one eliminates them at source.

10.3.1 Optical glass fibres

It has been realised from the time of Griffith that strengths approaching the theoretical cohesive limit (sects. 1.5, 6.1) can be obtained on pristine glass fibres and crystal whiskers. The demand of the communications industry in the 1970s for high-transmission optical fibres with strengths > 1 GPa over lengths > 1 km stimulated a renaissance in this area. Several techniques for achieving defect-free fibres have been investigated, including post-drawing etching and flame polishing, but the most effective is fresh drawing from a heated rod pre-form in a clean atmosphere.

The Weibull diagram in fig. 10.7 shows inert strength distributions for freshly drawn silica fibres at two gauge lengths (Maurer 1985). Observe the distinctive bimodal character of the distributions: an intrinsic steep region with modulus $m > 50$ and central strength ≈ 3.5 GPa (i.e. $E/20$, not far below that theoretically achievable); an extrinsic region, more pronounced at the longer gauge length. Strengths in the central intrinsic region correspond to $c_f \approx 50$ nm for penny flaws in (10.1), which is probably of the order of surface fluctuations in these fibres. The extrinsic tail, usually attributed to spurious degradation from contacts with atmospheric dust, etc., becomes more prominent with aging in unprotected fibres. With prolonged exposure and handling the strengths ultimately degrade to the low levels characteristic of ordinary glass surfaces in fig. 10.2.

Pristine glass is subject to fatigue from moisture. Static fatigue plots for silica fibres correspond to $n = 15$–20 in (10.5), which compares with 35–40 from conventional tests on long-crack specimens. A comparable increase in susceptibility is observed in the subthreshold–postthreshold transition with indentation flaws (Dabbs & Lawn 1985), where residual contact fields exert a strong modifying influence on the flaw stability (sect. 8.5).

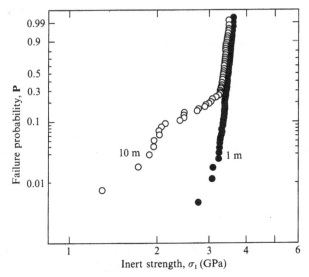

Fig. 10.7. Weibull inert strength distributions for freshly drawn fused silica fibres, two gauge lengths. (After Maurer, R. D. (1985), in *Strength of Inorganic Glass*, ed. C. R. Kurkjian, Plenum, New York, p. 291.)

Modern fibre production lines meticulously guard against contamination of the glass composition and apply a protective coating on the emergent fibre as it is drawn from the melt. Considerable research has gone into optimising the coating conditions, and to incorporating surface compressive stresses (recall sect. 8.6.2), to inhibit evolution of intrinsic flaws. Current technology routinely produces fibre products over 3 km long capable of surviving stresses in excess of 1 GPA for several years.

10.3.2 Heterogeneity-free ceramics

Lange (1984, 1989) and others have embarked on a concerted program of refining powder processing to eliminate microstructural flaws from brittle ceramics and thereby produce materials at the top end of the strength spectrum. Their approach is to identify the primary sources of failure from test specimens at each stage of production, and then to implement steps to eliminate the offending flaws (or at least to reduce their severity) in an iterative processing sequence. A schematic diagram indicating a typical

◄――― Strength

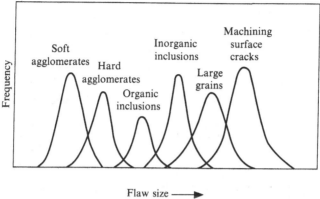

Flaw size ―――►

Fig. 10.8. Potential spectrum of flaw populations in fine-grained ceramics to be eliminated by iterative refinement in processing for high-strength defect-free ceramics. (After Lange, F. F. (1989) *J. Amer. Ceram. Soc.* **72** 3.)

spectrum of flaw populations that might be encountered *en route* to the fabrication of an ultra-high-strength ceramic is indicated in fig. 10.8.

The best investigated ceramic to date is a fine-grained (1–2 μm) alumina/zirconia system (Lange 1984). Using conventional dry powder methods, moderate strengths ≈ 600 MPa were obtained on polished surfaces. These strengths were determined to be limited by cracks around occasional, hard agglomerates. Elimination of the agglomerates by sedimentation and subsequent consolidation of the powders from colloidal solution improved the strength to ≈ 1 GPa. Finally, elimination of voids from organic binder contaminants using a burnout/isopress treatment resulted in strengths > 2 GPa. Unfortunately, as yet little definitive Weibull analysis has been done to confirm the intimation that refined processing should simultaneously diminish the variability in strength.

The promise of ultra-high strength is indeed seductive. But the flaw elimination route has its drawbacks. First is the issue of economy. The processing is exacting, requiring ultra-clean conditions to avoid powder contamination. The expense and effort may be warranted only in special high-technology applications. Then, if the allowable risk of failure is tight, screening to eliminate occasional weak components may be necessary. We have already pointed out the inadequacies of NDE in ordinary ceramics: the reduced flaw size in the refined materials serves only to magnify that inadequacy. Finally, there is a tendency for pristine surfaces to develop bimodal flaw distributions with long tails on aging (cf. fig. 10.7): the introduction of a single service flaw can degrade the elevated strength to its ordinary level in a stroke. It may then be necessary to resort to surface protection or encapsulation. For polycrystalline ceramics, coating with passive films (cf. optical fibres), introducing surface compressive stresses (e.g. thermal tempering, ion implantation, transformation toughening), are possible sources of protection that have received some attention in the literature.

It is interesting to reflect that this philosophy of flaw elimination has been used with some success in the development of 'macro-defect-free' cements. By adding surface-active chemicals to control the rheology of the cement paste in mixing and by adopting novel extrusion procedures, one is able to effect dramatic reductions in the porosity of the finished product, increasing the strength from a traditionally low 5–15 MPa to a more respectable 50–70 MPa.

10.4 Flaw tolerance

Of all the philosophies for improved reliability, that of designing flaw tolerance into the starting material is the most far reaching. It retains the conceptual shift away from component evaluation to material refinement advocated in the preceding section, with a singular difference in the processing strategy: instead of *eliminating* heterogeneities one *incorporates* them, albeit in a controlled manner. The goal is to stabilise crack extension by microstructural shielding (chapter 7) and so reduce the sensitivity of strength to flaw size. Failure is governed by a resistance-curve (*R*-curve, G_R-curve) or toughness-curve (*T*-curve, K_R-curve) (sect. 3.6).

10.4.1 Strength of materials with toughness-curves

Consider the mechanics of failure from microstructural penny-like flaws in a material with moderate shielding. Assume that the flaws are of the most deleterious *intrinsic* kind that evolve from inception through the full shielding field, i.e. including the tensile zone prior to first bridge intersection, fig. 7.27(c). The requirements for spontaneous (Griffith)

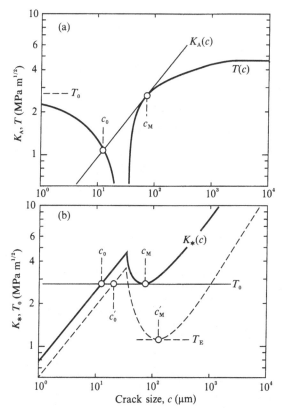

Fig. 10.9. Failure in alumina of grain size 35 μm (fig. 7.29), from
natural flaws. (a) T-curve construction showing $K_A(c)$ for inert strength
$\sigma_A = \sigma_I = 250$ MPa at critical point of tangency to $T(c)$. (b) Equivalent
construction showing $K_*(c)$ at $\sigma_A = 250$ MPa (solid curve) intersecting
$T = T_0$ at $c = c_M$ (inert strength), and at $\sigma_A = 110$ MPa (dashed curve)
intersecting $T = T_E$ (environment) at $c = c'_M$ (fatigue limit). Strength is
independent of flaw size within $c_0 \leqslant c_f \leqslant c_M$.

instability are now inappropriate: the flaws lie in the short-crack domain
of the T-curve, and the modified stability conditions of sect. 3.6 apply.

Fig. 10.9 illustrates with K-field data on a high-density alumina of grain
size 35 μm from chapters 7 and 8. The solid curves represent *inert strength*
configurations, plotted for $\sigma_A = \sigma_M = \sigma_I = 250$ MPa (cf. shaded area for
$l = 35$ μm, fig. 8.24): (a) is the 'global' construction for failure at the
tangency point, i.e. $K_A(c) = T(c)$, $dK_A/dc = dT/dc$, with $T(c) = T_0 + T_\mu(c)$
from fig. 7.29 and $K_A(c) = \psi\sigma_A c^{1/2}$ for penny cracks from (2.20) and (2.21d);
(b) is the corresponding 'enclave' failure construction (cf. fig. 8.19(c))
at $K_*(c) = T_0$, $dK_*/dc = 0$, with $K_*(c) = K_A(c) + K_\mu(c)$ $(-K_\mu = T_\mu)$.

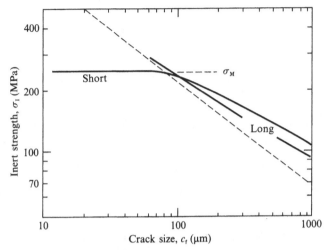

Fig. 10.10. Predicted inert strength as function of initial penny-crack flaw size for same alumina as in fig. 10.9. Curve for intrinsic *short-crack* flaws (microstructural flaws, annealed indentation or machining flaws) that have evolved completely within bridging field, i.e. $c_0 = d$ in fig. 7.27, shows characteristic flaw-tolerance plateau. Comparison curve for extrinsic *long-crack* flaws (initially stress-free notch-like defects) bridged only after extension, $c_0 = c_f$ in fig. 7.27, reflects reduced tolerance. Dashed curve is for (Griffith) flaws in hypothetical material with no bridging (zero friction and residual thermal expansion anisotropy stress). (Courtesy S. J. Bennison.)

By virtue of precursor stable crack growth, from c_f to c_M at $K_A(c) = T(c)$ or $K_* = T_0$, the strength is independent of starting flaw size within $10 \ \mu\text{m} \leqslant c_f \leqslant 80 \ \mu\text{m}$. This characteristic is demonstrated more directly as the plateau at $\sigma_I = \sigma_M$ in the curve for short-crack flaws in fig. 10.10. The notion of a critical defect size is no longer universally applicable: the material is *flaw tolerant*.

This flaw tolerance extends to component *lifetime* under static stress $\sigma_A < \sigma_I$ in reactive environments. Included in fig. 10.9(b) is a plot of $K_*(c)$ at $\sigma_A = 110$ MPa, representing the *fatigue limit*. The crack velocity $v(K_*)$ diminishes as the flaw extends down this curve toward the minimum at threshold $K_* = T_E$ in (5.14), so (contrary to the condition that pertains in (10.5)) it is c'_M and not c_f that is the rate-determining crack length. Note that any strengthening of the shielding component $-K_\mu = T_\mu$ will depress the minimum in $K_*(c)$ still further, with consequent enhancement in the threshold stress.

It is highly significant that materials with strong T-curves continue to exhibit variability in strength and lifetime, despite their professed flaw insensitivity. We note for instance the error bars in the strength data of fig. 8.22 for our alumina material, even at those points representing breaks exclusively from well-controlled indentation flaws. These error bars are considerably greater than those in fig. 8.25 for homogeneous glass, confirming that the scatter is not attributable to variation in the indentation-strength testing procedure itself. One concludes that the scatter must be due to microstructural fluctuation, in the case of alumina to fluctuation in bridging parameters (bridge grain spacing, internal residual stress, pullout friction, etc., sect. 7.5). Thus, while heterogeneities are a vital component of the processing, there should be a certain uniformity in their distribution within the microstructure. And whereas the *Weibull* probability functions (10.2) and (10.5) may be reasonably retained as indicators of strength and lifetime variability, it is the distribution of microstructural shielding elements, not of intrinsic flaws, that dictates the route to a high modulus.

A major appeal of the T-curve route is the potential for tailoring microstructures for optimal flaw insensitivity. With alumina-based ceramics, say, the objective is to strengthen microstructural bridging (sect. 7.5). As seen in fig. 8.24, a simple way of doing this is to increase the grain size. However, there is a limit to any such benefits obtainable with monophase materials. Increasing grain size scales up the crack-opening displacements over which the pullout friction tractions remain intact, enhancing the shielding; but it also scales up the bridge spacing, expanding the distance flaws may initially grow before bridges are activated. The net result is the compromise seen in fig. 8.24: the curves cross each other, so flaw tolerance is bought at the expense of a lower strength in the small-flaw (low-load) region. Ideally, one seeks a means of strengthening bridges without simultaneously reducing their density.

One such means is to augment existing bridging tractions by including a second phase with high thermal expansion mismatch relative to the matrix, so as to increase the residual stress σ_R in (7.29). An extreme example is the mismatch between aluminium titanate and alumina, sufficient to intensify the internal stresses in an alumina matrix by more than an order of magnitude. Indentation-strength data in fig. 10.11 for an alumina-based composite of matrix grain size 6 μm with 20 vol.% aluminium titanate particles confirm the effectiveness of this approach. We see a marked enhancement of flaw insensitivity relative to a comparative curve for the base alumina (by theoretical interpolation from fig. 8.24), without the same severe depression in plateau strength associated with microstructural coarsening. As a quantitative measure of the improvement in flaw

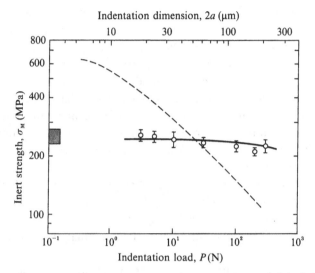

Fig. 10.11. Inert strength for composite material of alumina matrix (6 μm grain size) and aluminium titanate second phase (20 vol.%) as function of Vickers indentation load. All data points represent breaks from indentation sites. Error bars are standard deviation limits. Solid curve is empirical fit through data. Dashed curve is theoretical prediction for base alumina of same matrix grain size. (Data courtesy S. J. Bennison & J. L. Runyan.)

insensitivity, we observe that while the indentation flaw size increases from ≈ 15 to 250 μm over the data range in fig. 10.11, the corresponding strength diminishes by less than 10%.

10.4.2 Design implications and misconceptions

Flaw tolerance offers the prospect of a well-defined *design stress*, e.g. the plateau σ_M in fig. 10.10 for inert conditions. Whereas flaw-tolerant materials may have only moderate laboratory strengths relative to their more 'refined' flaw-free counterparts in sect. 10.3, these same strengths are far less susceptible to degradation from the kind of in-service impact damage described in sect. 8.7.1. We have just highlighted the extreme flaw insensitivity of the alumina-based composite in fig. 10.11: from that figure we may deduce that a particle-impact event which results in a strength loss of ≈ 10% in the composite would cause a comparative loss of ≈ 400% in the base alumina. It is interesting to reflect that such a state of flaw-invariant strength is realised only with non-invariant toughness. The notion of a well-defined design stress extends to failure in chemically active environments: recall again from sect. 8.3 (e.g. fig. 8.29) the tendency for microstructural shielding to manifest itself as an enhanced fatigue limit, i.e. an increase in the threshold stress below which the time to failure is effectively infinite.

Flaw tolerance bears on other consequential issues in design with ceramics:

(i) *Mechanical evaluation.* (a) *NDE.* As a result of pre-failure stable crack growth in the shielding K-field (further enhanced by any local residual nucleation field, sect. 9.5), near-critical flaws stand a much greater chance of detection by conventional means. Moreover, because the stabilising field originates from an underlying discreteness in the microstructure, crack growth is inevitably 'noisier' than in homogeneous materials, raising the possibility of 'early warning' to failure. (b) *Proof testing.* Proofing cycles may be employed in the usual way to eliminate tails in strength distributions. Substantial stable growth without failure may occur during these tests. Such extension will not significantly reduce the strength or subsequent time to failure, since the still-intact system remains to the left of the tangency point in fig. 10.9; again, it is not c_f but c_M that is the limiting crack size.

(ii) *Processing strategy.* Prescriptions for design must now be formulated in terms of microstructural characteristics rather than flaw populations. We reiterate the advocated processing route of incorporating shielding heterogeneities rather than eliminating them, with its attendant advantages of simplicity and economy: to enhance tolerance, increase the *density* of heterogeneities; to increase Weibull modulus, control the *dispersion* of heterogeneities. With this philosophy we move closer to the realm of the composite and complex materials like concrete. Some caution is necessary to ensure that the processes which give us flaw tolerance do not manifest themselves in spontaneous microcracking (sect. 9.4) or other potentially adverse properties (see sect. 10.5).

The notion of 'microstructural design' offers much scope to the ceramics engineer for the development of reliable components. As yet, the attention devoted to the reliability of T-curve materials has been little more than perfunctory.

We complete this section by discussing some common misrepresentations of the T-curve (R-curve) construction.

(i) *Long-crack vs short-crack regions of T-curve.* The strength of materials with intrinsic penny-like flaws is sensitive to the shape of the T-curve in the short-crack region (e.g. fig. 10.9), which for bridging ceramics is governed by the microstructural discreteness in the shielding K-field. As noted in fig. 7.24, long-crack measurements from conventional *notched* specimens with straight crack fronts inevitably smooth over this discreteness. Indentation flaws would appear most suitable for evaluating the requisite short-crack T-curve, since they closely simulate the essential scale and shape of intrinsic flaws, under controlled testing conditions.

(ii) *Validity of alternative T-curve constructions.* A common representation of the failure condition for a material with microstructural shielding is one that replots the toughness curve as crack resistance R against extension Δc from a stress-free notch of length c_0; as before, instability is determined by a tangency condition, but after displacing the origin of the applied stress function $G_A(c)$ along the extension axis from $\Delta c = 0$ ($c = c_0$) to $\Delta c = -c_0$ ($c = 0$) (Broek 1982). The illustrative plot in fig. 10.12 is for a long-crack R-curve appropriate to the alumina material of fig. 10.9. Strength, although still governed by stable growth from $c = c_0$ to $c = c_M$, reverts to dependence on initial crack size. This dependence holds for initially unbridged penny-like flaws, i.e. extrinsic stress-free voids,[2] identifying $c_0 = c_f$. An appropriately moderated flaw tolerance for such long-crack defects is evident in fig. 10.10. However, the construction in fig. 10.12 is restrictive, in that it is representative *only* of initially unbridged long-crack flaws, and certainly not of the wider spectrum of intrinsic short-crack flaws which evolve entirely through the bridging field.

(iii) *Power-law T-curves.* Some analysts employ a power-law function to represent toughness, e.g. $T \propto c^q$. There are two objections to this function. First, it is unequivocally *empirical*. It contains no provision for incorporating a fundamental constitutive function for the underlying shielding process, and hence for predicting dependence on microstructural parameters (grain or inclusion size or shape, residual stress, etc.). Nor does it have provision for distinguishing between discrete short-crack and continuum long-crack features in the T-curve. (In this context it suffers

___

[2] And then with appropriate modification to the $K_A(c)$ function, e.g. (9.10).

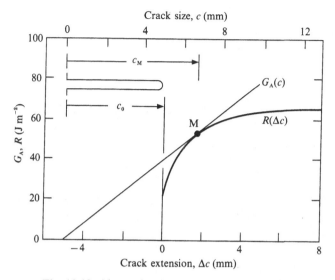

Fig. 10.12. Alternative *R*-curve construction, for extension Δc of a straight-fronted crack from a notch of length c_0. $R(\Delta c)$ (origin at $c = c_0$) is long-crack resistance function for alumina. $G_A(c)$ line (origin at $c = 0$) is for notch of length $c_0 = 5$ mm. (Cf. fig. 3.10(a).)

from the same restrictions as the power-law velocity function used in (10.4) for lifetime predictions.) Second, it is *unphysical*. On a logarithmic plot the $K_A(c)$ and $T(c)$ functions are straight lines with just one mutual intersection point: if $q < \frac{1}{2}$ one has failure at $c = c_f$ without stable growth (i.e. spontaneous failure); if $q > \frac{1}{2}$ one has stable growth from $c = c_f$ to $c = \infty$ without failure. There is no tangency point, so the essential quality of stabilised failure is lost.

10.5 Other design factors

Strength *per se* is not the only property that must be considered when designing with ceramics. In certain cases other factors may take precedence, depending on the specific function of the component part. We identify some of these in the superficial accounts below, leaving the more interested reader to pursue the detailed ceramics engineering literature.

(i) *Thermal shock.* Brittle components in high-temperature applications may not survive the thermal stresses generated in rapid heating or cooling cycles (Hasselman 1969; Davidge 1979). For the ideal case of an isotropic, infinite plate with instantaneous heat transfer from the surrounds in infinitesimal quench time, the surface tensile stress is

$$\sigma_{\mathrm{TS}} = E\alpha\Delta\mathrm{T}/(1-\nu) \tag{10.8}$$

with α the expansion coefficient, $\Delta\mathrm{T}$ the temperature differential, E Young's modulus and ν Poisson's ratio. A countervailing compressive stress exists at the centre plane of the plate. For real, non-ideal systems the stresses are smaller and are strongly time-dependent, in accordance with conventional heat-transfer equations. Nonetheless, the gradient from surface tension to subsurface compression prevails, so that a surface crack, once initiated at critical $\Delta\mathrm{T}_{\mathrm{C}}$, pops-in to a subsurface arrest configuration (cf. sect. 4.2.2). Thus the thermal shock K-field $K_{\mathrm{TS}}(c/d)$ passes through a maximum at some crack depth $c_{\mathrm{f}} < c < d$, d the plate half-thickness. The strength in subsequent mechanical applied loading is thereby degraded.

Strength data for thermally shocked high-density alumina ceramics at four grain sizes are shown in fig. 10.13. Three regions may be distinguished: $\Delta\mathrm{T} < \Delta\mathrm{T}_{\mathrm{C}}$, $K_{\mathrm{TS}}(c_{\mathrm{f}}) < T(c_{\mathrm{f}})$, insufficient driving force to extend the flaw and thus no strength loss; $\Delta\mathrm{T} = \Delta\mathrm{T}_{\mathrm{C}}$ ($\approx 200\ ^\circ\mathrm{C}$), $K_{\mathrm{TS}}(c_{\mathrm{f}}) = T(c_{\mathrm{f}})$, flaw pops-in and arrests, with corresponding abrupt strength drop; $\Delta\mathrm{T} > \Delta\mathrm{T}_{\mathrm{C}}$, $K_{\mathrm{TS}}(c_{\mathrm{f}}) > T(c_{\mathrm{f}})$, enhanced stable crack extension after pop-in, with monotonically increasing strength degradation.[3] Whereas the strength drop at $\Delta\mathrm{T}_{\mathrm{C}}$ is most precipitous at the smallest grain size, it is almost imperceptible at the largest grain size. This latter is the realm of strong toughness-curve enhancement (fig. 7.29), suggesting that the shocked crack in the coarser materials is constrained to arrest on the T-curve prior to the tangency point (fig. 10.9). The strength loss in subsequent applied loading then shows no discontinuity. This accounts in part for the utility of low-grade refractory ceramics in furnace applications.

(ii) *Wear.* Surface removal processes are an important consideration where an application constrains surfaces or interfaces to be in sustained sliding contact. As foreshadowed in sect. 8.7.2, this is especially the case with brittle ceramics, because of the potential severity of wear by contact-induced microfracture from particulate debris. Intuition has it that good wear resistance requires high toughness, cf. (8.26) for homogeneous

[3] Note a striking similarity between the strength degradation curves of fig. 10.13 for thermal shock and fig. 8.14 for contact damage, in which critical temperature differential $\Delta\mathrm{T}_{\mathrm{C}}$ is replaced by critical contact load P_{C}.

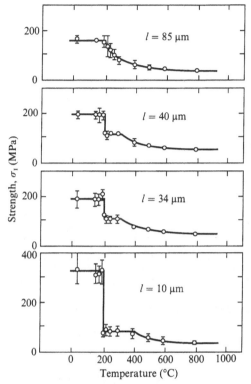

Fig. 10.13. Thermal shock of high-density aluminas of different grain sizes. Data are inert strengths after quench from temperature T. (After Gupta, T. K. (1972) *J. Amer. Ceram. Soc.* **55** 249.)

materials, and this is broadly supported by the experimental evidence in the literature. However, the potential for degradation is greater in materials with pronounced T-curves because of the superposition of intrinsic, microstructural residual forces onto the extrinsic, contact-induced driving forces for local fracture. Design is then no longer a simple matter of optimising a single toughness parameter, but demands a deeper appreciation of material properties at the microstructural level.

The wear data in fig. 10.14 for high-density aluminas at three values of grain size l illustrate the point. In each alumina the wear rate increases monotonically with time, initially slowly and subsequently, after an 'incubation' period, abruptly. Of the three aluminas it is the coarsest that shows the greatest susceptibility to degradation in fig. 10.14. Whereas the initial increase is microstructure-invariant, the transition shifts markedly to smaller incubation times at larger grain sizes. Recall the crossover in T-

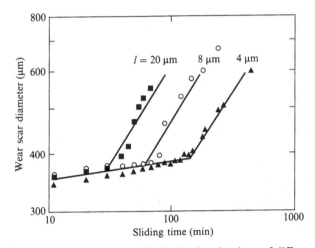

Fig. 10.14. Wear of high-density aluminas of different grain sizes, *l*. Data show scar diameter as function of time in contact with rotating hard (silicon nitride) sphere. Initial slow increase in scar diameter represents region of deformation-controlled wear. Subsequent rapid increase at critical sliding time indicates transition to fracture-controlled wear, where progressive build-up in contact sliding deformation stresses is sufficient to initiate grain boundary microcracking. Note strongly diminishing transition time with increasing grain size. (Data courtesy S-J. Cho.)

curves in fig. 7.29: the toughness of the coarser aluminas, enhanced in the long-crack domain, is conversely depressed in the short-crack domain; and it is this latter domain that is pertinent to wear. When augmented by extrinsic stresses σ_D from local contact deformation, the same intrinsic stresses σ_R that play such a vital role in crack-tip shielding can induce premature microcracking from pre-existing sub-facet flaws (sects. 7.3.2, 9.4). Suppose that the deformation stresses accumulate uniformly with time, i.e. $\dot{\sigma}_D = \sigma_D/t = $ const. Replacing $\bar{\sigma}_{ii}$ by σ_D in the relation (7.16) for critical flaws and invoking the limiting condition $l = l_C$, $\bar{\sigma}_{ii} = 0$ for spontaneous microfracture in (7.17) yields the incubation time

$$t_* = (\sigma_R/\dot{\sigma}_D)[(l_C/l)^2 - 1], \quad (l \leqslant l_C). \tag{10.9}$$

Then as *l* decreases below l_C, t_* increases from zero. The observed shifts in fig. 10.14 are quantitatively consistent with a stress cumulation rate $\dot{\sigma}_D = 5$ MPa s^{-1}.

Hence in aluminas and other *T*-curve materials flaw tolerance and wear resistance may be countervailing qualities.

(iii) *Cyclic fatigue.* In many potential applications ceramic components are subject to repeated loading and unloading. In sect. 10.2 we gave explicit consideration to fatigue properties under static stress, assuming lifetime to be governed exclusively by chemically enhanced rate-dependent crack growth. In metals and polymers *mechanical* fatigue from cumulative irreversibility in frontal plastic zones can lead to failures at maximum stresses well below the static fatigue limits. The question is only now being raised as to whether similar mechanical fatigue exists in brittle ceramics (Ritchie 1988; Suresh 1991). Whereas ceramics may be immune to true crack-tip plasticity, other modes of microstructural shielding outlined in chapter 7 may well be susceptible to mechanical degradation.

Data confirming cyclic fatigue effects have been reported on phase-transforming zirconias and bridging aluminas. Although some kind of progressive weakening of the shielding elements in these materials does appear to be involved, specific micromechanics of the degradation remain obscure. The effects reported thus far are not as severe as those in metals and polymers. Nevertheless, the potential for cyclic degradation has been forewarned, and further investigation, especially in composite systems, appears warranted.

(iv) *Damage accumulation.* In materials where fracture is highly stabilised, a stressed component may develop dense arrays of contained microcracks rather than a single dominant extending macrocrack. The material then responds in an increasingly nonlinear fashion, in extreme cases somewhat like a yielding metal, due to a reduced compliance. The 'yield' stress itself may be low, yet the corresponding strain to failure inordinately high. Under such circumstances the entire basis of linear fracture mechanics is suspect. Damage accumulation is desirable in applications where energy absorption is a critical factor, e.g. impact with projectiles.

An obvious route to enhanced damage accumulation and ensuing stress–strain nonlinearity is via reinforced shielding, to promote the essential multiple-crack stability. This kind of strong nonlinearity is exhibited by zirconias and two-phase materials (e.g. the alumina-based material in fig. 10.11). Again, this takes us toward the realm of the composite.

(v) *Creep.* Above a temperature of $\approx 1000\ °C$ the mechanical responses of ceramics are far more diverse. This is the domain of high strain-rate and temperature dependence in the failure mechanics, where the effects of thin grain boundary phases and environmental chemistry can be dramatic

(Evans 1985; Wiederhorn & Fuller 1985). Above a threshold stress-intensity factor pre-existent flaws may lead to failure by traditional crack propagation, according to some high-temperature velocity function with relatively weak stress dependence (power-law exponent $n \approx 2$). Below the threshold the cracks blunt out in the viscous flow field, and entirely new populations of cavitation flaws may be generated by deformation and chemical or physical change. Such cavities coalesce to produce strain-rate-limited failure, as embodied in the empirical Monkman–Grant lifetime relation

$$t_F = \beta / \dot{\varepsilon}_A^\alpha \tag{10.10}$$

where α is an exponent of order unity and β is a weak function of applied stress. In this subthreshold region the concept of damage accumulation is especially pertinent (Ashby & Dyson 1984).

Creep rupture is potentially a vast area of study, as yet not all that well-understood. Its complexity is reflected in the empirical codes that currently exist for high-temperature design.

Balancing strength and toughness against the factors outlined above in designing with ceramics is an exercise in compromise: 'specific materials for specific applications'. We have seen evidence that what is good for one property is not necessarily so for another, depending on details of the material microstructure. The future of design with brittle solids rests with a proper dialogue between structural engineer and materials scientist.

References and reading list

Chapter 1

Gordon, J. E. (1976) *The New Science of Strong Materials*. Penguin, Harmonds-worth.
Griffith, A. A. (1920) The phenomena of rupture and flow in solids. *Phil. Trans. Roy. Soc. Lond.* **A221** 163.
Griffith, A. A. (1924) The theory of rupture. In *Proc. First Internat. Congr. Appl. Mech.* (ed. C. B. Biezeno & J. M. Burgers). J. Waltman Jr., Delft, p. 55.
Inglis, C. E. (1913) Stresses in a plate due to the presence of cracks and sharp corners. *Trans. Inst. Naval Archit.* **55** 219.
Obreimoff, J. W. (1930) The splitting strength of mica. *Proc. Roy. Soc. Lond.* **A127** 290.

Chapter 2

Adamson, A. W. (1982) *Physical Chemistry of Surfaces*. John Wiley, New York.
Atkins, A. G. & Mai, Y-M. (1985) *Elastic and Plastic Fracture*. Ellis Horwood, Chichester.
Barenblatt, G. I. (1962) The mathematical theory of equilibrium cracks in brittle fracture. *Adv. Appl. Mech.* **7** 55.
Cotterell, B. & Rice, J. R. (1980) Slightly curved or kinked cracks. *Int. J. Fract.* **16** 155.
Gell, M. & Smith, E. (1967) The propagation of cracks through grain boundaries in polycrystalline 3% silicon-iron. *Acta Metall.* **15** 253.
Gurney, C. & Hunt, J. (1967) Quasi-static crack propagation. *Proc. Roy. Soc. Lond.* **A299** 508.
Hutchinson, J. W. (1990) Mixed-mode fracture mechanics of interfaces. In *Metal-Ceramic Interfaces* (ed. M. Rühle, A. G. Evans, M. F. Ashby & J. P. Hirth). Acta-Scripta Metall. Proceedings Series, Vol. 4, p. 295.
Hutchinson, J. W. & Suo, Z. (1991) Mixed-mode cracking in layered structures. *Adv. Appl. Mech.* **29** 64.

Irwin, G. R. (1958) Fracture. In *Handbuch der Physik*. Springer-Verlag, Berlin, Vol. 6, p. 551.

Maugis, D. (1985) Subcritical crack growth, surface energy, fracture toughness, stick–slip and embrittlement. *J. Mater. Sci.* **20** 3041.

Paris, P. C. & Sih, G. C. (1965) Stress analysis of cracks. In *Fracture Toughness Testing and its Applications*. A.S.T.M. Spec. Tech. Publ. 381, p. 30.

Rooke, D. P. & Cartwright, D. J. (1976) *Compendium of Stress-Intensity Factors*. Her Majesty's Stationery Office, London.

Tada, H., Paris, P. C. & Irwin, G. R. (1985) *The Stress Analysis of Cracks Handbook*. Del Research Corporation, St Louis.

Chapter 3

Atkins, A. G. & Mai, Y-M. (1985) *Elastic and Plastic Fracture*. Ellis Horwood, Chichester.

Barenblatt, G. I. (1962) The mathematical theory of equilibrium cracks in brittle fracture. *Adv. Appl. Mech.* **7** 55.

Broek, D. (1982) *Elementary Engineering Fracture Mechanics*. Martinus Nijhoff, Boston.

Budiansky, B., Hutchinson, J. W. & Lambropoulus, J. C. (1983) Continuum theory of dilatant transformation toughening in ceramics. *Int. J. Solids Structs.* **19** 337.

Dugdale, D. S. (1960) Yielding of steel sheets containing slits. *J. Mech. Phys. Solids* **8** 100.

Elliot, H. A. (1947) An analysis of the conditions for rupture due to Griffith cracks. *Proc. Phys. Soc. Lond.* **59** 208.

Hertzberg, R. W. (1988) *Deformation and Fracture Mechanics of Engineering Materials*. Wiley, New York.

Irwin, G. R. (1958) Fracture. In *Handbuch der Physik*. Springer-Verlag, Berlin, Vol. 6, p. 557.

Knott, J. F. (1973) *Fundamentals of Fracture Mechanics*. Butterworths, London.

Mai, Y-W. & Lawn, B. R. (1986) Crack stability and toughness characteristics in brittle materials. *Ann. Rev. Mater. Sci.* **16** 415.

Orowan, E. (1955) Energy criteria of fracture. *Weld. Res. Supp.* **34** 157-s.

Rice, J. R. (1968a) A path-independent integral and the approximate analysis of strain concentration by notches and cracks. *J. Appl. Mech.* **35** 379.

Rice, J. R. (1968b) Mathematical analysis in the mechanics of fracture. In *Fracture* (ed. H. Liebowitz). Academic, New York, Vol. 2, chapter 3.

Thomson, R. M. (1986) Physics of fracture. *Solid State Physics* **39** 1.

Weertman, J. (1978) Fracture mechanics: a unified view for Griffith–Irwin–Orowan cracks. *Acta Metall.* **26** 1731.

Chapter 4

Berry, J. P. (1960) Some kinetic considerations of the Griffith criterion for fracture. I. Equations of motion at constant force. II. Equations of motion at constant deformation. *J. Mech. Phys. Solids* **8** 194, 207.

Dickinson J. T. (1990) Fracto-emission. In *Non-Destructive Testing of Fibre-Reinforced Plastics Composites* (ed. J. Summerscales). Elsevier, London, Vol. 2, chapter 10.

Dickinson, J. T., Donaldson, E. E. & Park, M. K. (1981) The emission of electrons and positive ions from fracture of materials. *J. Mater. Sci.* **16** 2897.

Erdogan, F. (1968) Crack-propagation theories. In *Fracture* (ed. H. Liebowitz). Academic, New York, Vol. 2, chapter 5.

Field, J. E. (1971) Brittle fracture: its study and application. *Contemp. Phys.* **12** 1.

Freund, L. B. (1990) *Dynamic Fracture Mechanics*. Cambridge University Press, Cambridge.

Kerkhof, F. (1957) Ultrasonic fractography. In *Proceedings Third Internat. Congress High-Speed Photography*. Butterworths, London, p. 194.

Kolsky, H. (1953) *Stress Waves in Solids*. Clarendon, Oxford.

Mott, N. F. (1948) Brittle fracture in mild steel plates. *Engineering* **165** 16.

Roberts, D. K. & Wells, A. A. (1954) The velocity of brittle fracture. *Engineering* **24** 820.

Schardin, H. (1959) Velocity effects in fracture. In *Fracture* (ed. B. L. Averbach, D. K. Felbeck, G. T. Hahn & D. A. Thomas). Wiley, New York, p. 297.

Yoffe, E. H. (1951) The moving crack. *Phil. Mag.* **42** 739.

Chapter 5

Adamson, A. W. (1982) *Physical Chemistry of Surfaces*. John Wiley, New York.

Bailey, A. I. & Kay, S. M. (1967) Direct measurement of the influence of vapour, of liquid and of oriented monolayers on the interfacial energy of mica. *Proc. Roy. Soc. Lond.* **A301** 47.

Burns, S. J. & Lawn, B. R. (1968) A simulated crack experiment illustrating the energy balance criterion. *Int. J. Fract. Mech.* **4** 339.

Charles, R. J. & Hillig, W. B. (1962) The kinetics of glass failure by stress corrosion. In *Symposium sur la Resistance Mechanique du Verre et les Moyens de L'Ameliorer*. Union Sciences Continentale du Verre, Charleroi, Belgium, p. 511.

Glasstone, S., Laidler, K. J. & Eyring, H. (1941) *The Theory of Rate Processes*. McGraw-Hill, New York.

Hart, E. (1980) A theory for stable crack extension rates in ductile materials. *Int. J. Solids Struct.* **16** 807.

Johnson, H. H. & Paris, P. C. (1968) Subcritical flaw growth. *Eng. Fract. Mech.* **1** 3.

Lawn, B. R. (1974) Diffusion-controlled subcritical crack growth. *Mater. Sci. Eng.* **13** 277.

Lawn, B. R. (1983) Physics of fracture. *J. Amer. Ceram. Soc.* **66** 83.

Maugis, D. (1985) Subcritical crack growth, surface energy, fracture toughness, stick–slip and embrittlement. *J. Mater. Sci.* **20** 3041.

Orowan, E. (1944) The fatigue of glass under stress. *Nature* **154** 341.

Pollett, J-C. & Burns, S. J. (1977) Thermally activated crack propagation – theory. *Int. J. Fract.* **13** 667.

Rice, J. R. (1978) Thermodynamics of the quasi-static growth of Griffith cracks. *J. Mech. Phys. Solids* **26** 61.

Stavrinidis, B. & Holloway, D. G. (1983) Crack healing in glass. *Phys. and Chem. Glasses* **24** 19.

Wiederhorn, S. M. (1967) Influence of water vapour on crack propagation in soda-lime glass. *J. Amer. Ceram. Soc.* **50** 407.

Wiederhorn, S. M. & Bolz, L. H. (1970) Stress corrosion and static fatigue of glass. *J. Amer. Ceram. Soc.* **53** 543.

Chapter 6

Adamson, A. W. (1982) *Physical Chemistry of Surfaces*. John Wiley, New York.

Chan, D. C. & Horn, R. H. (1985) The drainage of thin liquid films between solid surfaces. *J. Chem. Phys.* **83** 5311.

Derjaguin, B. V., Churaev, N. V. & Muller, V. M. (1987) *Surface Forces*. Consultants Bureau (Plenum), New York.

Fuller, E. R., Lawn, B. R. & Thomson, R. M. (1980) Atomic modeling of crack-tip chemistry. *Acta Metall.* **28** 1407.

Fuller, E. R. & Thomson, R. M. (1978) Lattice theories of fracture. In *Fracture Mechanics of Ceramics* (ed. R. C. Bradt, A. G. Evans, D. P. H. Hasselman & F. F. Lange). Plenum, New York, Vol. 4, p. 507.

Gilman, J. J. (1960) Direct measurements of the surface energies of crystals. *J. Appl. Phys.* **31** 2208.

Glasstone, S., Laidler, K. J. & Eyring, H. (1941) *The Theory of Rate Processes*. McGraw-Hill, New York.

Hockey, B. J. (1983) Crack healing in brittle materials. In *Fracture Mechanics of Ceramics* (ed. R. C. Bradt, A. G. Evans, D. P. H. Hasselman & F. F. Lange). Plenum, New York, Vol. 6, p. 637.

Horn, R. G. & Israelachvili, J. N. (1981) Direct measurement of structural forces between two surfaces in a nonpolar liquid. *J. Chem. Phys.* **75** 1400.

Israelachvili, J. N. (1985) *Intermolecular and surface forces*. Academic, London.

Kanninen, M. F. & Gehlen, P. C. (1972) A study of crack propagation in α-iron. In *Interatomic Potentials and Simulation of Lattice Defects* (ed. P. C. Gehlen *et al.*). Plenum, New York, p. 713.

Kelly, A., Tyson, W. R. & Cottrell, A. H. (1967) Ductile and brittle crystals. *Phil. Mag.* **15** 567.

Lawn, B. R. (1975) An atomistic model of kinetic crack growth in brittle solids. *J. Mater. Sci.* **10** 469.

Lawn, B. R. (1983) Physics of fracture. *J. Amer. Ceram. Soc.* **66** 83.

Lawn, B. R., Hockey, B. J. & Wiederhorn, S. M. (1980) Atomically sharp cracks in brittle solids: an electron microscopy study. *J. Mater. Sci.* **15** 1207.

Lawn, B. R., Jakus, K. & Gonzalez, A. C. (1985) Sharp vs blunt crack hypotheses in the strength of glass: a critical study using indentation flaws. *J. Amer. Ceram. Soc.* **68** 25.

Lawn, B. R., Roach, D. H. & Thomson, R. M. (1987) Thresholds and reversibility in brittle cracks: an atomistic surface force model. *J. Mater. Sci.* **22** 4036.

Michalske, T. A. & Freiman, S. W. (1981) A molecular interpretation of stress corrosion in silica. *Nature* **295** 511.

Michalske, T. A. & Bunker, B. (1987) The fracturing of glass. *Scientific American* **257** 122.

Orowan, E. (1949) Fracture and strength of solids. *Rep. Progr. Phys.* **12** 48.

Rice, J. R. & Thomson, R. M. (1974) Ductile vs brittle behaviour of crystals. *Phil. Mag.* **29** 73.

Sinclair, J. E. (1975) The influence of the interatomic force law and of kinks on the propagation of brittle cracks. *Phil. Mag.* **31** 647.

Sinclair, J. E. & Lawn, B. R. (1972) An atomistic study of cracks in diamond-structure crystals. *Proc. Roy. Soc. Lond.* **A329** 83.

Slater, J. C. (1939) *Introduction to Chemical Physics.* McGraw-Hill, New York, chapter 10.

Thomson, R. M. (1973) The fracture crack as an imperfection in a nearly perfect solid. *Ann. Rev. Mater. Sci.* **3** 31.

Thomson, R. M., Hsieh, C. & Rana, V. (1971) Lattice trapping of fracture cracks. *J. Appl. Phys.* **42** 3154.

Chapter 7

Aveston, J., Cooper, G. A. & Kelly, A. (1971) Single and multiple fracture. In *The properties of fibre composites.* Guildford IPC Science and Technology Press, Surrey, p. 15.

Becher, P. F. (1991) Microstructural design of toughened ceramics. *J. Amer. Ceram. Soc.* **74** 255.

Bennison, S. J. & Lawn, B. R. (1989) Role of interfacial grain-bridging sliding friction in the crack-resistance and strength properties of nontransforming ceramics. *Acta Metall.* **37** 2659.

Budiansky, B., Hutchinson, J. W. & Lambropoulus, J. C. (1983) Continuum theory of dilatant transformation toughening in ceramics. *Int. J. Solids Structs.* **19** 337.

Burns, S. J. & Webb, W. W. (1966) Plastic deformation during cleavage of LiF. *Trans. Met. Soc. A.I.M.E.* **236** 1165.

Burns, S. J. & Webb, W. W. (1970) Fracture surface energies and dynamical cleavage of LiF. I. Theory. II. Experiments. *J. Appl. Phys.* **41** 2078, 2086.

Clarke, D. R. & Faber, K. T. (1987) Fracture of ceramics and glasses. *J. Phys. Chem. Solids* **11** 1115.

Dörre, E. & Hübner, H. (1984) *Alumina: Processing, Properties and Applications.* Springer-Verlag, Berlin, chapter 3.

Evans, A. G. (1990) Perspective on the development of high-toughness ceramics. *J. Amer. Ceram. Soc.* **73** 187.

Evans, A. G. & Faber, K. T. (1984) Crack growth resistance of microcracking brittle materials. *J. Amer. Ceram. Soc.* **67** 255.

Faber, K. T. & Evans, A. G. (1983) Crack deflection processes: I. Theory; II. Experiment. *Acta Metall.* **31** 565, 577.

Garvie, R. C., Hannink, R. H. J. & Pascoe, R. T. (1975) Ceramic steel? *Nature* **258** 703.

Green, D. J., Hannink, R. H. J. & Swain, M. V. (1989) *Transformation toughening of ceramics*. CRC, Boca Raton, Florida.

Hart, E. (1980) A theory for stable crack extension rates in ductile materials. *Int. J. Solids Struct.* **16** 807.

Hutchinson, J. W. (1990) Mixed-mode fracture mechanics of interfaces. In *Metal–Ceramic Interfaces* (ed. M. Rühle, A. G. Evans, M. F. Ashby & J. P. Hirth). Acta-Scripta Metall. Proceedings Series, Vol. 4, p. 295.

Hutchinson, J. W. & Suo, Z. (1991) Mixed-mode cracking in layered structures. *Adv. Appl. Mech.* **29** 64.

Kelly, A. (1966) *Strong Solids*. Clarendon, Oxford, chapter 5.

Knehans, R. & Steinbrech, R. (1982) Memory effect of crack resistance during slow crack growth in notched Al$_2$O$_3$ bend specimens. *J. Mater. Sci.* **1** 327.

Mai, Y-W. (1988) Fracture resistance and fracture mechanisms of engineering materials. *Mater. Forum* **11** 232.

Mai, Y-W. & Lawn, B. R. (1987) Crack-interface grain bridging as a fracture-resistance mechanism in ceramics: II. Theoretical fracture mechanics. *J. Amer. Ceram. Soc.* **70** 289.

Majumdar, B. S. & Burns, S. J. (1981) Crack-tip shielding – an elastic theory of dislocations and dislocation arrays near a sharp crack. *Acta Metall.* **29** 579.

Marshall, D. B., Cox, B. N. & Evans, A. G. (1985) The mechanics of matrix cracking in brittle-matrix fibre composites. *Acta Metall.* **23** 2013.

Marshall, D. B., Drory, M. D. & Evans, A. G. (1983) Transformation toughening in ceramics. In *Fracture Mechanics of Ceramics* (ed. R. C. Bradt, A. G. Evans, F. F. Lange & D. P. H. Hasselman). Plenum, New York, Vol. 6, p. 289.

McMeeking, R. M. & Evans, A. G. (1982) Mechanics of transformation toughening in brittle materials. *J. Amer. Ceram. Soc.* **65** 242.

Swanson, P. L. (1988) Crack-interface traction: a fracture-resistance mechanism in brittle polycrystals. In *Advances in Ceramics*. American Ceramic Society, Columbus, Vol. 22, p. 135.

Swanson, P. L., Fairbanks, C. J., Lawn, B. R., Mai, Y-W. & Hockey, B. J. (1987) Crack-interface grain bridging as a fracture-resistance mechanism in ceramics: I. Experimental study on alumina. *J. Amer. Ceram. Soc.* **70** 279.

Thomson, R. M. (1978) Brittle fracture in a ductile material with application to hydrogen embrittlement. *J. Mater. Sci.* **13** 128.

Thomson, R. M. (1986) Physics of fracture. *Solid State Physics* **39** 1.

Weertman, J. (1978) Fracture mechanics: a unified view for Griffith–Irwin–Orowan cracks. *Acta Metall.* **26** 1731.

Chapter 8

Anstis, G. R., Chantikul, P., Marshall, D. B. & Lawn, B. R. (1981) A critical evaluation of indentation techniques for measuring fracture toughness: I. Direct crack measurements. II. Strength method. *J. Amer. Ceram. Soc.* **64** 533, 539.

Auerbach, F. (1891) Measurement of hardness. *Ann. Phys. Chem.* **43** 61.

Cook, R. F., Lawn, B. R. & Fairbanks, C. J. (1985) Microstructure–strength properties in ceramics: I. Effect of crack size on toughness. *J. Amer. Ceram. Soc.* **68** 604.

Cook, R. F., Fairbanks, C. J., Lawn, B. R. & Mai, Y-W. (1987) Crack resistance by interfacial bridging: its role in determining strength characteristics. *J. Mater. Research* **2** 345.

Cook, R. F. & Pharr, G. M. (1990) Direct observation and analysis of indentation cracking in glasses and ceramics. *J. Amer. Ceram. Soc.* **73** 787.

Dabbs, T. P. & Lawn, B. R. (1985) Strength and fatigue properties of optical glass fibres containing microindentation flaws. *J. Amer. Ceram. Soc.* **68** 563.

Evans, A. G. & Wilshaw, T. R. (1976) Quasi-static solid particle damage in brittle solids. *Acta Metall.* **24** 939.

Frank, F. C. & Lawn, B. R. (1967) On the theory of Hertzian fracture. *Proc. Roy. Soc. Lond.* **A299** 291.

Hagan, J. T. (1980) Shear deformation under pyramidal indentations in soda-lime glass. *J. Mater. Sci.* **15** 1417.

Hertz, H. H. (1896) *Hertz's Miscellaneous Papers.* Macmillan, London, chapters 5, 6.

Johnson, K. L. (1985) *Contact Mechanics.* Cambridge University Press, Cambridge.

Johnson, K. L., Kendall, K. & Roberts, A. D. (1971) Surface energy and the contact of elastic solids. *Proc. Roy. Soc. Lond.* **A324** 301.

Lawn, B. R. (1983) The indentation crack as a model indentation flaw. In *Fracture Mechanics of Ceramics* (ed. R. C. Bradt, A. G. Evans, D. P. H. Hasselman & F. F. Lange). Plenum, New York, Vol. 5, p. 1.

Lawn, B. R., Dabbs, T. P. & Fairbanks, C. J. (1983) Kinetics of shear-activated indentation crack initiation in soda-lime glass. *J. Mater. Sci.* **18** 2785.

Lawn, B. R. & Evans, A. G. (1977) A model for crack initiation in elastic/plastic indentation fields. *J. Mater. Sci.* **12** 2195.

Lawn, B. R., Evans, A. G. & Marshall, D. B. (1980) Elastic/plastic indentation damage in ceramics: the median/radial crack system. *J. Amer. Ceram. Soc.* **63** 574.

Lawn, B. R., Jakus, K. & Gonzalez, A. C. (1985) Sharp vs blunt crack hypotheses in the strength of glass: a critical study using indentation flaws. *J. Amer. Ceram. Soc.* **68** 25.

Lawn, B. R. & Marshall, D. B. (1979) Hardness, toughness, and brittleness. *J. Amer. Ceram. Soc.* **62** 347.

Lawn, B. R., Marshall, D. B., Chantikul, P. & Anstis, G. R. (1980) Indentation fracture: applications in the assessment of strength of ceramics. *J. Austral. Ceram. Soc.* **16** 4.

Lawn, B. R. & Wilshaw, T. R. (1975) Indentation fracture: principles and applications. *J. Mater. Sci.* **10** 1049.

Mai, Y-W. & Lawn, B. R. (1986) Crack stability and toughness characteristics in brittle materials. *Ann. Rev. Mater. Sci.* **16** 415.

Marshall, D. B. (1984) An indentation method for measuring matrix–fibre frictional stresses in ceramic composites. *J. Amer. Ceram. Soc.* **67** C–259.

Marshall, D. B., Lawn, B. R. & Chantikul, P. (1979) Residual stress effects in sharp-contact cracking. I. Indentation fracture mechanics. II. Strength degradation. *J. Mater. Sci.* **14** 2001, 2225.

Marshall, D. B. & Lawn, B. R. (1980) Flaw characteristics in dynamic fatigue: the influence of residual contact stresses. *J. Amer. Ceram. Soc.* **63** 532.

Maugis, D. & Barquins, M. (1978) Fracture mechanics and the adherence of viscoelastic bodies. *J. Phys. D: Appl. Phys.* **11** 1989.

Puttick, K. (1980) The correlation of fracture transitions. *J. Phys. D: Appl. Phys.* **13** 2249.

370   *References and reading list*

Roesler, F. C. (1956) Brittle fractures near equilibrium. *Proc. Phys. Soc. Lond.* **B69** 981.
Swain, M. V., Williams, J. S., Lawn, B. R. & Beek, J. J. H. (1973) A comparative study of the fracture of various silica modifications using the Hertzian test. *J. Mater. Sci.* **8** 1153.

Chapter 9

Chantikul, P., Bennison, S. J. & Lawn, B. R. (1990) Role of grain size in the strength and *R*-curve properties of alumina. *J. Amer. Ceram. Soc.* **73** 2419.
Cottrell, A. H. (1958) Theory of brittle fracture in steel and similar metals. *Trans. Met. Soc. A.I.M.E.* **212** 192.
Davidge, R. W. (1979) *Mechanical Behaviour of Ceramics*. Cambridge University Press, London, chapter 6.
Dörre, E. & Hübner, H. (1984) *Alumina: Processing, Properties and Applications*. Springer-Verlag, Berlin, chapter 3.
Ernsberger, F. M. (1960) Detection of strength-impairing flaws in glass. *Proc. Roy. Soc. Lond.* **A257** 213.
Green, D. J. (1983) Microcracking mechanisms in ceramics. In *Fracture Mechanics of Ceramics* (ed. R. C. Bradt, A. G. Evans, D. P. H. Hasselman & F. F. Lange). Plenum, New York, Vol. 5, p. 457.
Lange, F. F. (1978) Fracture mechanics and microstructural design. In *Fracture Mechanics of Ceramics* (ed. R. C. Bradt, A. G. Evans, D. P. H. Hasselman & F. F. Lange). Plenum, New York, Vol. 4, p. 799.
Lange, F. F. (1989) Powder processing science and technology for increased reliability. *J. Amer. Ceram. Soc.* **72** 3.
Petch, N. J. (1968) Metallographic aspects of fracture. In *Fracture* (ed. H. Liebowitz). Academic, New York, vol. 1, chapter 5.
Puttick, K. (1980) The correlation of fracture transitions. *J. Phys. D: Appl. Phys.* **13** 2249.
Zener, C. (1948) Micromechanism of fracture. In *Fracturing of Metals*. A.S.M., Cleveland, p. 3.

Chapter 10

Ashby, M. F. & Dyson, B. F. (1984) Creep damage mechanisms and micromechanisms. In *Advances in Fracture Research* (ed. S. R. Valluri, D. M. R. Taplin, P. Rama Rao, J. F. Knott & R. Dubey). Pergamon, Oxford, p. 3.
Broek, D. (1982) *Elementary Engineering Fracture Mechanics*. Martinus Nijhoff, Boston, chapter 5.
Creyke, W. E. C., Sainsbury, I. E. J. & Morrell, R. (1982) *Design With Non-Ductile Materials*. Applied Science Publishers, London.
Dabbs, T. P. & Lawn, B. R. (1985) Strength and fatigue properties of optical glass fibers containing microindentation flaws. *J. Amer. Ceram. Soc.* **68** 563.
Davidge, R. W. (1979) *Mechanical Behaviour of Ceramics*. Cambridge University Press, London, chapters 8, 9.

Davidge, R. W., McLaren, J. R. & Tappin, G. (1973) Strength–probability–time (SPT) relationships in ceramics. *J. Mater. Sci.* **8** 1699.

Evans, A. G. (1985) Engineering property requirements for high performance ceramics. *Mater. Sci. Eng.* **71** 3.

Evans, A. G. & Wiederhorn, S. M. (1974) Proof testing of ceramic materials – an analytical basis for failure prediction. *Int. J. Fract.* **10** 379.

Hasselman, D. P. H. (1969) Unified theory of thermal shock fracture initiation and crack propagation in brittle ceramics. *J. Amer. Ceram. Soc.* **52** 600.

Lange, F. F. (1984) Structural ceramics: a question of fabrication. *J. Mater. Energy System* **6** 107.

Lange, F. F. (1989) Powder processing science and technology for increased reliability. *J. Amer. Ceram. Soc.* **72** 3.

Mai, Y-W. & Lawn, B. R. (1986) Crack stability and toughness characteristics in brittle materials. *Ann. Rev. Mater. Sci.* **16** 415.

Marshall, D. B. & Ritter, J. E. (1987) Reliability of advanced structural ceramics and ceramic matrix composites – a review. *Ceram. Bull.* **66** 309.

Maurer, R. D. (1985) Behavior of flaws in fused silica fibers. In *Strength of Inorganic Glass* (ed. C. R. Kurkjian). Plenum, New York, p. 291.

Ritchie, R. O. (1988) Mechanisms of fatigue crack propagation in metals, ceramics, composites: role of crack-tip shielding. *Mater. Sci. Eng.* **103A** 15.

Ritter, J. E. (1978) Engineering design and fatigue failure of brittle materials. In *Fracture Mechanics of Ceramics* (ed. R. C. Bradt, D. P. H. Hasselman & F. F. Lange). Plenum, New York, Vol. 4, p. 667.

Suresh, S. (1991) *Fatigue of Materials.* Cambridge University Press, Cambridge.

Weibull, W. (1951) A statistical distribution function of wide applicability. *J. Appl. Mech.* **18** 293.

Wiederhorn, S. M. (1972) Subcritical crack growth in ceramics. In *Fracture Mechanics of Ceramics* (ed. R. C. Bradt, D. P. H. Hasselman & F. F. Lange). Plenum, New York, Vol. 2, p. 613.

Wiederhorn, S. M. (1978) A probabilistic framework for structural design. In *Fracture Mechanics of Ceramics* (ed. R. C. Bradt, A. G. Evans, D. P. H. Hasselman & F. F. Lange). Plenum, New York, Vol. 5, p. 197.

Wiederhorn, S. M. & Fuller, E. R. (1985) Structural reliability of ceramic materials. *Mater. Sci. Eng.* **71** 169.

Index

acoustic detection of cracks, 99, 103, 196, 341–2
activated crack propagation, 107–8, 128–32, 138–9, 157–62, 165–72, 183
activated failure, 266–8, 285, 294–5, 330–1
activation area, 130–1, 160–1, 168, 171
activation energy (*see under* energy)
adhesion, contact (*see under* contact adhesion)
adhesion zone (*see* Barenblatt cohesion zone)
adsorption, 107–12, 128–31, 134–5, 169–75, 178, 180, 182
aging
  of flaws, 286–7, 307, 321, 325, 346, 350
  of zirconia, 122–7
alumina (*see under* materials)
anisotropy, in crack systems, 28, 41, 50, 98, 255–6
atomic aspects of fracture, 12, 15, 53–4, 143–93
atomic bond (*see* cohesive bond)
Auerbach's law, 249, 285–6

Barenblatt cohesion zone, 59–66, 69–73, 80–2, 112, 117, 129, 143, 165, 176, 179, 180–4
bathtub curve, 335–6
bi-material interfacial-crack specimen (*see under* fracture test specimens)
blunt crack (*see under* crack-tip)
blunt indenter (*see under* indenter)
bond (*see* cohesive bond)
bond-rupture concept of fracture, 12, 41, 51, 54, 105–7, 113, 117, 128–9, 137–8, 144–62, 165–8, 169–74, 317
branching, crack, 93, 95–9, 206
bridging, crack-interface, 80–2, 194, 202, 205, 209–10, 225, 227, 230–48, 274–6, 278, 282, 325–6, 332–3, 350–4, 356, 361

brittleness, 14–15, 187, 188, 291–3

capillary, crack-interface, 112, 132, 136, 141
ceramics (*see also under* materials), 51, 55–6, 99, 125–6, 194–248, 261–2, 271–6, 292–3, 302–4, 307–8, 314–18, 321, 324–7, 332–3, 335–62
chemistry, in fracture, 11, 106–42, 165–85, 320, 343–4
cleavage, 49–50, 55, 93, 98–9, 183, 196–9, 207–8, 255–6
  of mica, 10–12, 23, 50, 140–1, 163–4, 176–9, 182–3
closure, crack (*see under* crack)
coating, protective, 14, 103, 314, 348–9
cohesion zone (*see* Barenblatt cohesion zone)
cohesive bond, 6, 12, 15, 40, 53–4, 143–92
cohesive energy- and force-separation function, 53–4, 59–66, 69, 110, 144–7, 149–57, 161–3, 166–9, 175–6, 178–81, 185
cohesive strength (*see* theoretical cohesive strength)
compaction, 261, 272, 288
compliance, crack specimen, 20–3, 30, 70, 82, 89, 91, 361
composites, ceramic-matrix, 55–6, 202, 210, 242–8, 299–300, 337, 353–5, 361
computer simulation, crack-tip (*see under* crack-tip)
cone crack (*see* Hertzian cone crack)
contact
  adhesion, 304–6
  damage, 300–4, 309–12
  elastic field, 249–50, 251–6, 283–7, 309–12
  elastic–plastic field, 247–50, 259–61, 288–90, 309
  residual field (*see under* residual stress)

continuum descriptions of cracks, 16–85
controlled flaws (*see under* flaws;
    indentation fracture)
covalent–ionic solids (*see also under*
    materials), 50, 103–4, 163–5, 174,
    185–92, 200, 211–12, 215, 249, 287,
    307, 314–15, 355
crack
    arrest, 42, 79, 91–2, 99, 196, 259, 290,
        323, 325, 358
    closure, 19, 28–9, 39–40, 178–80, 253
    healing (*see* healing, crack-interface;
        hysteresis)
    initiation (*see* initiation, crack)
    irreversibility (*see* irreversibility, crack)
    propagation (*see* propagation, crack)
    reversibility (*see* reversibility, crack)
crack-extension force, or 'motive', 41, 58,
    76, 114, 154
crack-interface bridging (*see* bridging,
    crack-interface)
crack-opening displacement, 27–8, 64, 80,
    134–5, 143, 152, 235–6, 240, 353
crack path, 17, 44–50, 95–7, 196–206, 252
crack-resistance energy (*see also R*-curve),
    52, 55–8, 72–85, 194–248
crack stability (*see also* crack path), 17, 23,
    37, 41–4, 78–9, 99, 187, 233, 240–1,
    249–50, 257–8, 266, 277, 283–5, 290,
    296, 307–9, 328–34, 350–2, 356–7
    neutral, 37–8
    quasi-, 107, 113–16, 132, 155–6
    stable, 10–11, 37, 42–3, 78–9, 249, 251,
        253–4, 265–70, 275, 284–5, 290, 341,
        355–6, 358, 361
    unstable, 9, 37, 42–3, 78–9, 86–99, 155,
        253, 265–6, 277, 290, 299
crack-tip
    blunting, 136–8, 185, 187–8, 192, 296–7,
        362
    chemistry (*see* chemistry, in fracture)
    cohesion, 52–4, 144–7, 185–6
    computer simulation, 144, 162–5, 185
    corrosion, 128, 133, 136–8, 141, 188
    deflection (*see* deflection, crack-front)
    deformation (*see* plasticity, *this entry*)
    field, 4, 18–19, 23–8, 44–5, 48, 51, 59–64,
        72–7, 89–90, 95–8, 117–18, 162–5,
        181–2, 195, 210, 212–17, 228
    kink, 45, 157–62, 165, 171, 188
    plasticity, 56–8, 132–3, 136–8, 162,
        185–93, 211–16, 229, 361
    profile (*see* structure, *this entry*)
    radius, 3–5, 12, 137–8, 147, 164–5,
        181–2, 188
    reaction, 128–31, 138–9, 165–9, 172–4,
        183

sharpness, 137, 188–93, 195, 296–7
    shielding (*see* shielding)
    singularity, 26–8, 41, 51–4, 59–66, 129,
        164
    stress field (*see* field, *this entry*)
    structure, 2–5, 14–15, 28, 52, 59–66, 128,
        163–5, 173, 180–3, 305–6
crack velocity (*see* velocity, crack)
creep, 321, 361–2
cyclic fatigue (*see under* fatigue)

damage accumulation, 361–2
dead-weight (constant-force) loading, 20–3,
    89–92
deflection, crack-front (*see also* branching,
    crack), 44–50, 99, 194–208, 223, 227,
    235–6, 244
diamond-structure crystals (*see under*
    materials)
diffusion, crack-interface, 128–9, 131–6,
    182–4, 321
discreteness, in brittle-crack models, 54, 65,
    72, 143–95, 209, 233, 288, 355–6
dislocation, 23, 185, 187–92, 194, 211–16
    cloud, 195, 211–16
    emission, 211
    friction, or Peierls stress, 141, 213
    network, 189–92, 200
    pile-up, 314–18, 328
    source, 190, 211, 213–15, 315
double-cantilever beam test specimen (*see
    under* fracture test specimens)
double-torsion test specimen (*see under*
    fracture test specimens)
ductile–brittle transition, 190, 215, 291, 315
Dugdale model, 59, 188
Dupré work of adhesion (*see* work of
    adhesion)
dynamic fracture, 86–105
dynamical loading, 99–102

elastic contact field (*see under* contact)
elastic–plastic contact field (*see under*
    contact)
electron microscopy, of crack tips, 188–92,
    212
Elliot crack, 65–6, 143–4, 165, 176
elliptical
    cavity, 2–5, 7, 14–15, 320, 326
    crack front, 32–3
energy-balance concept (*see* Griffith
    energy-balance concept)
energy, of crack system, 5–11, 13, 17–23,
    39–42, 45, 51–8, 86–8, 112–16, 150–62,
    304–6
    activation, 113, 116, 130–1, 158–60,
        130–1, 139, 154, 157, 160, 165, 267

crack-resistance (*see* crack-resistance energy)
dissipative, 76–7, 83, 118, 132–3, 157, 196, 206, 211, 213, 220, 236, 238, 243, 247–8
kinetic, 86–92, 95, 105
mechanical, 6, 8, 10–11, 18–23, 29, 37, 57, 67–8, 70, 154, 163
surface, 6, 8, 10–11, 40, 50, 54–5, 58, 107–12, 132, 141, 145–7, 151, 154–7, 163, 170, 179, 184, 200, 203, 211
entropy production rate, in crack growth, 113–15
environment, influence on fracture, 1, 11, 106–42, 165–85, 276–82, 296, 320, 343–7, 352
equilibrium crack systems (*see also* crack stability), 2–12, 16–19, 39–44, 51–66, 70–2, 73–9, 86, 106–16, 139, 143–85, 209
erosion (*see* wear and erosion)

fatigue, and time-dependent failure, 108, 276–82, 295, 343–6, 348
cyclic 361
lifetime, 280–2, 335, 343–6, 352, 361–2
limit, 280–2, 344–5, 351–2, 355
static, 280–2, 343–5
fibres, 13–14, 55, 210, 242–7, 299–300, 307, 346–8
field (*see under* contact; crack-tip)
fixed-grips (constant-displacement) loading, 21–3, 89–92
flaw characteristics, in relation to design
detection, 307, 312–14, 336–47
elimination, 321, 327, 336–7, 340, 344, 347–50
tolerance, 79, 230, 273, 275–8, 303, 332–3, 337, 350–7, 360
flaws, 2, 49, 95, 118, 253, 301, 307–34, 335–62, 358
chemically-induced, 318, 320
controlled, 249, 263–98, 309–12, 341–2
critical, 16, 308, 338
Griffith, 13–14, 16, 79, 102, 115, 307–9, 319, 335–6, 343
microstructural (and processing), 206, 216–18, 233, 235, 240, 323–7, 348–50
radiation-induced, 321–5
stability of, 79, 283, 328–34, 350–4
thermally-induced, 319–22
fluid, at crack interface, 9, 107–11, 114–16, 128–9, 133–6, 144, 176–80
force–separation function, for bond rupture, 12, 145, 149–51, 164, 166–8
fracto-emission, 103–5, 157
fracture mechanics, linear, 16–50
fracture mechanics, nonlinear, 51–85
fracture steps, 4–5, 49–50, 198–9, 206–8

fracture surface energy (*see* crack-resistance energy)
fracture test specimens
bi-material interfacial-crack, 38–9
double-cantilever beam, 36–7
double-torsion, 38
flexure, 35–6, 267
indentation-strength (*see under* indentation-strength test)
tensile, 30–3
free molecular flow, of gas along crack interface, 128–9, 133–6
friction, in fracture systems, 101, 129, 132–3, 213, 215, 231, 234–7, 241, 244–5, 255–6, 286, 289, 299–300, 309, 315, 352–3
frontal zone, 57, 75, 80, 83–5, 194–5, 209–30, 239, 243, 248, 361

glass, silica (*see under* materials)
grains and grain boundaries (*see also under* size effects, in fracture; strength), 194–5, 196–202, 216–19, 225, 230, 232–41, 262, 275–6, 302–4, 315–16, 320, 323, 325–6, 332–4, 353, 358–61
Griffith energy-balance concept, 1–19, 39–41, 51–2, 54, 56–8, 64, 70–2, 86–9, 106–13, 125, 130, 132, 139, 147, 153–7, 160, 195
Griffith flaws (*see under* flaws)

hackle (*see* mirror, mist and hackle)
hardness, 137, 250–1, 257, 259–60, 271–3, 292–3, 303–4
healing, crack-interface, 12, 19, 29, 40, 106, 111, 115–16, 139–42, 152–3, 170, 175, 180–4, 189–93, 306, 319, 321
Hertzian cone crack, 249, 253–7, 64–5, 283–7, 301, 304–5, 309–13
hydrolysis, of silicon-oxygen bonds, 172–4, 178
hysteresis, in fracture processes, 83–4, 106, 116, 139–42, 144, 175, 178–80, 183–5, 221, 223, 237, 242, 342

impact loading, 86, 99–102, 300–4, 309–12, 354, 361
identation fracture (*see also* Hertzian cone crack, radial crack, lateral crack), 233, 249–306, 309
as controlled flaws, 263–82, 293–5
postthreshold, and crack propagation, 253–82
special applications, 296–300
subthreshold, and crack initiation, 282–93
threshold, 282–93

indentation-strength test, 263–82, 293–5, 301–3, 331, 352–4, 356
indenter
blunt, 250–1, 253–7, 264–5, 282–7, 309
sharp, 250–1, 257–63, 287–95, 341–2
Inglis analysis, of elliptical cavity, 2–5, 7–8, 12, 14, 18–19, 320
initiation, of crack, 14, 95, 98, 138, 216–17, 249–50, 265, 282–93, 307–34, 323, 328–32, 337, 358
integral, path-independent (*see* J-integral)
interatomic bond (*see* cohesive bond)
interatomic potential and force function (*see* cohesive energy– and force–separation function)
interface energy (*see* surface or interface energy)
interfaces (*see* grains and grain boundaries; weak interfaces; *see also under* microstructure)
intergranular fracture, 196, 199–202, 204, 211, 216, 233, 235, 262, 279
interphase boundaries (*see under* microstructure)
ionic bonding (*see* covalent–ionic bonding)
irreversibility, crack, 12, 15, 51–8, 70, 72–86, 88, 91, 95, 112–17, 132–3, 137, 141, 157, 194, 196, 207, 216, 229, 236, 239, 242–3, 245, 247–8, 251, 257, 341–2, 361
Irwin–Orowan small-scale zone model, 56–8, 72, 77, 86

J-integral, 66–71, 74–6, 80–4, 248
Joffe effect, 320

kinetic energy, of crack system (*see under* energy, of crack system)
kinetic fracture, 106–42, 157–62, 169–74, 184, 276–82, 296–7, 343–5
kink, crack-front (*see under* crack-tip)
Knoop indenter, 257, 259, 266–7, 269, 341–2

lateral crack, 257–9, 261, 274, 296–7, 303
lattice trapping, 144, 149–72
lifetime (*see under* fatigue)
limit, fatigue (*see under* fatigue)
linear elastic fracture mechanics (*see* fracture mechanics)

materials (*see also* ceramics; composites, ceramic-matrix; polycrystalline ceramics; two-phase ceramics)
alumina, 55, 126–7, 199, 219, 221, 230, 232–4, 239–40, 242–3, 247–8, 262,
272–6, 278–82, 302, 323–6, 332–3, 338, 340, 344–5, 349, 351–4, 356–61
body-centred cubic metals, 162, 185–6, 188
cement, 243, 246–7, 349
concrete, 55, 210–11, 230, 307, 355
diamond-structure crystals, 50, 55, 162–5, 185–6, 255–6, 292, 310, 312
face-centred cubic metals, 186, 188
glass (*see* silica glass, *this entry*)
glass-ceramics, 242–3, 271–2
hexagonal-close-packed metals, 188
ionic solids (potassium chloride, sodium chloride), 200, 318, 320
lithium fluoride, 55, 198, 213, 321–2
magnesium oxide, 55, 190, 192, 212, 292, 315–16
mica, 10–12, 50, 55, 103–4, 119, 121, 123, 133, 136, 139, 140–1, 176–8, 181–4, 213, 304–6
sapphire, 55, 120–1, 139, 142, 183, 190–1, 233, 262, 271–4, 292, 307, 319, 321
silica glass, quartz, 1, 7–9, 12–14, 50, 55, 93, 95–8, 102–3, 108, 119–25, 128, 133, 137–42, 161, 171–4, 183, 204–5, 242–3, 249, 253–6, 258–60, 262–5, 268–73, 277–81, 285–8, 292, 294–305, 307–8, 311–12, 314, 319–23, 338, 340, 346–8
silicon, 55, 162–5, 174, 189–90, 208, 272, 292, 307
silicon carbide, 55, 190, 243, 255–6, 272, 292, 299–300
silicon nitride, 55, 266–7, 268, 272, 292, 300, 302, 339–40, 341–2
steel (and iron), 15, 55, 162, 185, 292
tungsten carbide, 55, 247, 272, 292
zirconia, 55, 195, 221–30, 242–3, 272, 292, 349, 361
mechanical energy, of crack system (*see under* energy, of crack system)
mechanical-energy-release rate, definition, 17, 20–3, 29–41
metallic solids (*see also under* materials), 1, 23, 55–6, 58, 117, 119, 127–8, 185–8, 211, 215–16, 247–8, 291–3, 315, 318, 361
metastable interfacial states, 144, 170, 175–85, 192
mica (*see under* materials)
microcrack (*see also* frontal zone), 13–14, 210–11, 216, 232–3, 246, 289, 304, 308, 315–32, 360–1
cloud, activation, 216–18
cloud, and toughness, 217–21
microstructure, 194–248, 323–7
dislocations (*see* dislocation)

grain boundaries (*see*
    grains and grain boundaries)
interfaces (*see* weak interface)
interphase boundaries, 23, 50, 203, 216,
    235–6, 243, 325
pores (and voids), 194, 205, 318, 323–4,
    326–7, 349, 356
second-phase particles (and inclusions,
    precipitates), 194, 202–6, 210, 221,
    242, 315, 321, 326–7, 332, 353–4
slip bands, 315–16
twins, 194, 227, 230, 315
mirror, mist and hackle, on fracture
    surface, 95–8
mist (*see* mirror, mist and hackle)
modes of fracture, 23–4, 26–7, 29, 31–2, 39,
    44–50
    I, opening, 23–4, 25–8, 31–2, 39, 41, 43,
        46–50, 203, 213
    II, sliding, 23–4, 25, 27–8, 31–2, 38, 39,
        47–8, 196, 199, 201, 203, 300
    III, tearing, 23–4, 27–8, 31–2, 47–50,
        196, 199, 201, 203, 206
moisture (*see* water)
molecular theory of strength (*see*
    theoretical cohesive strength)
motive (*see* crack-extension force)
Mott dynamic crack, 86–92

neutral crack (*see under* crack stability)
notch, 2–5, 35–6, 39, 78–9, 97, 138, 211,
    223–4, 231, 235–6, 247, 356–7
nucleation, crack, 16, 307, 309–12, 314–18,
    325–6, 328

Obreimoff's experiment, 9–12, 37, 103
opening mode of fracture (*see under* modes
    of fracture)
Orowan generalisation of Griffith concept,
    108–12

path, crack (*see* crack path)
path-independent integral (*see* *J*-integral)
Peierls barrier (*see under* dislocation)
penny cracks, 31–3, 35–6, 217–18, 236–7,
    239–41, 254, 258–61, 321, 329–30, 341,
    350–2, 356
Petch relation, 317, 332–4
phase transformation, in zirconia
    cloud initiation, 225–8
    and toughness, 195, 210, 220–30, 328,
        349
pile-up model of crack nucleation (*see*
    under dislocation)
pinning and bowing, 205–6
plane strain and plane stress, 8–9, 25, 29,
    46–8, 61, 186

plasticity (*see under* crack-tip)
polycrystalline ceramics (*see also under*
    materials), 50, 55–6, 121, 125–7, 194,
    196–202, 216, 230, 233–5, 262, 271–4,
    276, 292, 302, 304, 307, 321, 323, 325,
    332
pop-in, 42, 79, 233, 240, 253, 257, 275, 284,
    285–6, 290–1, 295, 309, 312, 319, 327,
    330, 358
pores (*see under* microstructure)
postthreshold indentations (*see under*
    indentation fracture)
proof test, 336, 339–41, 344–7, 355
propagation, crack, 16–193

quartz (*see under* materials)
quasi-equilibrium crack (*see under* crack
    stability)

radial–median crack, 249–50, 255, 257–63,
    265–73, 279, 287–93, 296–8, 301, 321,
    326–9, 331–2
R-curve (*see also* *T*-curve), 55, 72–85,
    117–19, 195, 209, 216, 229, 230–3, 246,
    248–9, 337, 346, 350, 356–7
reaction, crack-tip (*see under* crack-tip)
reliability, 307, 335–62
residual stress (or strain)
    mismatch, 73, 83, 190–4, 195, 202,
        204–5, 211, 213, 216–17, 219, 221, 223,
        232, 235–7, 240–1, 244–5, 248,
        298–300, 304, 308, 318, 320–3, 325–32,
        346–7, 353, 355–6, 359
    contact, 258–66, 268, 273, 278, 280,
        288–91, 296–7, 308, 312, 341
reversibility, crack, 5–7, 11, 17–19, 29, 40,
    54, 67, 70, 106, 109–16, 139–42, 170,
    178–80
Rice integral (*see* *J*-integral)
Rice thermodynamical generalisation of
    Griffith concept, 107, 112–16, 132,
    139, 155, 175, 184

sapphire (*see under* materials)
second-phase particles (and inclusions,
    precipitates) (*see under* microstructure;
    see also two-phase ceramics)
sharp crack (*see under* crack-tip; *see also*
    slit crack)
sharp indenter (*see under* indenter)
shielding (*see also* bridging, crack-
    interface; frontal zone), 52, 54–6,
    72–85, 117–19, 126–7, 132–3, 194–5,
    208–48, 350–4
silica glass (*see under* materials)
silicon (*see under* materials)
silicon carbide (*see under* materials)

silicon nitride (*see under* materials)
size effects, in fracture, 13, 227, 293
    crack-initiation threshold, 219, 250, 282, 293, 328–32
    grain-, 237, 332–4
    indenter radius (and Auerbach's law), 286
    molecular, 143
sliding mode of fracture (*see under* modes of fracture)
slit crack, 17–28, 52, 59–66, 143, 148–50, 164, 173, 181
slip band (*see under* microstructure)
slow crack growth (*see* velocity, crack)
source–sink, in crack field, 56, 72–3, 79, 118, 209, 212
spontaneous fracture, 9, 217–18, 241, 265, 268, 273, 285, 290, 294, 327, 330–1, 332, 338, 357, 360
stability, crack (*see* crack stability)
stable crack (*see under* crack stability)
static fatigue (*see under* fatigue)
statistical distribution of flaws (*see under* flaw characteristics; *see also* Weibull statistics)
steady-state
    crack velocity, 92, 95, 102, 106–7, 111, 113–15, 132, 134, 215
    R-curve or *T*-curve, 57, 81–2, 84, 209–10, 214–15, 219, 223, 226, 228–30, 239, 246, 248
steel (*see under* materials)
steps, on fracture surface (*see* fracture steps)
straight cracks, 22, 31–6, 40, 42–3, 67, 78, 89, 154, 156, 238–40, 357
strain-energy-release rate (*see also* mechanical-energy-release rate), 22, 29
strength (*see also* indentation-strength test; theoretical cohesive strength), 1, 7–9, 15, 78–9, 226, 245, 249–50, 263–82, 293–5, 298–9, 314–18, 326, 328–34, 335–62
    and flaw size, 9, 13–14, 79, 248, 307–9, 336
    and grain (microstructure) size, 275–6, 315–17, 328–34
    degradation, 250, 265, 301–3
    inert, 106, 263–76, 295, 296–301, 337–42, 352
    fatigue, 108–9, 276–82, 343–7
    specimens, 35–6
    variability, 335–48
stress concentrators (*see* notches)
stress-intensity factor, definition, 17, 23–44
subcritical crack growth (*see under* velocity, crack)

subthreshold indentations (*see under* indentation fracture)
superposability, 17, 41–4
    of *G*-fields, 29
    of *K*-fields, 27–8, 43–4, 237, 246
surface or interface energy, 6–12, 39–41, 50, 55–6, 63, 106, 108–10, 119, 123, 132, 145–7, 150–1, 155–6, 163, 170, 180–4
    periodicity in, 154–6
surface forces, 40, 53, 109, 112, 175–85, 304–6
surface stresses, 298–300

T-curve (*see also R*-curve), 55, 72–85, 117, 126–7, 195, 209, 223–4, 232–5, 237, 239–41, 246–8, 249, 263–4, 269–70, 274–6, 278–9, 282, 304, 332–3, 337, 350–6, 358–60
tearing mode of fracture (*see under* modes of fracture)
temperature, effect on fracture, 106, 113–14, 121–3, 125, 127–8, 131, 138–9, 141–2, 189–90, 215, 285–7, 291, 296, 319–22, 357–9, 361–2
tensile test (*see under* fracture test specimens)
terminal velocity, 86, 89, 92–8
theoretical cohesive strength, 12, 144–7, 185–6, 287, 347
thermally-activated crack propagation (*see* activated crack propagation)
thermal expansion mismatch (*see under* residual stress)
thermodynamic basis of fracture
    first law (Griffith), 2, 5–7, 16–20, 29, 39–41, 44, 51, 56–8, 66, 70–1, 106, 108–10
    second law (Rice), 107, 112–16, 157, 180, 183
threshold (*see under* indentation fracture; size effects, in fracture; velocity, crack)
time-dependent failure (*see* fatigue)
toughness (*see also T*-curve), 52, 55–6, 61, 72–85, 194–248, 271–6
transformation toughening (*see under* phase transformation, in zirconia)
transgranular fracture, 195–9, 201, 211, 234
transition, ductile–brittle (*see* ductile–brittle transition)
twins (*see under* microstructure)
two-phase ceramics (*see also under* materials), 202–6, 210, 221, 353–4, 361

unstable crack (*see under* crack stability)

van der Waals forces, 176–7

velocity, crack (*see also* terminal velocity;
    see also under steady-state)
    'fast' (dynamic), 86–102
    'slow', or 'subcritical' (kinetic), 106–8,
        112–42, 157–62, 171, 174–5, 184, 215,
        276–82, 295, 343, 352, 362
    threshold, 106–8, 115–17, 123–5, 130,
        137–42, 170, 175, 181, 184, 262, 278–9,
        282, 296, 362
Vickers indenter, 190, 255, 257–9, 262–4,
    269–73, 275–82, 288–9, 291, 293–300,
    354
virgin crack interface, 40, 110, 115–16, 121,
    140–2, 179–84
viscosity, of crack-interface fluid, 128–9,
    132–3

Wallner lines, on fracture surface, 99
water, effect on crack propagation, 108,
    120–7, 134–6, 139, 140–2, 171–4,

177–8, 181–3, 192, 262, 277–82, 295–7,
    306, 346–7, 320
weak interface, 23, 49, 194–6, 200–1, 211,
    203, 231, 235, 242, 245
wear, and erosion, 250, 302–4, 314, 358–60
Weibull statistics, and Weibull modulus,
    335, 337–41, 343–9, 355
work of adhesion (*see also* surface or
    interface energy), 40, 54, 107, 108–12,
    141, 170, 179–80, 200–1, 203–4, 304

yield stress, 59, 188, 248, 315, 361
Yoffe hypothesis, for dynamic crack, 95–8

zirconia (*see under* materials)
zone models of cracks (*see* Barenblatt
    cohesion zone; bridging, crack-
    interface; frontal zone; Irwin–Orowan
    small-scale zone model)